Space Manifold Dynamics

Ettore Perozzi · Sylvio Ferraz-Mello
Editors

# Space Manifold Dynamics

## Novel Spaceways for Science and Exploration

Springer

*Editors*
Ettore Perozzi
Telespazio SpA
Via Tiburtina, 965
00156 Roma
Italy
ettore.perozzi@telespazio.com

Sylvio Ferraz-Mello
Universidade de São Paulo
Instituto de Astronomia,
Geofísica e
Ciência Atmosféricas (IAG)
Rua do Matão 1226
São Paulo-SP
Cidade Universitaria
Brazil
sylvio@astro.iag.usp.br;
sylvio@usp.br

ISBN 978-1-4419-0347-1 ISBN 978-1-4419-0348-8 (eBook)
DOI 10.1007/978-1-4419-0348-8
Springer New York Dordrecht Heidelberg London

Library of Congress Control Number: 2009940710



Printed on acid-free paper

Springer is part of Springer Science+Business Media (www.springer.com)

# Preface

*One cannot predict how knowledge will be applied – only that it often is*

(Charles Conley, 1968)

The advances in the field of dynamical systems led to the development of innovative methods and techniques for investigating stable and chaotic dynamics. The application of these findings to the classical three- and N-body problems provided a novel approach to studying the dynamical evolution of the celestial bodies and finding novel spaceways and orbital configurations for artificial satellites and spacecrafts. The exploitation of quasi-periodic orbits around the collinear Lagrangian points of the Earth-Sun system for solar and astronomy missions is a well known example of this kind.

In the last years the renewed interest of the major space agencies in the exploration of Solar System bodies, foreseeing also manned missions to the Moon and Mars, widened the potential benefits of the dynamical systems approach to spaceflight dynamics. Moreover, mission profiles are becoming more and more complex, often requiring a multi-disciplinary approach, where the contribution of operations, the impact of the space environment, and the possibility of in-situ resource exploitation play an increasingly important role.

Within this framework, the Space OPS Academy promoted by Telespazio (a Finmeccanica-Thales Company) to foster the development of scientific and professional skills on flight dynamics, ground system management, and Earth observation, organized in October 2007 the workshop "Novel Spaceways for Scientific and Exploration Missions – a dynamical systems approach to affordable and sustainable space applications." The aim was to work out a coherent picture of the possibilities offered by the stable/unstable manifold approach to space mission design, by analyzing the advantages and the drawbacks of using these novel trajectories when faced to the requirements imposed by the scientific payload, the operational aspects, and the industrial approach.

The three-day meeting (from 15 to 17 October 2007) was honored by the participation of more than 70 experts coming from widely different institutions and countries. The opening session was hosted at the Telespazio Fucino Space Center (Avezzano, Italy), one of the largest space communication facilities in the world, where an impressive collection of antennas (up to 32-m dishes) is displayed. The workshop moved then to the nearby "Scuola Superiore Guglielmo Reiss Romoli" (L'Aquila, Italy), where the other sessions and the final round-table discussion were held. This last event turned out to be particularly fruitful, and a list of recommendations for follow-up actions, as reported in a separate section after the preface, was compiled.

The need for a book somehow different from the usual "workshop proceedings" was also pointed out, an indication that resulted in what is now presented to you. The reason is that the workshop gathered together different communities, both scientific and technological, thus requiring an additional effort for clarity in the contributions. Each topic should be discussed in a proper context, articles written as far as possible self-consistent, and the use of introductory sections and extended explanations encouraged. The need for a reference publication highlighting the common dynamical ground underlying the "novel spaceways" was also felt, as witnessed by the different names under which they are referred to in the literature (e.g., Lagrangian trajectories, stable/unstable manifolds, weak stability boundary, etc.). We owe to Alessandra Celletti the term "Space Manifold Dynamics," which found immediate consensus among all participants. Our hope is of having succeeded in spreading the good news that there is an alternative way of thinking to spacecraft trajectories if one dares to abandon the "patched conics paradigm" when entering the realm of the three- and N-body problems.

Many people and institutions deserve acknowledgments. The Telespazio CEO Giuseppe Veredice and the top management of the company, with a special reference to Giorgio Dettori, HR Director of the company, provided the will, the contacts, and the funding scenario needed for making it all come true. The professional skills and the enthusiasm of the director of the SpaceOPS Academy, Francesco Perillo, have been a steady source of motivation for us all. We are in debt to Silvano Casini, former administrator of the Italian Space Agency and presently CEO of DdeB, for the original idea of a workshop devoted to the subject of space manifold dynamics. We would like to thank Roberto Battiston, Alessandra Celletti, Guido Di Cocco, Glauco Di Genova, Reno Mandolesi, Walter Pecorella, Piergiorgio Picozza, and Giovanni Valsecchi for trustfully joining the organizing committee and Livio Mastroddi for hosting us at the Fucino Space Center. To Gianna Fattore, Alessandra Gaetani, and Orietta Gagliano goes our sincere appreciation for taking care of logistics, administration, and positive thinking. Viviana Panaccia and Paolo Mazzetti have properly managed communication aspects. Maria Luisa Porciatti, Biagio Calicchio, and Alfredo Calzolaio ensured safe and timely car/bus/train/plane trips. A special thank to all participants and in particular to Letizia Stefanelli for her support in "contingency" situations.

Last but not least, we acknowledge the support of ASI (Agenzia Spaziale Italiana), SIMCA (Società Italiana di Meccanica Celeste e Astrodinamica), Regione Abruzzo, Provincia dell'Aquila, the BCC (Banca di Credito Cooperativo) Roma, of the Scuola Superiore Guglielmo Reiss Romoli and its highly professional personnel. The workshop "Novel Spaceways for Scientific and Exploration Missions" has greatly profited from the experience coming from the CELMEC meetings on celestial mechanics and from the Moon Base International Conference.

Rome, Italy — Ettore Perozzi
Sao Paulo, Brazil — Sylvio Ferraz-Mello

NOTE ADDED IN PROOF - While the book was in print the Space Academy has grown into a non profit foundation jointly promoted by Telespazio, Thales Alenia Space Italia and The University of L'Aquila. The aim is to bridge space-related issues between the scientific and the industrial communities and to promote high level training in space culture in a wide context.

# Recommendations

During the final panel discussion, the participants to the workshop "Novel Spaceways for Scientific and Exploration Missions" have produced a commonly agreed list of strategic considerations and of specific actions to be undertaken for fully exploiting the potential benefits for the scientific, exploration and technological programs of the major space agencies.

Introductory statements:

> **There is the need of establishing a *strong and continuous link* among the research, the industrial communities and the space agencies, even at a basic level (e.g. regular organization of workshops and schools).**
>
> **The terminology "*Space Manifold Dynamics*" (SMD) is adopted for referring to the dynamical systems approach to spaceflight dynamics, thus encompassing more specific definitions (stable/unstable manifolds, Lagrange trajectories, weak stability boundary, etc.);**

SMD topics deserving immediate attention:

- ***filling the gap* between industry and research on specific issues used to investigate space manifold dynamics (e.g. apply new methods, translating mission requirements into theory);**
- ***focus on the effect of dissipative systems* on SMD in terms of outcomes, methods and applications (e.g. low-thrust engines, non-gravitational forces, tethered systems);**
- ***explore the "geography" of the solar system* using novel dynamical approaches, such as mapping techniques and lagrangian trajectories;**
- ***build the "cartography"* associated to the utilization of the near-Earth space (e.g. periodic orbits, lunar transfers);**
- ***define the concept of "mixed approach"* to space mission design, i.e. dividing a complex trajectory into sections for which a given methods applies better (e.g. traditional ballistic + gravity assist for transfer trajectories, SMD for orbit insertion and satellite tour design);**
- ***develop a simulation environment* to investigate the new challenges posed by manned and unmanned exploration to numerical optimization methods (e.g. flight time vs. delta-V, electric vs chemical propulsion or other alternatives);**

- ○ ***study the impact of SMD trajectories*** **on human spaceflight and in particular on the radiation issue (e.g. finding "sheltered" mission profiles);**
- ○ ***evaluate the consequences*** **of adopting a SMD approach on both, ground support and spacecraft subsystems requirements (e.g. TLC, GNC, propulsion);**

Specific actions to be considered:

***Interplanetary trajectory orbit determination*** **needs a new start: novel methods are required for developing new generation high-precision orbit determination operational software**.

**The Workshop highlights the relevance of the generation of** ***high-quality planetary ephemeris*** **as well of other celestial and artificial bodies for both, astronomical research and space applications.**

*L'Aquila, 17 October 2007*

# Contents

# Contributors

**Mariano Andrenucci** Alta S.p.A., Pisa, Italy m.andrenucci@alta-space.com

**Edward Belbruno** Department of Astrophysical Sciences, Princeton University, Princeton, New Jersey 08544, USA belbruno@princeton.edu

**Miguel Belló** DEIMOS Space SL, Ronda de Poniente 19, Edificio Fiteni VI, 28760 Tres Cantos, Madrid, Spain miguel.bello@deimos-space.com

**Carlo Burigana** INAF-IASF Bologna, Via Gobetti 101, I-40129 Bologna, Italy burigana@iasfbo.inaf.it

**Cosmo Casaregola** Alta S.p.A., Pisa, Italy c.casaregola@alta-space.com

**Silvano Casini** DdeB – Domaine de Beauregard Sarl, Houston, TX, USA silvano.casini@gmail.com

**Alessandra Celletti** Dipartimento di Matematica, Università di Roma Tor Vergata, Via della Ricerca Scientifica 1, I-00133 Roma, Italy celletti@mat.uniroma2.it

**Christian Circi** Scuola di Ingegneria Aerospaziale, Università di Roma "La Sapienza", Italy christian.circi@uniroma1.it

**Bruce A. Conway** Department of Aerospace Engineering, University of Illinois, 104 South Wright Street, Urbana, 61801 IL, USA bconway@uiuc.edu

**Alessio Di Salvo** Rheinmetall Italia, Via Affile 102, 00131, Roma, Italy a.disalvo@rheinmetall.com

**Koen Geurts** Alta S.p.A., Pisa, Italy k.geurts@alta-space.com

**Jesús Gil-Fernández** GMV, Tres Cantos, Madrid 28760, Spain jesusgil@gmv.com

**Gerard Gómez** Departament de Matem`atica Aplicada i An`alis. Universitat de Barcelona, Gran Via 585, 08007 Barcelona, Spain gerard@maia.ub.es

**Raúl Cadenas Gorgojo** GMV, Tres Cantos, Madrid 28760, Spain rcadenas@gmv.com

**Mariella Graziano** GMV, Tres Cantos, Madrid 28760, Spain
mgraziano@gmv.com

**Massimiliano Guzzo** Dipartimento di Matematica Pura ed Applicata, Universit`a degli Studi di Padova, Via Trieste, 63 - 35121 Padova
guzzo@math.unipd.it

**Pablo Ibánez** ETSI Aeronáuticos, Technical University of Madrid (UPM), Pz Cardenal Cisneros 3, 28040 Madrid, Spain p.ibaneztarrago@gmail.com

**Nazzareno Mandolesi** INAF-IASF Bologna, Via Gobetti 101, I-40129 Bologna, Italy mandolesi@iasfbo.inaf.it

**Riccardo Marson** Telespazio, Via Tiburtina 965, 00156 Roma, Italy
guest468.marson@telespazio.com

**Christopher Martin** Department of Aerospace Engineering, University of Illinois, 104 South Wright Street, Urbana, 61801 IL, USA cmartin6@uiuc.edu

**Josep J. Masdemont** Departament de Matemàtica Aplicada I. ETSEIB, Universitat Politècnica de, Catalunya. Diagonal 647, 08028 Barcelona, Spain
josep@barquins.upc.edu

**Filippo Ongaro** ISMERIAN - Istituto di Medicina Rigenerativa e Anti-Aging
filippo.ongaro@ismerian.com

**Pierpaolo Pergola** Alta S.p.A., Pisa, Italy p.pergola@alta-space.com

**Ettore Perozzi** Telespazio, Via Tiburtina 965, 00156 Roma, Italy
ettore.perozzi@telespazio.com

**Vladimir M. Petrov** IMBP, Institute for Biomedical Research of the Russian Academy of Science, Moscow, Russia petrov@imbp.ru

**Franco Rossitto** Moon Base Working Group – former head of ESA Astronaut Center, Moscow, Russia franco.rossitto@gmail.com

**Paolo Teofilatto** Scuola di Ingegneria Aerospaziale, Università di Roma “La Sapienza”, Italy paolo.teofilatto@uniroma1.it

**Luca Valenziano** INAF-IASF Bologna, Via Gobetti 101, I-40129 Bologna, Italy valenziano@iasfbo.inaf.it

**Carlos Corral Van Damme** GMV, Tres Cantos, Madrid 28760, Spain
cfcorral@gmv.com

# About the Editors

Ettore Perozzi is at Telespazio, a Finmeccanica/Thales Company, in the Scientific Missions Department. He has long-standing experience in celestial mechanics and space mission analysis after spending many years in research institutions, space agencies, and industry.

Sylvio Ferraz-Mello has been Director of the National Observatory, Rio de Janeiro, received the doctor "honoris causa" degree from the Observatory of Paris, and is at present editor-in-chief of the international journal *Celestial Mechanics and Dynamical Astronomy*

# Invariant Manifolds, Lagrangian Trajectories and Space Mission Design

**Miguel Belló, Gerard Gómez, and Josep J. Masdemont**

## 1 Introduction

The last 30 years have produced an explosion in the capabilities of designing and managing libration point missions. The starting point was the ground-breaking mission of the third International Sun-Earth Explorer spacecraft (ISEE–3). The ISEE-3 was launched August 12, 1978 to pursue studies of the Earth–Sun interactions, in a first step of what now is known as Space Weather. After a direct transfer of the ISEE-3 to the vicinity of the Sun-Earth $L_1$ Lagrange point, it was inserted into a nearly-periodic halo orbit, in order to monitor the solar wind about 1 h before it reached the Earth's magneto-sphere as well as the ISEE–1 and 2 spacecraft (which where in an elliptical orbit around the Earth). After completing some revolutions around the halo orbit, the spacecraft visited the vicinity of the $L_2$ libration point to explore the magneto-tail of the Earth. Finally, and after making use of a double lunar swing-by the spacecraft was renamed as the International Cometary Explorer (ICE) and had a close encounter with the comet Giacobini-Zinner on September 11, 1985. The spacecraft later flew between the Sun and comet Haley, and now is in a 355 day heliocentric orbit with $a = 0.98$ AU and $e = 0.051$. Since "old spacecraft" never die, ISEE-3 will return to the Earth's vicinity on August 10, 2014, 36 years after its launch (Fig. 1).

Interest in the scientific advantages of the Lagrange libration points for space missions has continued to increase and to inspire even more challenging objectives that are reflected, in part, in some of the current missions, such as SOHO, MAP and Genesis, and of the most challenging future ones, such as Darwin and TPF. Also, increasing understanding of the available mission options has emerged due to the theoretical, analytical, and numerical advances in many aspects of libration point mission design.

In this paper we will try to show how advanced analytical and numerical techniques can be used in the mission design of spacecraft moving in the vicinity of libration points.

---

M. Belló (✉)
DEIMOS Space SL, Ronda de Poniente 19, Edificio Fiteni VI, 28760, Tres Cantos, Madrid, Spain
e-mail: miguel.bello@deimos-space.com

E. Perozzi, S. Ferraz-Mello (eds.), *Space Manifold Dynamics*,
DOI 10.1007/978-1-4419-0348-8_1, 

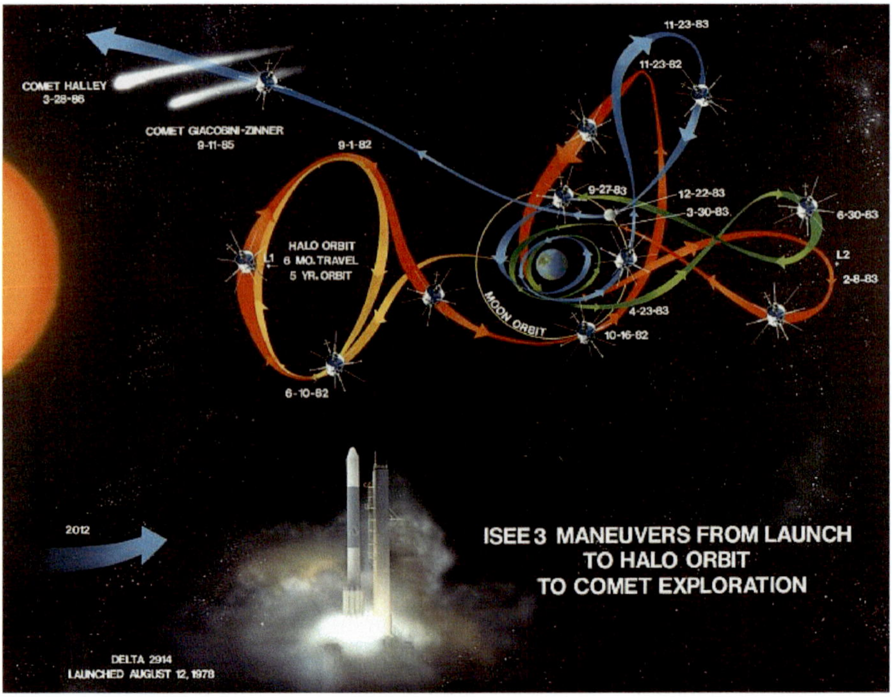


**Fig. 1** The ISEE–3 extended mission trajectory. (http://heasarc.gsfc. nasa.gov)

## *1.1 The CRTBP and the Libration Points*

The Lagrange libration points are the equilibrium solutions of the Circular Restricted Three Body Problem (CRTBP), which describes the motion of a particle, of very small mass, under the gravitational attraction of two massive bodies (usually called primaries, or primary and secondary). It is assumed that the particles are in a circular (Keplerian) motion around their centre of masses. For space missions, the particle is the spacecraft and the two primaries can be taken, for example, as the Sun and the Earth-Moon barycentre, or the Earth and the Moon. In all the cases, both the small mass and the primaries will be considered point masses and not rigid bodies.

Since Euler and Lagrange, some relevant solutions of both the General Three Body Problem and the CRTBP are known. For one of these solutions, the three bodies are in the edges of an equilateral triangle, with the centre of masses at the origin, that can rotate with an angular velocity that depends on the masses of the bodies and the size of the side of the triangle. Aside from these triangular configurations, the bodies can also rotate aligned, if the ratio of the relative distances of one body to the other two verifies some algebraic quintic equation. For the CRTBP, suitable rotating coordinates can be introduced to keep both primaries fixed. In this reference system, the solutions found by Euler and Lagrange become equilibrium solutions. They are the so-called Lagrange libration points.

Three of the five libration points lie on the line joining both primaries: one, that is usually denoted by $L_1$, is between the primaries, and the other two at both sides of the two primaries, the one closest to the smaller primary is called $L_2$ and the third one $L_3$. The two remaining equilibrium points, $L_4$ and $L_5$, are in the plane of motion of the primaries and they form an equilateral triangle with the two primaries (see Fig. 2).

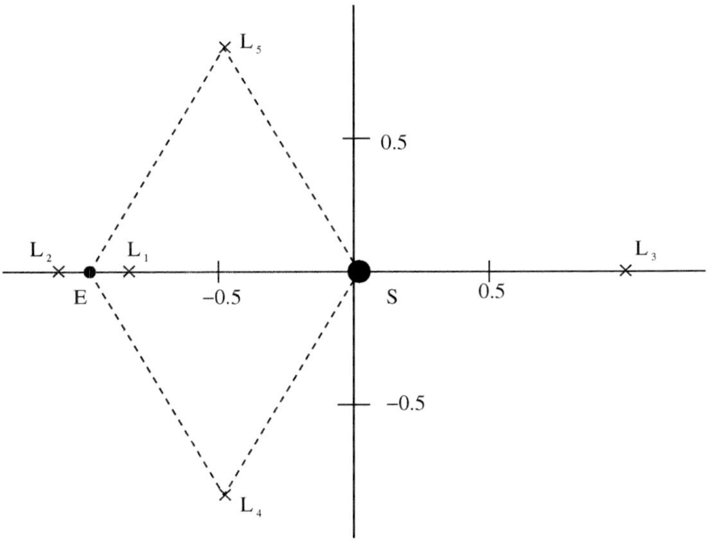


**Fig. 2** The Lagrange libration points in the usual CRTBP synodic reference system and units

## *1.2 Properties of the Libration Point Orbits*

The Lagrange points offer many new orbits and applications. Around the triangular equilibrium points, $L_4$ and $L_5$, there are large regions with good stability properties that could be used as parking regions at which no station keeping is needed. The collinear points, $L_1$, $L_2$ and $L_3$, generate and control many trajectories with interesting applications to space missions and planetary science due to several reasons:

- They are easy and inexpensive to reach from Earth.
- They provide good observation sites of the Sun.
- For missions with heat sensitive instruments, orbits around the $L_2$ point of the Sun Earth system provide a constant geometry for observation with half of the

entire celestial sphere available at all times, since the Sun, Earth and Moon are always behind the spacecraft.

- The communications system design is simple and cheap, since the libration orbits around the $L_1$ and $L_2$ points of the Sun–Earth system always remain close to the Earth, at a distance of roughly 1.5 million km with a near-constant communications geometry.
- The $L_2$ environment of the Sun–Earth system is highly favourable for non-cryogenic missions requiring great thermal stability, suitable for highly precise visible light telescopes.
- The libration orbits around the $L_2$ point of the Earth–Moon system, can be used to establish a permanent communications link between the Earth and the hidden part of the Moon, as was suggested by A.C. Clark in 1950 and proposed by R. Farquhar in 1966.
- The libration point orbits can provide ballistic planetary captures, such as for the one used by the Hiten mission.
- The libration point orbits provide Earth transfer and return trajectories, such as the one used for the Genesis mission.
- The libration point orbits provide interplanetary transport which can be exploited in the Jovian and Saturn systems to design a low energy cost mission to tour several of their moons (Petit Grand Tour mission).
- Recent work has shown that even formation flight with a rigid shape is possible using libration point orbits.

The fundamental breakthrough that has given the theoretical and numerical framework for most of the mission concepts of the list above is the use of Dynamical Systems tools. Classical methods can be used only for ordinary halo orbit missions, but all the new concepts require the more powerful Dynamical Systems methods, in order to get qualitative and quantitative insight into the problem.

Dynamical Systems Theory, founded by Poincaré by the end of the nineteenth century, has used the CRTBP as one of the paradigmatic models for its application. Following Poincaré's idea, that it is better to study the full set of orbits rather than individual ones, the Dynamical Systems approach looks at these models from a global point of view. Its procedures are both qualitative and quantitative and have as their final goal to get a picture, as accurate as possible, of the evolution of all the states of the system. This full set of states constitutes the phase space. So, Dynamical Systems tries to get the dynamic picture of the phase space of a given model.

Although the application of Dynamical Systems Theory to space mission design is very recent, it has already been used in various missions, starting with SOHO and followed by Genesis, MAP and Triana. In the case of Genesis this approach not only provided a $\Delta v$ saving of almost 100 m/s but also a systematic and fast way to perform the mission analysis, with which it was possible to easily redefine the nominal trajectory when the launch date was delayed.

Another relevant example of these advantages are the transfers to libration point orbits, for which previous efforts in its design relied on a manual trial-and-error search followed by and optimisation procedure. The process can now be addressed in a more meaningful and insightful manner by introducing the concept, as well as the explicit calculations, of invariant manifolds as a means to describe the phase space. The result is not only the efficient determination of the desired transfers, but also the emergence of other trajectory and mission options. By understanding the geometry of the phase space and the solution arcs that populate it, the mission designer is free to creatively explore concepts and ideas that previously may have been considered intractable, or even better, had not yet been envisioned. This has been evidenced recently as studies, ranging from flying formations of spacecraft near libration points to sending humans further into space, have been initiated and in fact show great promise.

Beyond baseline trajectory design, of course, other analyses required for any mission can also benefit from studies of motion in this regime, for example, station-keeping strategies for various mission scenarios. The techniques, developed using a variety of approaches, have helped establishing many options that provide robust control scenarios for many or all of the current mission scenarios. Some station-keeping methods have also been shown to be applicable for a more general class of trajectories, i.e., not just libration point trajectories. The availability of these methods has played an important part in establishing more confidence in mission designers and managers alike regarding potential real world problems that may arise and the ability to effectively handle them.

The final goal of this paper is to present a study on the implementation of the tools derived from the Dynamical Systems Theory, taking into account the performance of nowadays computers. It has been framed in the context of present and future missions, as well as in the current state of supporting mathematical tools.

## 2 Libration Point Orbits and Their Synthetic Representation

### 2.1 Libration Point Orbits

The reference model that will be used is the CRTBP. As it is well known[61], this problem studies the behaviour of a particle with infinitesimal mass moving under the gravitational attraction of two primaries revolving around their centre of masses in circular orbits.

Using a suitable reference system and a dimensionless set of units, the equations of motion can be written as

$$\begin{aligned}\ddot{x} - 2\dot{y} &= x - \frac{(1-\mu)}{r_1^3}(x-\mu) - \frac{\mu}{r_2^3}(x+1-\mu),\\ \ddot{y} + 2\dot{x} &= y - \frac{(1-\mu)}{r_1^3}y - \frac{\mu}{r_2^3}y,\\ \ddot{z} &= -\frac{(1-\mu)}{r_1^3}z - \frac{\mu}{r_2^3}z,\end{aligned}$$

where $r_1 = [(x-\mu)^2 + y^2 + z^2]^{\frac{1}{2}}$ and $r_2 = [(x+1-\mu)^2 + y^2 + z^2]^{\frac{1}{2}}$ are the distances from the infinitesimal mass particle to the two primaries.

By introducing momenta as $p_x = \dot{x} - y$, $p_y = \dot{y} + x$ and $p_z = \dot{z}$, the equations of the CRTBP can be written in Hamiltonian form with Hamiltonian function

$$H = \frac{1}{2}\left(p_x^2 + p_y^2 + p_z^2\right) - xp_y + yp_x - \frac{1-\mu}{r_1} - \frac{\mu}{r_2}.$$

The Hamiltonian is related to the well known Jacobi first integral through

$$C = -2H + \mu(1-\mu).$$

As it has already been said, the above CRTBP equations have five equilibrium points, the so called libration points. If $x_{L_i}$ $(i = 1,2,3)$ denotes the abscissa of the three collinear points, we will assume that

$$x_{L_2} < \mu - 1 < x_{L_1} < \mu < x_{L_3},$$

and we will focus our attention in $L_1$ and $L_2$.

Using a linear symplectic change of coordinates, it is easy to cast the second order part of the Hamiltonian into its real normal form,

$$H_2 = \lambda x p_x + \frac{\omega_p}{2}(y^2 + p_y^2) + \frac{\omega_v}{2}(z^2 + p_z^2), \tag{1}$$

where, for simplicity, we have kept the same notation for the variables. Here, $\lambda$, $\omega_p$ and $\omega_v$ are positive real numbers given by

$$\lambda^2 = \frac{c_2 - 2 + \sqrt{9c_2^2 - 8c_2}}{2}, \qquad \omega_p^2 = \frac{2 - c_2 + \sqrt{9c_2^2 - 8c_2}}{2}, \qquad \omega_v^2 = c_2.$$

In the above equations $c_2$ is a constant given by

$$c_2 = \frac{1}{\gamma_j^3}\left(\mu - \frac{(1-\mu)\gamma_j^3}{(1 \mp \gamma_j)^3}\right), \quad \text{for } L_j, j = 1,2,$$

and $\gamma_j$ is the distance from the libration point $L_j$ to the closest primary.

From (1) it is clear that the linear behaviour near the collinear equilibrium points is of the type

$$\textbf{saddle} \times \textbf{centre} \times \textbf{centre},$$

so, the motion in the vicinity of the collinear equilibrium points can be seen as the composition of two oscillators and some "hyperbolic" behaviour. This means that the oscillations are not stable and that very small deviations will be amplified as time increases. One of the oscillations takes place in the plane of motion of the primaries and the other orthogonal to this plane. These two periodic motions are known as the planar and vertical Lyapunov periodic orbits. The frequencies of the oscillations vary with the amplitudes (since the problem is not linear), and for a suitable amplitude both frequencies become equal. At this point the well known halo–type periodic orbits appear. When the frequencies of the two oscillations (vertical and planar) are not commensurable, the motion is not periodic and it remembers a Lissajous orbit. Then we say that we have a quasi-periodic orbit. This kind of motion can be found both around the vertical periodic orbit and around the halo orbits. Of course, "between" Lissajous orbits, and associated to other conmensurabilities between the two natural frequencies, there are periodic orbits. Some of these resonances have been identified and computed in [31] for the Sun–Earth and Earth–Moon mass ratios. Some of these different kinds of orbits have been represented in Fig. 3.

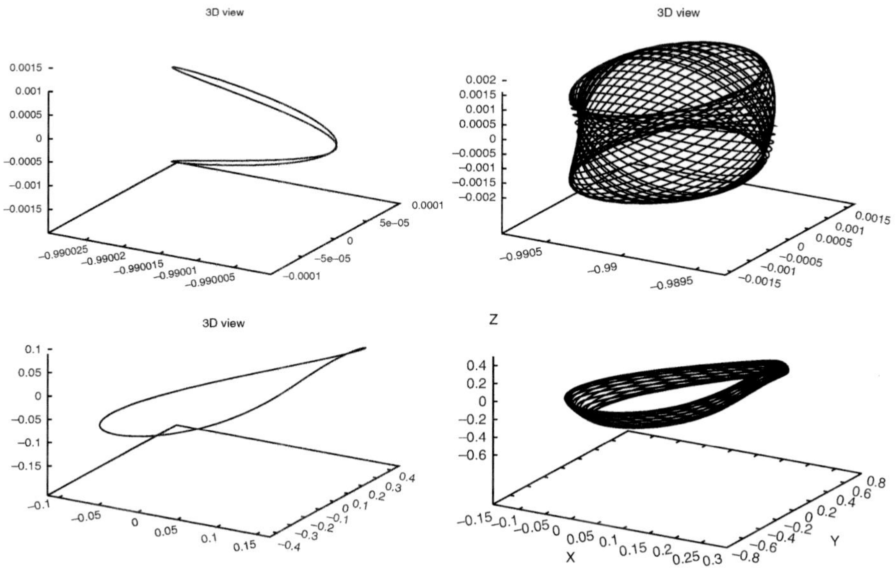


**Fig. 3** Several types of orbits around $L_1$. *Upper left*: vertical Lyapunov periodic orbit. *Upper right*: Quasi-periodic orbit around a vertical periodic orbit (Lissajous orbit). *Lower left*: halo periodic orbit. *Lower right*: quasi-halo orbit (quasi-periodic orbit around a halo orbit)

## 2.2 Synthetic Representations

### 2.2.1 Poincaré Map Representation

A more synthetic way of displaying all this zoo of orbits consists in representing only their intersection with the $z = 0$ plane. This is what is usually called a Poincaré map representation. A planar orbit will appear as a closed curve on the plane and

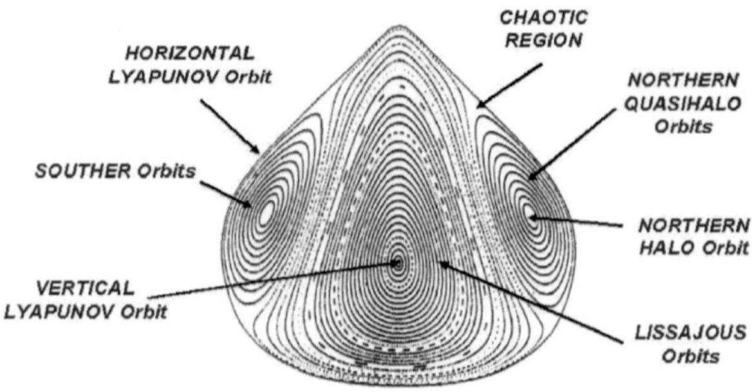


**Fig. 4** Poincaré map representation of the orbits near the libration point $L_1$ for the value of the Jacobi constant 3.00078515837634. The CRTBP mass parameter corresponds to the Earth+Moon–Sun system

a quasi-periodic orbit as a set of points lying, more or less, on a smooth closed curve. Figure 4 shows one of these representations. Near the centre of the figure one can see a fixed point. It corresponds to a vertical periodic orbit that crosses the $z = 0$ plane just at this point. It (and so, the corresponding orbit) is surrounded by quasi-periodic motions that take place on invariant tori. The external curve of the figure is the planar Lyapunov orbit (corresponding to a given value of the Jacobi constant). The two other fixed points are associated to the two halo orbits, which are symmetrical to one another with respect to $z = 0$. They are, in turn, surrounded by invariant 2D tori. Between the 2D tori around the vertical orbit and the ones around the halo orbit there is the trace of the stable and unstable manifolds of the planar Lyapunov orbit, which acts as a separatrix between two different kinds of motion: the ones around the vertical periodic orbits and the ones around the halo orbits.

Due to the unstable behaviour of the collinear libration points, this Poincaré map representation cannot be obtained by direct numerical integration of the CRTBP equations of motion. Figure 4 was obtained after performing a normal form reduction of the Hamiltonian of the CRTBP and removing from the reduced Hamiltonian its unstable terms. Of course, this 2-dimensional figure corresponds to a fixed energy level (fixed value of the Jacobi constant or, equivalently, of the Hamiltonian). To get a global representation of the 4-dimensional central manifold around the libration point (recall that for the representation we have fixed the value of the $z$-coordinate equal to zero, which reduces in one unit the total dimension) we need to vary the value of the energy and for each energy level get the associated Poincaré map representation of the flow. This 3-dimensional picture can also be obtained by a more direct method. For this purpose we must compute the periodic orbits and invariant 2D tori of the centre manifolds of the libration points, using either Lindstedt-Poincaré or purely numerical procedures, and once they have been determined, represent their intersections with the $z = 0$ section for each energy level. To avoid the convergence problems of the Lindstedt-Poincaré method, we have selected numerical procedures.

### 2.2.2 Energy vs Rotation Number Representation

Consider a 2-dimensional invariant torus of the CRTBP with frequencies $\omega = (\omega_1,\omega_2)\mathbb{R}^2$, and let $T_i$ be the period corresponding to the $\omega_i$ frequency, that is

$T_i = 2\pi/\omega_i$, and $\theta = (\xi,\eta)$. We consider a curve in the torus invariant under the flow of the CRTBP after $T_2$ time units: $\phi_{T_2}$.

If $\varphi : \mathbb{T}^1 \to \mathbb{R}^n$ is the parametrisation of this curve, then we require

$$\varphi(\xi + \omega_1 T_2) = \phi_{T_2}(\varphi(\xi)), \quad \text{for all} \quad \xi \in \mathbb{T}^1, \tag{2}$$

where $\rho = \omega_1 T_2 = 2\pi\omega_1/\omega_2$ is the so called *rotation number* of the curve we are looking for. Note that the rotation number $\rho$ uniquely identifies a torus at a given energy level.

The dynamical interpretation of the rotation number is clear: it represents the average variation of the angle $\xi$ when the angle $\eta$ has done one revolution, this is, has increased $2\pi$ units (see Fig. 5).

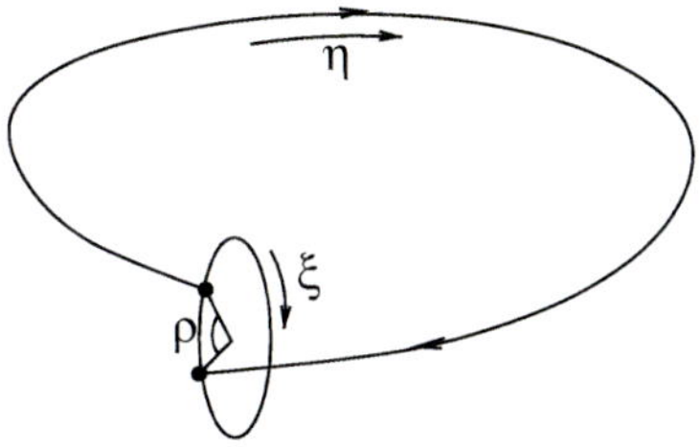


**Fig. 5** Qualitative representation of the rotation number $\rho$

In Fig. 6 we have displayed, for the $L_1$ point the region, in the energy-rotation number plane, covered by the two-parametric family of tori computed starting from the vertical Lyapunov family of periodic orbits.

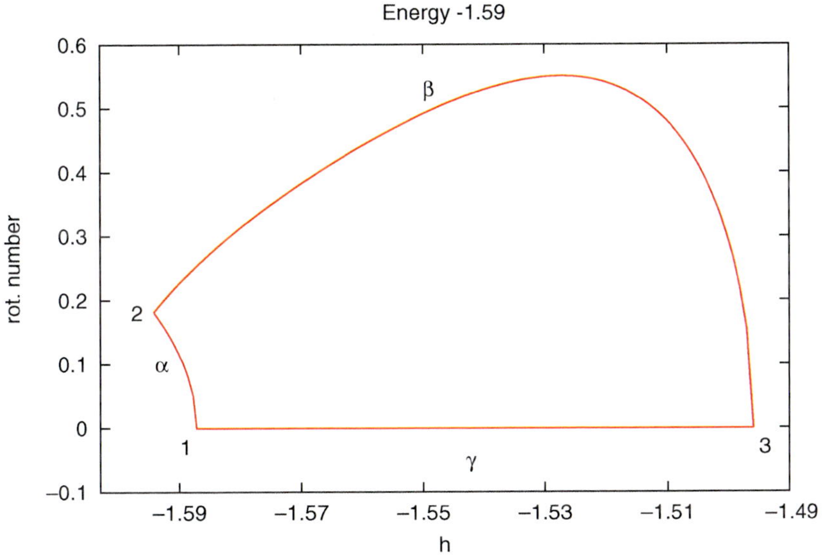


**Fig. 6** Synthetic representation of the 2-parametric family of Lissajous orbits around the $L_1$ point, using as parameters the value of the Hamiltonian $h$ and the rotation number $\rho$

In the $(h, \rho)$ plane, the region of existence of Lissajous orbits shown in Fig. 6 is bounded by three different curves:

- The lower-left curve $\alpha$ (from vertex 1 to 2) is related to the planar Lyapunov family of periodic orbits. The orbits of this family represented in the curve are only those with central part, which are the only ones surrounded by tori. They are the "first" orbits of the family generated from the libration point.
- The upper piece $\beta$ (from vertex 2 to 3) is strictly related to the vertical Lyapunov family of periodic orbits.
- The bottom boundary $\gamma$ (from vertex 3 to 1) that corresponds to $\rho = 0$, begins at the value of the energy where the halo families are born. It is related to a separatrix between the tori around the vertical Lyapunov families and the halo ones.

Next, we can see how this synthetic representation of the 2-parameter family of Lissajous orbits is related to their Poincaré map representation at different energy levels. This is shown in Fig. 7. If in the $(h,\rho)$ diagram we fix a value of the energy, we get a vertical segment connecting the $\beta$ curve with the $\alpha$ or $\gamma$ curve (depending on the value of $h$). For values of $h$ lower than the one associated to the bifurcation of the halo families of periodic orbits (for instance, $h = -1.59$), the vertical segment connects $\beta$ and $\alpha$, which are associated, respectively, to the vertical and planar Lyapunov families of periodic orbits. This means that there is a 1-parameter family of Lissajous orbits (tori) connecting these two periodic orbits. The intersection of

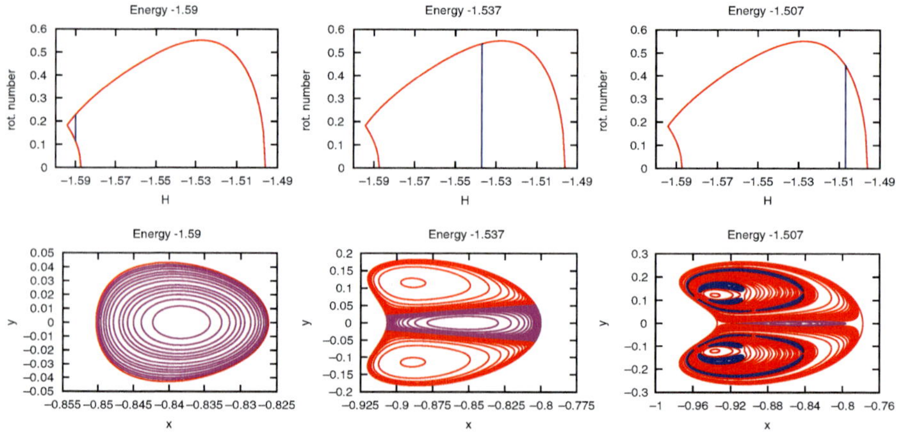


**Fig. 7** Two synthetic representations of the 2-parametric family of Lissajous orbits around the $L_1$ point. In each $(h,\rho)$ representation the vertical (*blue*) line represents the family of Lissajous orbits around the vertical periodic orbit with a fixed value of the energy. In the Poincaré map representation, which is displayed below, these Lissajous orbits are represented by their intersection with $z = 0$ and are the (*pink*) closed curves in the centre of each plot, around the central fixed point associated to the vertical periodic orbits

these Lissajous orbits with the $z = 0$ plane is displayed in the Poincaré map representation as a set of "concentric circles" with centre at the fixed point associated to the vertical periodic orbit and having as outer boundary the planar Lyapunov orbit. For any other value of $h$ the vertical segment in the $(h,\rho)$ goes from the $\beta$ curve, associated to the vertical Lyapunov family, to the $\gamma$ curve, for which $\rho = 0$. As it has been said, this last curve represents the separatrix between the tori around the vertical Lyapunov families and the halo ones, so now the 1-parameter family of Lissajous orbits will start also close to the vertical Lyapunov periodic orbit and will have a natural termination when it reaches the separatrix. The intersection of these Lissajous orbits with $z = 0$ are the closed curves around the central fixed point in the Poincaré map representation.

# 3 Computation of LPO's

## *3.1 Lindstedt–Poincaré Computation of Tori and Their Stable and Unstable Manifolds*

The planar and vertical Lyapunov periodic orbits as well as the Lissajous, halo and quasi-halo orbits can be computed using Lindstedt-Poincaré procedures and ad hoc algebraic manipulators. In this way one obtains their expansions, in CRTBP coordinates, suitable to be used in a friendly way. In this section we will give the main ideas used for their computation.

We will start with the computation of the Lissajous trajectories (2D tori) and halo orbits (1D tori or periodic orbits). The CRTBP equations of motion can be written as

$$
\begin{aligned}
\ddot{x} - 2\dot{y} - (1 + 2c_2)x &= \frac{\partial}{\partial x} \sum_{n\geq 3} c_n \rho^n P_n \left(\frac{x}{\rho}\right), \\
\ddot{y} + 2\dot{x} + (c_2 - 1)y &= \frac{\partial}{\partial y} \sum_{n\geq 3} c_n \rho^n P_n \left(\frac{x}{\rho}\right), \\
\ddot{z} + c_2 z &= \frac{\partial}{\partial z} \sum_{n\geq 3} c_n \rho^n P_n \left(\frac{x}{\rho}\right),
\end{aligned} \tag{3}
$$

where $c_n$ are constant depending on the equilibrium point and the mass ratio, $\rho^2 = x^2 + y^2 + z^2$ and $P_n$ is the Legendre polynomial of degree $n$. The bounded solution of the linear part of these equations is

$$
\begin{aligned}
x(t) &= \alpha \cos(\omega_p t + \phi_1), \\
y(t) &= \kappa\alpha \sin(\omega_p t + \phi_1), \\
z(t) &= \beta \cos(\omega_v t + \phi_2),
\end{aligned} \tag{4}
$$

where $\omega_p$ and $\omega_v$ are the planar and vertical frequencies and $\kappa$ is a constant. The parameters $\alpha$ and $\beta$ are the in-plane and out-of-plane amplitudes of the orbit and $\phi_1$,

$\phi_2$ are the phases. These linear solutions are already Lissajous trajectories. When we consider the nonlinear terms, we look for formal series solutions in powers of the amplitudes $\alpha$ and $\beta$ of the type

$$\begin{Bmatrix} x \\ y \\ z \end{Bmatrix} = \sum_{i,j=1}^{\infty} \left( \sum_{|k|\le i,|m|\le j} \begin{Bmatrix} x \\ y \\ z \end{Bmatrix}_{ijkm} \begin{Bmatrix} \cos \\ \sin \\ \cos \end{Bmatrix} (k\theta_1 + m\theta_2) \right) \alpha^i \beta^j, \qquad (5)$$

where $\theta_1 = \omega t + \phi_1$ and $\theta_2 = \nu t + \phi_1$. Due to the presence of nonlinear terms, the frequencies $\omega$ and $\nu$ cannot be kept equal to $\omega_p$ and $\omega_v$, and they must be expanded in powers of the amplitudes

$$\omega = \omega_p + \sum_{i,j=1}^{\infty} \omega_{ij} \alpha^i \beta^j, \quad \nu = \omega_v + \sum_{i,j=1}^{\infty} \nu_{ij} \alpha^i \beta^j.$$

The goal is to compute the coefficients $x_{ijkm}$, $y_{ijkm}$, $z_{ijkm}$, $\omega_{ij}$, and $\nu_{ij}$ recurrently up to a finite order $N = i + j$. Identifying the coefficients of the general solution (5) with the ones obtained from the solution of the linear part (4), we see that the non zero values are $x_{1010} = 1$, $y_{1010} = \kappa$, $z_{1010} = 1$, $\omega_{00} = \omega_p$ and $\nu_{00} = \omega_v$. Inserting the linear solution (4) in the equations of motion, we get a reminder for each equation, which is a series in $\alpha$ and $\beta$ beginning with terms of order $i + j = 2$. In order to get the coefficients of order two, this known order 2 terms must be equated to the unknown order 2 terms of the left hand side of the equations. The general step is similar. It assumes that the solution has been computed up to a certain order $n - 1$. Then it is substituted in the right hand side of the CRTBP equations, producing terms of order $n$ in $\alpha$ and $\beta$. This known order $n$ terms must be equated with the unknown terms of order $n$ of the left hand side.

The procedure can be implemented up to high orders. In this way we get, close to the equilibrium point, a big set of KAM tori. In fact, between these tori there are very narrow stochastic zones (because the resonances are dense). Hence we will have divergence everywhere. However, small divisors will show up only at high orders (except the one due to the 1:1 resonance), because at the origin $\omega_p/\omega_v$ is close to 29/28. The high order resonances have a very small stochastic zone and the effect is only seen after a big time interval.

Halo orbits are periodic orbits which bifurcate from the planar Lyapunov periodic orbits when the in plane and out of plane frequencies are equal. This is a 1:1 resonance that appears as a consequence of the nonlinear terms of the equations and, in contrast with the Lissajous orbits, they do not appear as a solution of the linearised equations. Of course, we have to look for these 1-D invariant tori as series expansion with a single frequency. In order to apply the Lindstedt-Poincaré procedure, following [56], we modify the equations of motion (3) by adding to the third equation a term like $\Delta \cdot z$, where $\Delta$ is a frequency type series

$$\Delta = \sum_{i,j=0}^{\infty} d_{ij} \alpha^i \beta^j,$$

that must verify the condition $\Delta$=0. We start looking for the (non trivial) librating solutions with frequency $\omega_p$

$$\begin{aligned} x(t) &= \alpha \cos(\omega_p t + \phi_1), \\ y(t) &= \kappa\alpha \sin(\omega_p t + \phi_1), \\ z(t) &= \beta \cos(\omega_p t + \phi_2). \end{aligned} \tag{6}$$

We note that after this step, halo orbits are determined up to order 1, and $\Delta$=0 is read as $d_{00} = 0$. Halo orbits depend only on one frequency or one amplitude since they are 1-D invariant tori, so we have not two independent amplitudes $\alpha$ and $\beta$. The relation between $\alpha$ and $\beta$ is contained in the condition $\Delta$=0 which implicitly defines $\alpha = \alpha(\beta)$.

When we consider the full equations, we look for formal expansions in powers of the amplitudes $\alpha$ and $\beta$ of the type

$$\begin{Bmatrix} x \\ y \\ z \end{Bmatrix} = \sum_{i,j=1}^{\infty} \left( \sum_{|k|\le i+j} \begin{Bmatrix} x \\ y \\ z \end{Bmatrix}_{ijk} \begin{Bmatrix} \cos \\ \sin \\ \cos \end{Bmatrix} (k\theta) \right) \alpha^i \beta^j,$$

where $\theta = \omega t + \phi$ and, as in the case of 2-D invariant tori, the frequency $\omega$ must be expanded as $\omega = \sum_{i,j=0}^{\infty} \omega_{ij}\alpha^i\beta^j$. The procedure for the computation of the unknown coefficients $x_{ijk}$, $y_{ijk}$, $z_{ijk}$, $\omega_{ij}$ and $d_{ij}$ is close to the one described for the Lissajous trajectories.

Quasi-halo orbits are quasi-periodic orbits (depending on two basic frequencies) on two dimensional tori around a halo orbit. Given a halo orbit of frequency $\omega$, the series expansions for the coordinates of the quasi-halo orbits around it will be of the form

$$\begin{Bmatrix} x \\ y \\ z \end{Bmatrix} = \sum_{i=1}^{\infty} \left( \sum_{|k|<i,|m|<i} \begin{Bmatrix} x \\ y \\ z \end{Bmatrix}_i^{km} \begin{Bmatrix} \cos \\ \sin \\ \cos \end{Bmatrix} (k(\omega t + \phi_1) + m(\nu t + \phi_2)) \right) \gamma^i.$$

These expansions depend on two frequencies ($\omega$, $\nu$) and one amplitude, $\gamma$ (related to the size of the torus around the halo orbit). The frequency $\nu$ is the second natural frequency of the torus, and it is close to the normal frequency around the base halo orbit. The amplitude, $\gamma$, is related to the size of the torus around the "base" halo orbit which is taken as backbone.

In order to apply the Lindstedt-Poincaré method to compute the quasi-halo orbits, it is convenient to perform a change of variables transforming the halo orbit to an equilibrium point of the equations of motion. Then, orbits librating around the equilibrium point in the new coordinates correspond to orbits librating around the halo orbit in the original ones. The details of the procedure for their computation can be found in [34].

All the different kinds of orbits displayed in Fig. 3 have been computed using the Lindstedt-Poincaré procedure according to the previous explanations.

## 3.2 Numerical Computation of Tori and Their Stable and Unstable Manifolds

Numerical methods have been widely used in the past to compute fixed points and periodic orbits. The computation of invariant tori is not so extended. The procedure that has been used is based on the computation of the Fourier series of an invariant curve on the torus [31]. This strategy is combined with a multiple shooting procedure in order to deal with the unstable behaviour of the flow.

Consider a 2-dimensional invariant torus of the CRTBP with frequencies $\omega = (\omega_1,\omega_2) \in \mathbb{R}^2$. If $\phi_t$ denotes the flow associated to the CRTBP, a parametrisation $\Psi : \mathbb{T}^2 = [0,2\pi]^2 \to \mathbb{R}^6$ of the torus satisfies the following invariant relation:

$$\Psi\begin{pmatrix} \xi + \omega_1 t \\ \eta + \omega_2 t \end{pmatrix} = \phi_t\left(\Psi\begin{pmatrix} \xi \\ \eta \end{pmatrix}\right), \qquad \text{for all} \quad \theta = (\xi,\eta) \in \mathbb{T}^2, \quad t \in \mathbb{R}. \tag{7}$$

Let us denote by $T_i$ the period corresponding to the $\omega_i$ frequency, that is $T_i = 2\pi/\omega_i$, and $\theta = (\xi,\eta)$. In order to reduce dimensions, instead of considering a parametrisation of the whole torus, we can consider the parametrisation of a curve in the torus which is invariant under $\phi_{T_2}$. Such a curve is given by $\{\eta = \eta_0\}$, since from equation (7) we have

$$\Psi\begin{pmatrix} \xi + \omega_1 T_2 \\ \eta_0 \end{pmatrix} = \phi_{T_2}\left(\Psi\begin{pmatrix} \xi \\ \eta_0 \end{pmatrix}\right), \qquad \text{for all} \quad \xi \in \mathbb{T}^1. \tag{8}$$

As we have already said, if $\varphi : \mathbb{T}^1 \to \mathbb{R}^n$ is the parametrisation of this curve, then we require

$$\varphi(\xi + \omega_1 T_2) = \phi_{T_2}(\varphi(\xi)), \qquad \text{for all} \quad \xi \in \mathbb{T}^1, \tag{9}$$

where $\rho = \omega_1 T_2 = 2\pi\, \omega_1/\omega_2$ is called the rotation number of the curve we are looking for.

It is convenient to assume for $\varphi$ a truncated Fourier series representation

$$\varphi(\xi) = A_0 + \sum_{k=1}^{N_f} (A_k \cos(k\xi) + B_k \sin(k\xi)), \tag{10}$$

in which the unknowns $A_k, B_k \in \mathbb{R}^n$ depend on the value of the energy $h$ (given by the value of the Hamiltonian) and the rotation number $\rho$. Then, by taking $1 + 2N_f$ values of $\xi$, equation (9) can be turned into a finite–dimensional non–linear system of equations which can be solved for $A_k, B_k$. Concretely, we solve

$$\varphi(\xi_i + \rho) = \phi_{T_2}(\varphi(\xi_i)), \tag{11}$$

where $\xi_i = 2\pi/(1 + 2N_f)$ for $i = 0, \ldots, 2N_f$.

Once we have a parametrisation of an invariant curve, a parametrisation of the whole torus can be recovered by numerical integration, as

$$\Psi(\xi,\eta) = \phi_{\frac{\eta}{2\pi}T_2}\left(\varphi\left(\xi - \frac{\eta}{2\pi}\rho\right)\right). \tag{12}$$

For fixed values of $h$ and $\rho$, this representation of the torus $\{\Psi(\theta)\}_{\theta\in\mathbb{T}^2}$ is non unique for two reasons: (a) For each choice of $\eta_0$ we have a different $\varphi$ in (10), i.e., a different invariant curve on the torus. (b) Given the parametrisation (10), for each $\xi_0 \in \mathbb{T}^1$, $\varphi(\xi - \xi_0)$ is a different parametrisation with a different Fourier expansion of the same invariant curve of the torus. A way to overcome both indeterminations is to fix some components of the Fourier coefficients $A_k$.

Finally, and in order to deal with high instabilities, a multiple shooting procedure is used. It consists of looking for several invariant curves on the torus $\{\Psi(\theta)\}_{\theta\in\mathbb{T}^2}$ instead of just one, in order to reduce the maximum time of integration to a fraction of $T_2$. Concretely, we will look for $m$ parametrisations $\varphi_0,\varphi_1,\ldots,\varphi_{m-1}$ satisfying $\varphi_{j+1}(\xi) = \phi_{T_2/m}\left(\varphi_j(\xi)\right)$ for all $\xi \in \mathbb{T}^1$ and $j = 0,\ldots,m-2$, and $\varphi_0(\xi+\rho) = \phi_{T_2/m}\left(\varphi_{m-1}(\xi)\right)$ for all $\xi \in \mathbb{T}^1$.

The details of the computational aspects (implementation, computing effort, parallel strategies, etc.) of this procedure are given in Gómez–Mondelo [31]. As a sample of the tori that can be computed with this procedure, in Fig. 8 we display families around bifurcated halo-type orbits of $L_1$ and $L_2$ with central part.

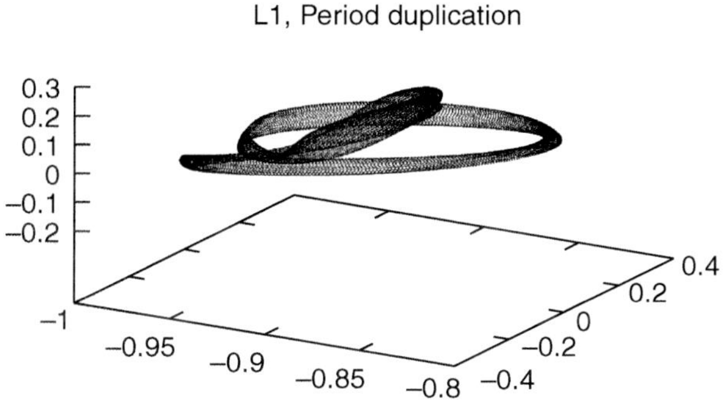


**Fig. 8** Torus around the bifurcated halo-type orbit around the $L_1$ point

# 4 Homoclinic and Heteroclinic Orbits

In the preceding sections we have shown how to use the local dynamics around a halo orbit for "local" purposes. In this one we will study the global behaviour of the invariant stable/unstable manifolds, of the central manifolds of $L_1$ and $L_2$, to perform some acrobatic motions connecting libration orbits around those equilibrium points.

In order to obtain heteroclinic trajectories between libration orbits around $L_1$ and $L_2$, we have to match an orbit of the unstable manifold of a libration orbit around one point, with another orbit, in the stable manifold of a libration orbit around the other point. This is, both orbits have to be the same one. Since these orbits, when looked in the $X$ coordinate of the CRTBP system, have to go from one side of the Earth to the other one, the place where we look for the connection is the plane $X = \mu - 1$, this is, the plane orthogonal to the $X$ axis that cuts it at the point where the centre of the Earth is located.

Although the technical details are much more complex, the main idea is similar to the computations introduced in [28] for $L_{4,5}$ connections. Once a Jacobi constant is fixed, we take initial conditions in the linear approximation of the unstable manifold of all the libration orbits inside the level of energy. Since the energy is fixed, we have three free variables (usually $q_1$, $q_2$ and $p_2$). A scanning procedure in these variables is done. Since the selected orbits will leave the neighbourhood of the libration point, each initial condition in the variables $(q,p)$ is changed into CRTBP coordinates and then propagated forward in time until it crosses the plane $X = \mu - 1$. We apply the same procedure to the orbits in the stable manifold, where all the propagation is done backward in time.

We have to remark that as usual, the unstable and stable manifolds have two branches. In the process we select only the branches that, at the initial steps of the propagations approach the $X = \mu - 1$ plane.

Since the Jacobi constant is fixed, the set of all CRTBP values $\mathcal{C} = \{(Y,\dot{Y},Z,\dot{Z})\}$ obtained, characterise the branch of the manifold of all the libration orbits around the selected equilibrium point for the particular section. Let us denote these sets by $\mathcal{C}_i^{+sj}$, where + or $-$ denote the branch of the $s$ (stable) or $u$ (unstable) manifold of the $L_j$, $j = 1,2$ libration orbits at the $i$-th intersection with the $X = \mu - 1$ plane.

Looking at the above mentioned branches of the manifolds, the simplest heteroclinic orbits will be obtained from $\mathcal{I}_{1^-} = \mathcal{C}_1^{-s1} \cap \mathcal{C}_1^{-u2}$ and $\mathcal{I}_{1^+} = \mathcal{C}_1^{-u1} \cap \mathcal{C}_1^{-s2}$. Both sets give transfer orbits that cross the plane $X = \mu - 1$ once. We will denote by $\mathcal{I}_{k^-}$ (respectively $\mathcal{I}_{k^+}$) the set of heteroclinic trajectories from $L_2$ to $L_1$ (resp. from $L_1$ to $L_2$) that cross the plane $X = \mu - 1$ $k$ times, following the above mentioned branches of the manifold. We note that, due to the symmetries of the CRTBP equations, for any heteroclinic orbit from $L_1$ to $L_2$ we have a symmetrical one from $L_2$ to $L_1$ and so just one exploration must be done.

Unhopefully, it has been found (see [29]) that $\mathcal{I}_{1^+}$ is empty and so one must look for connections crossing at least twice the plane $X = \mu - 1$. In this case many possibilities of connections appear. As an example, in Fig. 9 a connection between a Lissajous orbits around $L_2$ and quasi-halo orbit around $L_2$ is displayed. Both the 3-D representation of the homoclinic orbits and the intersections with the surface of section $Z = 0$, around both equilibrium points, are given.

As another kind of connection, the homoclinic transition inside the central manifold, that differences the central Lissajous orbits from the quasi-halo ones, is computed in [29]. These kind of solutions are interesting because they perform a transition from a planar motion (close to a Lyapunov orbit) to an inclined orbit

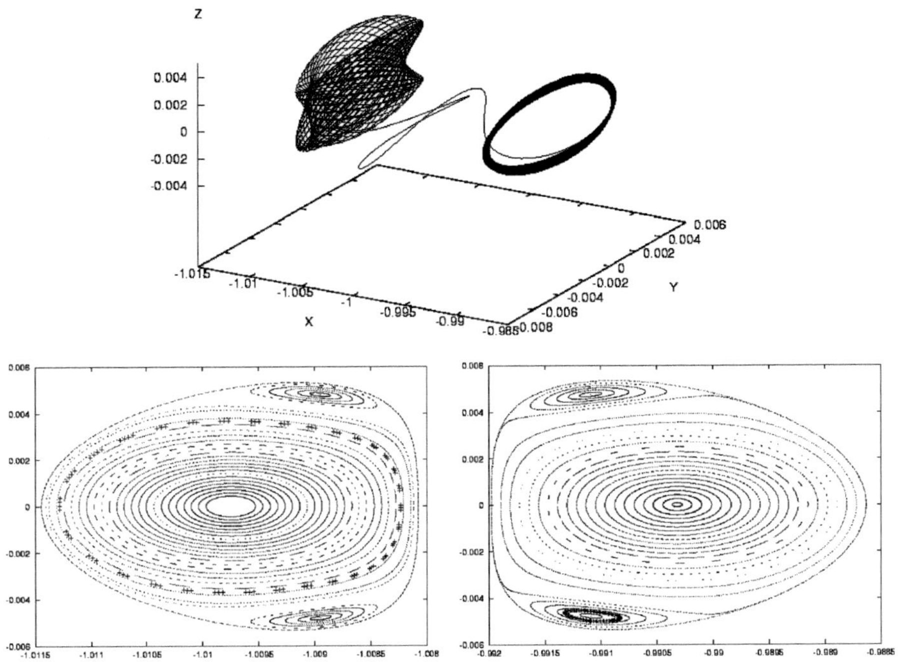


**Fig. 9** $L_1$-$L_2$ heteroclinic connection between a Lissajous orbit around $L_2$ and a quasi-halo orbit around $L_1$. In the lower pictures the intersections of the orbits with the surface of section ($Z = 0$) for $L_2$ (*left*) and for $L_1$ (*right*) are displayed with crosses

(close to the quasi-halo orbits) without any $\Delta v$. Figure 10 shows one of these orbits in central manifold ($q,p$) variables. Unfortunately, the transition is very slow but probably, with very small $\Delta v$, it could be possible to accelerate the transition from planar to inclined motion.

# 5 Applications of Invariant Manifolds

## *5.1 Transfers from the Earth*

In this and in the next sections the role of the invariant manifolds as natural channels of motion for spacecraft missions will be explained. Here we would like to mention that these manifolds, which play an important role in the evolution of the solar system and that when used for the design of spacecraft missions enhance the set of opportunities and can save precious fuel, are often seen as celestial mechanics "tricks" with no theory supporting them.

More recently, the heteroclinic connections in the region of space near $L_1$ and $L_2$ have also been successfully exploited for mission design. As scientific goals become increasingly ambitious and, therefore, mission trajectory requirements become more complex, innovative solutions become necessary. Thus, new directions emerge, such as the studies of dynamical channels and transport in and through this region of

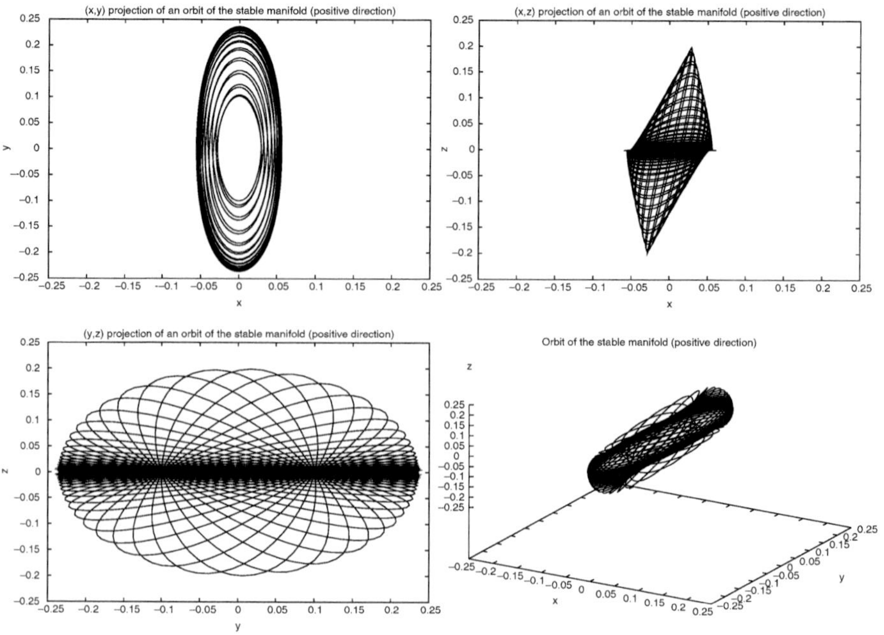


**Fig. 10** Homoclinic transition between Lyapunov orbits inside the central manifold (in central manifold coordinates)

space that offers options for transit-type orbits and low energy captures. When coupled with numerical techniques, frequently based on a differential correction process, the results of all these investigations are powerful and effective techniques for generating accurate and varied libration point trajectory options that may serve as a basis for many types of future mission opportunities. It is important to note that these various investigations are all ongoing and far from completion. Yet, even at this juncture, early insights and preliminary observations have yield useful results with immediate and significant impact.

## *5.2 Transfers using invariant manifolds*

There are two different approaches in the computation of transfer trajectories to a libration point orbit. One uses direct shooting methods (forward or backward) together with a differential corrector, for targeting and meeting mission goals. Proceeding in this way one can get direct or gravitational assisted transfers, such as the ones shown in Fig. 11. Since the libration point orbits, for values of the energy not too far from the ones corresponding to the libration point, have a strong hyperbolic character, its is also possible to use their stable manifold for the transfer. This is what is known in the literature as the dynamical systems approach to the transfer problem. Other ways to obtain transfer trajectories from the Earth to a libration

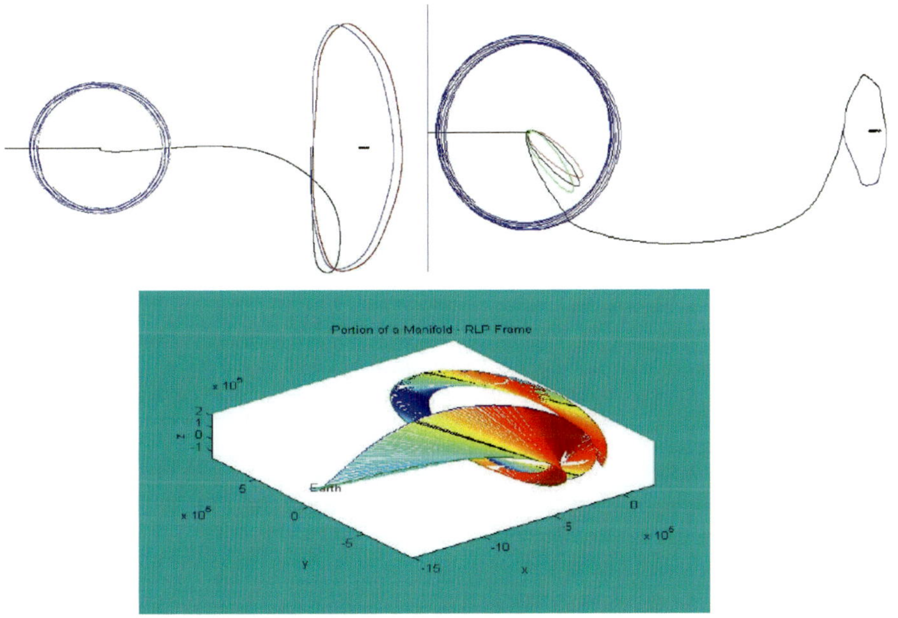


**Fig. 11** Three different kinds of transfer to a libration point orbit: direct transfer, gravitational assisted transfer and transfer using invariant manifolds ([27])

point orbit use optimisation procedures. These methods look for orbits between the Earth and the libration orbit maintaining some boundary conditions, subject to some technical constraints, and minimising the total fuel to be spent in manoeuvres during the transfer (see [40]). According to Masdemont, see [54] and [30], in the dynamical systems approach one can proceed as follows:

1. Take a local approximation of the stable manifold at a certain point of the nominal orbit. This determines a line in the phase space based on a point of the nominal orbit.
2. Propagate, backwards in time, the points in the line of the local approximation of the stable manifold until one or several close approaches to the Earth are found (or up to a maximum time span is reached). In this way some globalisation of the stable manifold is obtained.
3. Look at the possible intersections (in the configuration space) between the parking orbit of the spacecraft and the stable manifold. At each one of these intersections the velocities in the stable manifold and in the parking orbit have different values, $v_s$ and $v_p$. A perfect manoeuvre with $\Delta v = v_s - v_p$ will move the spacecraft from the parking orbit to an orbit that will reach the nominal orbit without any additional manoeuvre.
4. Then $|\Delta v|$ can be minimised by changing the base point of the nominal orbit at which the stable manifold has been computed (or any equivalent parameter).

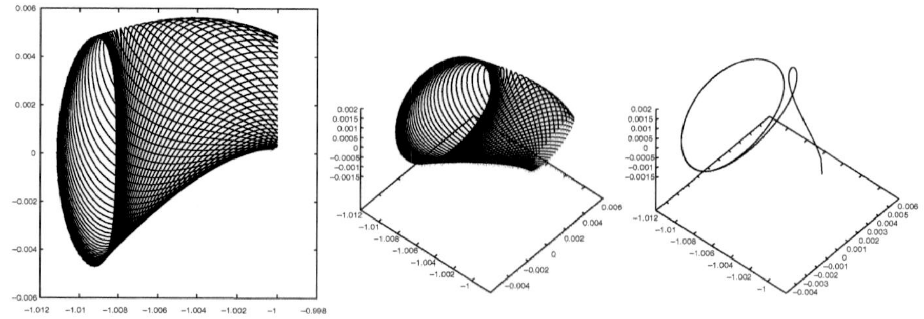


**Fig. 12** Illustration of the invariant manifold approach to the transfer to a halo orbit about $L_2$. The orbits in the stable manifold are represented in the *left*-hand side and *central* figures (*XY* and a 3D plot respectively). The key idea of the transfer is to insert the satellite in an orbit of the manifold having a close approach to the Earth (located near the point $(-1,0,0)$) as the one represented on the *right*-hand side figure

This idea is displayed in Fig. 12 and 13. Note that, depending on the nominal orbit and on the parking orbit, the intersection described in the third item can be empty, or the optimal solution found in this way can be too expensive. To overcome these difficulties several strategies can be adopted. One possibility is to perform a transfer to an orbit different from the nominal one and then, with some additional manoeuvres, move to the desired orbits. In a next section we will show how these last kind of transfers can be performed. Another possibility is to allow for some intermediate manoeuvres in the path from the vicinity of the Earth to the final orbit.

In the case the nominal orbit is a quasi-halo or Lissajous orbit and any phase can be accepted for the additional angular variable, the stable manifold has dimension 3. This produces, from one side, a heavier computational task than in the case of halo orbits, but from the other side it gives additional possibilities for the transfer. One should think that the stable manifolds of the full centre manifold (for a fixed value of $t$) have dimension 5 which offers a lot of possibilities.

Also low thrust trajectories can be considered for the transfers, as the ones illustrated in Fig. 14. In the examples of this figure the trust is of constant magnitude and always aligned with the inertial velocity of the satellite, computed in such a way that meets the stable manifold of the libration point zone at some point (labelled with a small box). At this point the thrust is stopped and the coast trajectory inside the manifold provides the rest of the transfer. For more details of these types of transfers see [21, 22].

## 5.3 The TCM Problem

The Trajectory Correction Manoeuvres (TCM) problem, deals with the manoeuvres to be done by a spacecraft in the transfer segment between the parking orbit and the

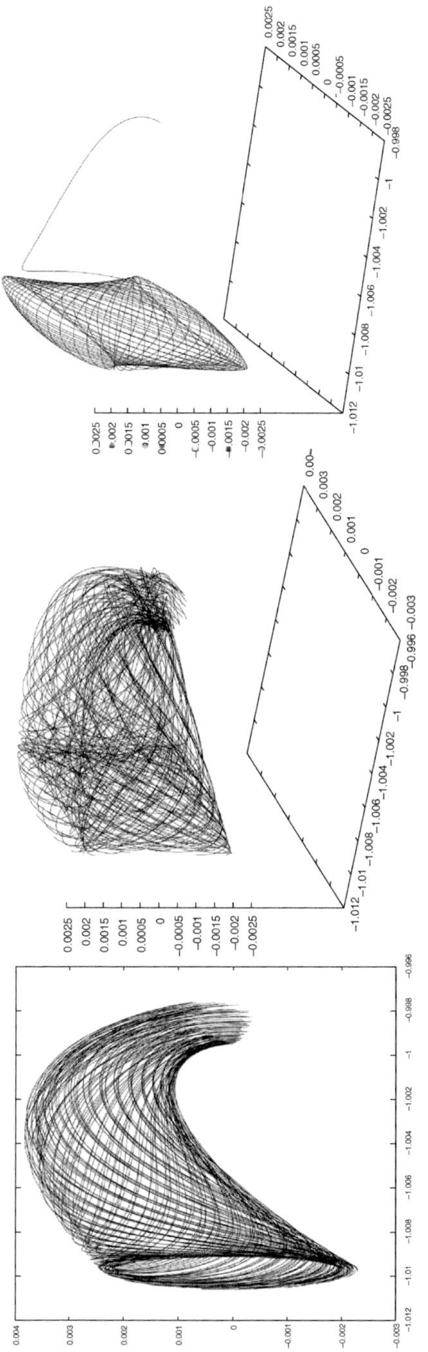

**Fig. 13** Illustration of the transfer approach used in Fig. 12 but targeting to a Lissajous orbit

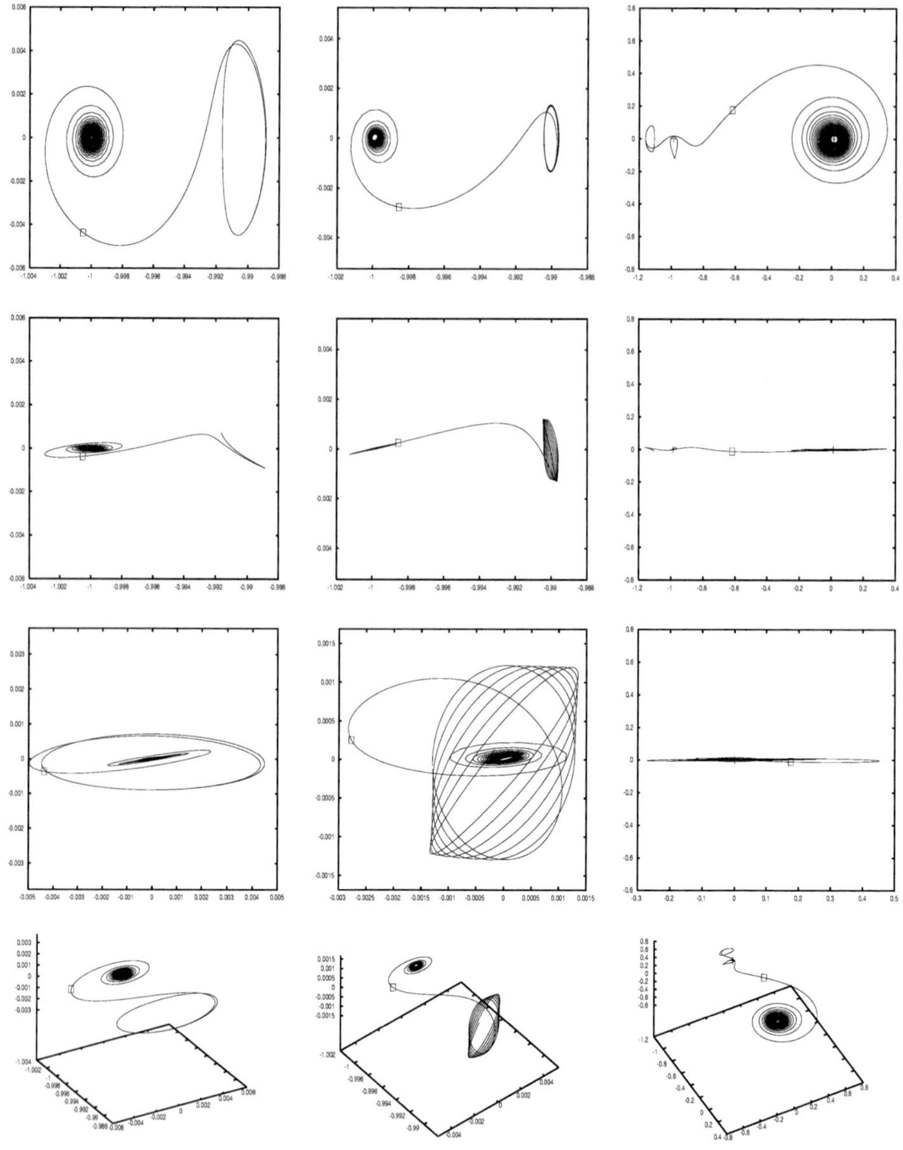

**Fig. 14** Examples of low thrust transfers to libration point orbits. By rows, *XY, XZ, YZ* and a 3D representation in the CRTBP coordinates are given. The first two columns display a transfer from the Earth to a halo and a Lissajous orbit about $L_1$ in the Sun-Earth system respectively. The last one displays a transfer from the Earth to a halo orbit about $L_2$ in the Earth-Moon system. The thrust is stopped when the orbit inserts into the manifold (the place is marked with a small box)

target nominal one. The purpose of the TCMs is to correct the error introduced by the inaccuracies of the injection manoeuvre.

In connection with the Genesis mission (see [53]), the TCM problem has been studied in [45] and [57]. For this mission a halo type orbit, around the $L_1$ point of the Earth–Sun system, is used as nominal orbit. The insertion manoeuvre, from the parking orbit around the Earth to the transfer trajectory, is a large one, with a $\Delta v$ of the order of 3000 m/s; for the Genesis mission, the error in its execution was expected to be about a 0.2 % of $\Delta v$ (1 sigma value) and a key point to be studied is how large the cost of the correction of this error is when the execution of the first correction manoeuvre is delayed.

In the paper by Serban et al. [57], two different strategies are considered in order to solve the problem, both using an optimisation procedure and producing very similar results. It is numerically shown that, in practice, the optimal solution can be obtained with just two TCMs and that the cost behaves almost linearly with respect to both TCM1 epoch and the launch velocity error.

The same results can be obtained without using any optimal control procedure. This is what is done in [37] and [38]. The quantitative results, concerning the optimal cost of the transfer and its behaviour as a function of the different free parameters, turn out to be the same as in [57]. Additionally, we provide information on the cost of the transfer when the correction manoeuvres cannot be done at the optimal epochs. These results are qualitatively very close to those obtained by Wilson et al. in [63] for the cost of the transfer to a Lissajous orbit around $L_2$, when the time of flight between the departure and the injection in the stable manifold is fixed but the target state (position and velocity) on the manifold is varied. For this problem it is found that the cost of the transfer can rise dramatically.

In our approach, the transfer path is divided in three different legs:

- The first leg goes from the fixed departure point to the point where the TCM1 is performed. Usually, this correction manoeuvre takes place few days after the departure.
- The second leg, between the two trajectory correction manoeuvres TCM1 and TCM2, is used to perform the injection in the stable manifold of the nominal orbit.
- The last path corresponds to a piece of trajectory on the stable manifold. Since both TCM1 and TCM2 are assumed to be done without errors, the spacecraft will reach the nominal halo orbit without any additional impulse.

Let $t_1$, $t_2$ and $t_3$ be the TCM1, TCM2 and arrival epochs, respectively, and $\Delta v_1$, $\Delta v_2$ the values of the correction manoeuvres at $t_1$ and $t_2$. In this way, given the departure state, $X_{dep}$, and the time $t_1$, we define $X_1 = \varphi_{t_1}(X_{dep})$, where $\varphi_t(X)$ denotes the image under the flow of the point $X$ after $t$. Then, the transfer condition is stated as

$$\varphi_{t_2-t_1}(X_1 + \Delta v_1) + \Delta v_2 = \varphi_{t_2-t_3}(X_a), \tag{13}$$

where $X_a$ represents the arrival state to the target orbit, which is chosen as $X_a = X_a^h + d \cdot V^s(X_a^h)$ in the linear approximation of the stable manifold based at the point $X_a^h$. In (13), a term like $X_1 + \Delta v_1$ has to be understood as: to the state $X_1$ (position and velocity) we add $\Delta v_1$ to the velocity. Note that for a given insertion error $\epsilon$ (which determines $X_{dep}$) we have six equality constraints, corresponding to the position and velocity equations (13), and ten parameters: $t_1$, $t_2$, $t_3$, $\Delta v_1$, $\Delta v_2$ and $X_a$ (given by the parameter along the orbit) which should be chosen in an optimal way within mission constraints.

The sketch of the exploration procedure is the following. To start with, we consider the error of the injection manoeuvre and $t_1$ fixed. Two types of explorations appear in a natural way: the fixed time of flight transfers for which $t_3$ is fixed, and the free time of flight transfers, where $t_3$ is allowed to vary. In both cases, we start the exploration by fixing an initial value for the parameter along the orbit, $X_a$. In the case of fixed time of flight, the problem then reduces to seven parameters ($t_2$, $\Delta v_1$, $\Delta v_2$) and the six constraints (13). Using $\Delta v_1$ and $\Delta v_2$ to match the constraints (13), the cost of the transfer, $\|\Delta v\| = \|\Delta v_1\| + \|\Delta v_2\|$, is seen as a function of $t_2$. In the case of free time of flight, $\|\Delta v\|$ is seen as a function of $t_2$ and $t_3$, or equivalently, as a function of $t_2$ and the parameter along the flow, $t_{ws} = t_3 - t_2$.

Once we have explored the dependence of the transfer cost on $t_2$ and $t_3$, we study the behaviour when moving the parameter along the orbit, $X_a$. Finally, the dependence with respect to the magnitude of the error (which is determined by the launch vehicle) and $t_1$ is studied (which, due to mission constraints, is enough to vary in a narrow and coarse range).

As an example, Fig. 15 shows the results obtained when: the magnitude of the error in the injection manoeuvre is $-3m/s$, the first manoeuvre is delayed 4 days after the departure ($t_1 = 4$), the total time of flight and $t_3$, is taken equal to 173.25 days.

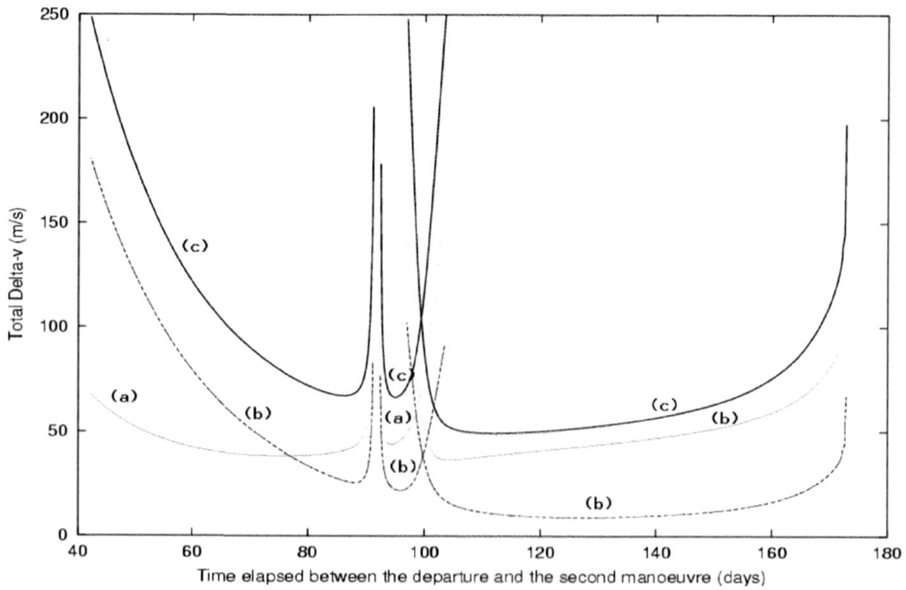


**Fig. 15** Cost of the trajectory correction manoeuvres when TCM1 is delayed 4 days after departure and the total time of flight is fixed to 173.25 days. The curves labelled with (**a**) correspond to $||\Delta v_1||$, those with (**b**) to $\|\Delta v_2\|$ and those with (**c**) to the total cost: $\|\Delta v_1\| + \|\Delta v_2\|$

Several remarks should be done concerning the figure:

- The solutions of equation (13) are grouped along three curves at least. For $t_2 = 99.5$ days there is a double point in the cost function, corresponding to two different possibilities.
- For $t_2 = 113$ days we get the optimum solution in terms of fuel consumption: $\|\Delta v_1\| + \|\Delta v_2\| = 49.31$ m/s. This value is very close to the one given in [57] for the MOI approach, which is 49.1817 m/s. The discrepancies can be attributed to slight differences between the two nominal orbits and the corresponding target points.
- When $t_2$ is small or very close to the final time, $t_3$, the total cost of the TCMs increases, as it should be expected.
- Around the values $t_2 = 92,97$ and 102 days, the total cost increases abruptly. This sudden growth is analogous to the one described in [45] in connection with the TCM problem for the Genesis mission. It is also similar to the behaviour found in [63] for the cost of the transfer to a Lissajous orbit around $L_2$, when the time of flight between the departure and the injection in the stable manifold is fixed. This fact can be explained in terms of the angle between the two velocity vectors at $t = t_2$. This is, when changing from the second to the third leg of the transfer path. This angle also increases sharply at the corresponding epochs.

## 5.4 Transfers Between Halo Orbits

The interest on this problem was initially motivated by the study of the transfer from the vicinity of the Earth to a halo orbit around the equilibrium point $L_1$ of the Earth-Sun system [35, 33, 54]. There, it was shown that the invariant stable manifolds of halo orbits can be used efficiently for the transfer from the Earth, if we are able to inject the spacecraft into these manifolds. This can be easily achieved when the orbits of the manifold come close to the Earth. But this is true only when the halo orbit is large enough, or when the effect of the Moon, bending some orbits of the manifold, is big enough to bring these orbits to the vicinity of the Earth. For small halo orbits, if a swing-by with the Moon is used, there are launch possibilities only during two or three days per month (see [23, 33, 30, 54]). These launching possibilities can be longer for halo orbits with larger $z$-amplitude. This is because they have a stable manifold coming closer to the Earth. After the transfer from the Earth to a large halo orbit has been done, we must be able to go from it to a smaller one in a not very expensive way, in terms of the $\Delta v$ consumption and time.

In what follows we will consider the problem of the transfer between halo type orbits and between Lissajous orbits, always around the same libration point. The method that we present is based on the local study of the motion around the halo orbits and uses the geometry of the problem in the neighbourhood of an orbit of this kind (see [54, 32]). The approach is different from the procedure developed by Hiday and Howell (see [41, 44]) for the same problem. In the Hiday & Howell approach, a departure and arrival states on two arbitrary halo orbits are selected, and a portion of a Lissajous orbit is taken as a path connecting these states. At the patch points there are discontinuities in the velocity which must be minimised. The primer vector theory (developed by Lawden [51] for the two body problem) is extended to the CRTBP and applied to establish the optimal transfers.

In a first step of our approach we study the transfer between two halo orbits which are assumed to be very close in the family of halo orbits. With this hypothesis, the linear approximation of the flow in the neighbourhood of the halo orbits, given by the variational equations, is good enough to have a better understanding of the transfer.

Assume that at a given epoch, $t_1$, we are on a halo orbit, $H_1$, and that at this point a manoeuvre, $\Delta v^{(1)}$, is performed to go away from the actual orbit. At $t = t_2 > t_1$, a second manoeuvre, $\Delta v^{(2)}$, is executed in order to get into the stable manifold of a nearby halo orbit $H_2$. Denoting by $\Delta\beta$ the difference between the $z$-amplitudes of these two orbits, the purpose of an optimal transfer is to perform both manoeuvres in such a way that the performance function

$$\frac{\Delta\beta}{\|\Delta v^{(1)}\|_2 + \|\Delta v^{(2)}\|_2},$$

is maximum.

Let $\varphi$ be, as usual, the flow associated to the differential equations of the CRTBP and $\varphi_\tau(y)$ the image of a point $y \in \mathbb{R}^6$ at $t = \tau$, so we can write

$$\varphi_\tau(y+h) = \varphi_\tau(y) + D\varphi_\tau(y)H + O(\mid h \mid^2) = \varphi_\tau(y) + A(\tau)h + O(\mid h \mid^2).$$

Let $\beta(0)$ be the initial point on a halo orbit, $H_1$, with $z$-amplitude $\beta$. The corresponding points in the phase space at $t = t_1, t_2$, and assuming that the time required to execute the manoeuvre can be neglected, will be, respectively,

$$\varphi_{t_1}(\beta(0)) + \begin{pmatrix} 0 \\ \Delta v^{(1)} \end{pmatrix}, \text{ and } \varphi_{t_2}(\beta(0)) + A(t_2)A(t_1)^{-1}\begin{pmatrix} 0 \\ \Delta v^{(1)} \end{pmatrix}.$$

At $t = t_2$, the insertion manoeuvre, $\Delta v^{(2)}$, into the stable manifold of the halo orbit of $z$-amplitude $\beta + \Delta\beta$ is done. So, denoting by $A_{t_1,t_2} = A(t_2)A(t_1)^{-1}$, we must have

$$\varphi_{t_2}(\beta(0)) + A_{t_1,t_2}\begin{pmatrix} 0 \\ \Delta v^{(1)} \end{pmatrix} + \begin{pmatrix} 0 \\ \Delta v^{(2)} \end{pmatrix} =$$
$$\varphi_{t_2}(\beta(0) + \Delta\beta) + \gamma_2 e_{2,\beta+\Delta\beta}(t_2) + \gamma_3 e_{3,\beta+\Delta\beta}(t_2),$$

where $e_{2,\beta+\Delta\beta}(t_2)$ and $e_{3,\beta+\Delta\beta}(t_2)$ are the eigenvectors related to the stable direction and to the tangent to the orbit direction, respectively, of the orbit of amplitude $\beta + \Delta\beta$ at $t = t_2$. The first term in the right hand side of the above equation can be written as

$$\varphi_{t_2}(\beta(0) + \Delta\beta) = \varphi_{t_2}(\beta(0)) + \frac{\partial\varphi_{t_2}(\beta(0))}{\partial\beta}\Delta\beta + O((\Delta\beta)^2).$$

We normalise taking $\Delta\beta = 1$, so the equation to be solved is

$$A_{t_1,t_2}\begin{pmatrix} 0 \\ \Delta v^{(1)} \end{pmatrix} + \begin{pmatrix} 0 \\ \Delta v^{(2)} \end{pmatrix} = \frac{\partial \phi_{t_2}(\beta(0))}{\partial \beta} + \gamma_2 e_{2,\beta}(t_2) + \gamma_3 e_{3,\beta}(t_2),$$

from which we can isolate $\Delta v^{(1)}$, $\Delta v^{(2)}$, getting

$$\Delta v^{(1)} = u_{10} + \gamma_2 u_{12} + \gamma_3 u_{13},$$
$$\Delta v^{(2)} = u_{20} + \gamma_2 u_{22} + \gamma_3 u_{23}.$$

All the magnitudes that appear in these two equations, except the scalars $\gamma_2$ and $\gamma_3$, are three-dimensional vectors. As $\Delta\beta = 1$ has been fixed, the maxima of the performance function corresponds to the minima of $\|\Delta v^{(1)}\|_2 + \|\Delta v^{(2)}\|_2$. Computing the derivatives of this function with respect to $\gamma_2$ and $\gamma_3$ and equating them to zero, we get a system of two polynomial equations of degree four in the two variables $\gamma_2$ and $\gamma_3$, that must be solved for each couple of values $t_1$, $t_2$ (which are the only free parameters).

The results of the numerical computations show that for a fixed value of $t_1$, there are, usually, two values of $t_2$ at which the performance function has a local maximum (for values of $t_1$ close to 90° and 240° there are three and four maxima). The difference between these two values of $t_2$ is almost constant and equal to 180°. That is, after the first manoeuvre has been done, the two optimal possibilities appear separated by a difference of 1/2 of revolution.

The cost of the transfer using the optimal $t_2$ is almost constant and the variation around the mean value does not exceed the 4%. As an example, if the $z$-amplitude of the departure orbit is $\beta = 0.1$, the optimum value is reached using the first maximum for $t_1 = 102°$ and $t_2 = 197°$. For these particular values, the cost of the transfer, per unit of $\Delta\beta$, is of 696 m/s. The cost increases with $\beta$: for $\beta = 0.15$, the optimal value is $\Delta v = 742$ m/s ($t_1 = 102°, t_2 = 193°$) and for $\beta = 0.2$ $\Delta v = 785$ m/s ($t_1 = 101°, t_2 = 187°$). It has been found that the variation with respect to $\beta$ of the optimal value of the cost is almost linear. The value of $t_1$ for the first manoeuvre is almost constant and equal to 100°, the corresponding point in the physical space being always very close to the $z = 0$ plane. For very small values of $\beta$, the second manoeuvre must be done after $t_2 = 270°$, but this value decreases quickly and for $\beta \in (0.1, 0.3)$ it is of the order of $t_2 = 190°$, approximately. That is, one has to wait, typically, 1/4 of revolution after the first manoeuvre, to do the second one.

The transfer computed with the above procedure is not optimal if the initial and final orbits are not close to each other. This is because the solution given by the linear analysis is not good enough when the orbits have very different $z$-amplitudes. Several possibilities are discussed in [32]. As a final conclusion we can say that the cost of a unitary transfer is of 756 m/s and the behaviour with the $z$–amplitude $\beta$ is almost linear. In this way, the cost of the transfer between two halo orbits of amplitudes $\beta = 0.25$ and 0.08 is $(0.25 - 0.08) \times 756$ m/s $= 128.5$ m/s. In Fig. 16 we show the three projections of a transfer trajectory that goes form $\beta = 0.25$ to $\beta = 0.08$

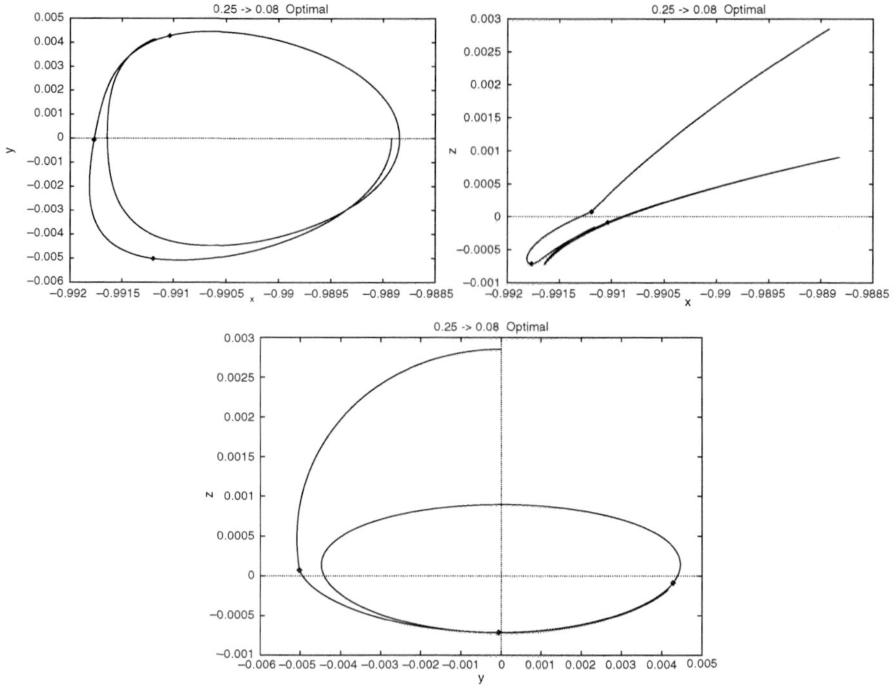


**Fig. 16** Projections of the transfer trajectory starting at a departure orbit of $z$-amplitude $\beta = 0.25$ and arriving at a final one with $z$-amplitude $\beta = 0.08$. The dotted points correspond to the epochs at which the manoeuvres have been done

## *5.5 Transfers Between Lissajous Orbits*

The study of the transfer between Lissajous orbits was first motivated by the missions Herschel/Plank and GAIA of the European Space Agency Scientific Program. Herschel is the cornerstone project in the ESA Science Program dedicated to far infrared Astronomy. Planck, renamed from COBRAS/SAMBA, is expected to map the microwave background over the whole sky and is now combined with GAIA for a common launch in 2009. Several options were considered during the orbit analysis work. The one finally adopted was the so-called "Carrier", where both spacecraft are launched by the same Ariane 5, but will separate after launch. For this option, the optimum solution is a free transfer to a large amplitude Lissajous orbit. Herschel will remain in this orbit whereas Plank, of much less mass, will perform a size reduction manoeuvres.

Although the methodology to transfer between halo orbits could also be applied to the transfer between Lissajous orbits, the method of this section is based in the dynamical study of the linearised CRTBP equations of motion about a collinear equilibrium point. The development was initiated during preliminary studies of the Herschel/Plank mission and is fully developed in [13].

Let us start with the solution of the linear part of the equations of motion (3) which can be written as,

$$\left.\begin{aligned} x(t) &= A_1 e^{\lambda t} + A_2 e^{-\lambda t} + A_3 \cos \omega t + A_4 \sin \omega t \\ y(t) &= cA_1 e^{\lambda t} - cA_2 e^{-\lambda t} - \bar{k} A_4 \cos \omega t + \bar{k} A_3 \sin \omega t \\ z(t) &= A_5 \cos \nu t + A_6 \sin \nu t \end{aligned}\right\} \tag{14}$$

where $A_i$ are arbitrary constants and $c$, $\bar{k}$, $\omega$, $\lambda$ and $\nu$ are constants depending only on $c_2$.

Introducing amplitudes and phases (14) can also be written as

$$\left.\begin{aligned} x(t) &= A_1 e^{\lambda t} + A_2 e^{-\lambda t} + A_x \cos(\omega t + \phi) \\ y(t) &= cA_1 e^{\lambda t} - cA_2 e^{-\lambda t} + \bar{k} A_x \sin(\omega t + \phi) \\ z(t) &= A_z \cos(\nu t + \psi) \end{aligned}\right\} \tag{15}$$

where the relations are $A_3 = A_x \cos\phi$, $A_4 = -A_x \sin\phi$, $A_5 = A_z \cos\psi$ and $A_6 = -A_z \sin\psi$.

The key point is that choosing, $A_1 = A_2 = 0$, we obtain periodic motions in the $xy$ components with a periodic motion in the $z$ component of a different period. These are the Lissajous orbits in the linearised restricted circular three-body problem, $A_x$, $A_z$ being the maximum in plane and out of plane amplitudes respectively. The first integrals $A_1$ and $A_2$ are directly related to the unstable and stable manifold of the linear Lissajous orbit. For instance, the relation $A_1 = 0$, $A_2 \neq 0$, defines a stable manifold . Any orbit verifying this condition will tend forward in time to the Lissajous (or periodic) orbit defined by $A_x$, $A_z$ since the $A_2$-component in (14) will die out. A similar fact happens when $A_1 \neq 0, A_2 = 0$, but now backwards in time. Then, this later condition defines a unstable manifold.

The analysis consists of computing the manoeuvres that keep the $A_1$ component equal to zero in order to prevent escape from the libration zone, and studying how the amplitudes change when a manoeuvre is applied. We note that for the linear problem the motion in the $z$-component is uncoupled from the motion in the $xy$ component and $z$-manoeuvres only change the $A_z$ amplitude but do not introduce instability. Assuming that the motion takes place in a Lissajous orbit with $A_z^{(i)}$ amplitude and phase $\psi_i$ and the desired final $z$-amplitude is $A_z^{(f)}$. The possible $z$-manoeuvres $\Delta\dot{z}$ which performs the transfer at time $t_m$ are given by,

$$\frac{\Delta\dot{z}}{\nu} = A_z^{(i)} \sin(\nu t_m + \psi_i) \pm \sqrt{A_z^{(f)2} - A_z^{(i)2} \cos^2(\nu t_m + \psi_i)} \tag{16}$$

We note that if,

- $A_z^{(f)} \geq A_z^{(i)}$ the transfer manoeuvre is possible at any time.

- $A_z^{(f)} < A_z^{(i)}$ the transfer manoeuvre is possible only if the expression inside the square root is positive; more precisely, when $t \in \left[\varepsilon, \frac{\pi}{\nu} - \varepsilon\right] \cup \left[\frac{\pi}{\nu} + \varepsilon, \frac{2\pi}{\nu} - \varepsilon\right]$, where $\varepsilon = \frac{1}{\nu}(\arccos(\frac{A_z^{(f)}}{A_z^{(i)}}) - \psi_i)$. This condition essentially says that it is not possible to reduce the amplitude with an impulsive manoeuvre in case that the actual position at time $t_m$ has a $z$ component bigger than $A_z^{(f)}$.

The change in the in-plane amplitude is a little more tricky since one must keep the unstable component equal to zero. Assuming that the motion takes place in a Lissajous orbit with $A_x^{(i)}$ amplitude and phase $\phi_i$ and the desired final in-plane amplitude is $A_x^{(f)}$, the possible manoeuvres at time $t_m$ are given by,

$$(\Delta\dot{x}, \Delta\dot{y}) = \alpha \frac{1}{\sqrt{c^2 + \bar{k}^2}}(d_2, -\bar{k}d_1), \quad \alpha \in \mathbb{R} \tag{17}$$

where, $\alpha$, indicating the size of the manoeuvre can be,

$$\alpha = A_x^{(i)} \sin(\omega t_m + \phi_i - \beta) \pm \sqrt{A_x^{(f)2} - A_x^{(i)2} \cos^2(\omega t_m + \phi_i - \beta)}.$$

Where $\beta$ is a fixed angle given by the direction of the vector $(c, \bar{k})$. Again we observe that if,

- $A_x^{(f)} \geq A_x^{(i)}$, the transfer manoeuvre is possible at any time.
- $A_x^{(f)} < A_x^{(i)}$, the transfer manoeuvre is possible only when the expression inside the square root is positive; more precisely, when $t \in \left[\delta, \frac{\pi}{\omega} - \delta\right] \cup \left[\frac{\pi}{\omega} + \delta, \frac{2\pi}{\omega} - \delta\right]$, where $\delta = \frac{1}{\omega}(\arccos(\frac{A_x^{(f)}}{A_x^{(i)}}) - \phi_i + \beta)$.

We also note that the manoeuvre (17) always has the same direction. This direction plays a similar role to the direction orthogonal to the $z$-plane in the case of the previous commented $z$-manoeuvres.

Once the target amplitudes are selected, the epochs of the manoeuvres essentially can be chosen according to the following possibilities,

- Select $t_m$ in such a way that the $\Delta v$ expended in changing the amplitude be a minimum.
- Select $t_m$ in such a way that you arrive at the target orbit with a selected phase.

Assuming that the amplitudes prior and after the manoeuvres are different, in the first case the optimal $t_m$ for changing the in-plane amplitude is when the angle $\omega t_m + \phi_i$ verifies $\omega t_m + \phi_i = \beta + \frac{\pi}{2} + k\pi$, $k \in \mathbb{Z}$. In this case the minimum fuel expenditure for the manoeuvre is $|A_x^{(f)} - A_x^{(i)}|$. In a similar way the optimal $t_m$ for $\nu t_m + \psi_i = \frac{\pi}{2} + k\pi, k \in \mathbb{Z}$ and the manoeuvre is given by $\Delta\dot{z} = \nu(A_z^{(f)} - A_z^{(i)})$.

In case that we decide to arrive at the selected Lissajous orbit with a certain phase the analysis proceeds considering the in-plane and out-of-plane amplitudes $A_x$ and $A_z$ written in term of its respective components $A_3, A_4$ and $A_5, A_6$ and studying the angle which they define.

### 5.5.1 Effective Phases and Eclipse Avoidance

Some interesting cases are the manoeuvres which maintain the amplitudes (the non trivial possibilities of (16) and (17)). In this case an in-plane manoeuvre (17) at time $t_m$ produces an in-plane change of phase given by,

$$\phi_f - \phi_i = -2(\omega t_m - \beta + \phi_i) \qquad (\text{mod } 2\pi), \tag{18}$$

and an out-of-plane manoeuvre (16) produces an out-of-plane change of phase given by,

$$\psi_f - \psi_i = -2(\nu t_m + \psi_i) \qquad (\text{mod } 2\pi). \tag{19}$$

These manoeuvres give two strategies for the avoidance of the exclusion zone needed in many missions (see, for instance, [25]). Besides the well known $z$-strategy given by (19), we have another $xy$ one given by (18) which for the Herschel/Plank mission implies only a delta-v expenditure of 15 m/s every 6 years. In the following we sketch the fundamentals of the procedures. More details can be found in [13] and in [17].

Effective Phases

Looking at the central part of (15) or equivalently, if the satellite is on a Lissajous orbit we have,

$$\begin{aligned} x(t) &= A_x \cos(\omega t + \phi), \quad z(t) = A_z \cos(\nu t + \psi), \\ y(t) &= \bar{k} A_x \sin(\omega t + \phi) \end{aligned} \tag{20}$$

We note that due to the autonomous character of the original system of differential equations, we can reset $t = 0$ at any time if we recompute the $A_i$ values of the solution (15). For the central part (20) mentioned above, due to the invariance of the amplitudes $A_x$ and $A_z$ it is even easier, since $t$ can be reseted to zero at time $t_0$ just changing the phases $\phi$ and $\psi$ by $\phi + \omega t_0$ and $\psi + \nu t_0$ respectively. This observation motivates the following definitions.

Let us define the effective phase $\Phi$ as all the epochs $t$ and all the phases $\phi$ such that $\Phi = \omega t + \phi$ (mod $2\pi$). In the same way we define the effective phase $\Psi$ as all the epochs $t$ and all the phases $\psi$ such that $\Psi = \nu t + \psi$ (mod $2\pi$). Although the effective phases are subsets in the space $\mathbb{R} \times [0,2\pi]$ by definition, they will be identified by the numbers $\Phi$ and $\Psi$ in $[0,2\pi]$ for convenience.

Using equation (20) and taking also into account the velocities, we note that there is a biunivocal correspondence between a pair of effective phases $(\Phi,\Psi)$ and a state $(x,y,z,\dot{x}\dot{y},\dot{z})$ on a Lissajous orbit of given amplitudes $A_x$ and $A_z$. In fact, from a dynamical systems point of view, this is a consequence that Lissajous orbits are 2D tori since $\Phi$ and $\Psi$ are identified (mod $2\pi$). We are just using the well known action-angle variables of the tori.

The convenience of using the effective phases becomes clear since in the space $(\Phi,\Psi)$ a trajectory such as (20) with initial phases $\phi_i$, $\psi_i$, is a straight line of slope $\nu/\omega$, starting at the point $(\phi_i,\psi_i)$, which propagates with constant velocity components $\omega$ and $\nu$ respectively in the directions $\Phi$ and $\Psi$. So, dynamics are much easier.

The space of effective phases, from now on the effective phases plane (EPP), can be used as a nice and general tool for mission design. An interesting application is the eclipse avoidance.

## Eclipse Avoidance. LOEWE

Usually a technical requirement for libration point satellites is to avoid an exclusion zone. For orbits about $L_1$ in the Sun–Earth system the exclusion zone is three degrees about the solar disk as seen from Earth[26]. For orbits about $L_2$ in the Sun–Earth system sometimes the Earth half-shadow has to be avoided. In both cases, since Sun and Earth are located in the $x$ axis, the exclusion zone is set as a disk in the $yz$ plane centred at the origin. As mentioned above, when it comes to orbiting $L_i$, Lissajous orbits suit much better than halo orbits in most of the cases. However, if the duration of the mission is long enough, the satellite will irremediable cross the exclusion zone. The time to enter eclipse depends on the initial phases $(\phi_i,\ \psi_i)$, and in the best case the time span between eclipses is about 6 years for an orbit of moderate size.

Based on the change of phase explained in the previous section we can design eclipse avoidance strategies. The strategies have been called LOEWE (Lissajous Orbit Ever Without Eclipse) and they are based on single impulses. Strategies with more than one impulse could also be studied but are not the purpose of this paper. Moreover, one of the nicest things of the single impulse strategies is their simplicity besides an affordable cost. As we will see, the idea is to perform the manoeuvre near the corners of the Lissajous figure corresponding to the $yz$ projection, where velocities are small. We are near these corners just when we are about to enter to or exit from an eclipse. This will provide us with maximum time without eclipse after having performed the manoeuvre.

Let us use the EPP to represent the LOEWE strategies. Assuming that the satellite is on a Lissajous orbit (20) of amplitudes $A_x$ and $A_z$, the exclusion zone appears in the $yz$ plane as a disk of radius $R$, $y^2+z^2<R^2$. Of course we must have $R<A_y=\bar{k}A_x$, and $R<A_z$. The border of the disk in the plane of effective phases satisfies the equation,

$$\bar{k}^2A_x^2\sin^2\Phi+A_z^2\cos^2\Psi=R^2, \tag{21}$$

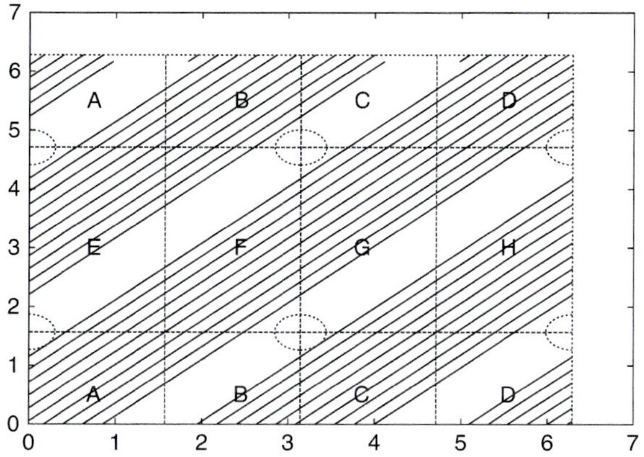


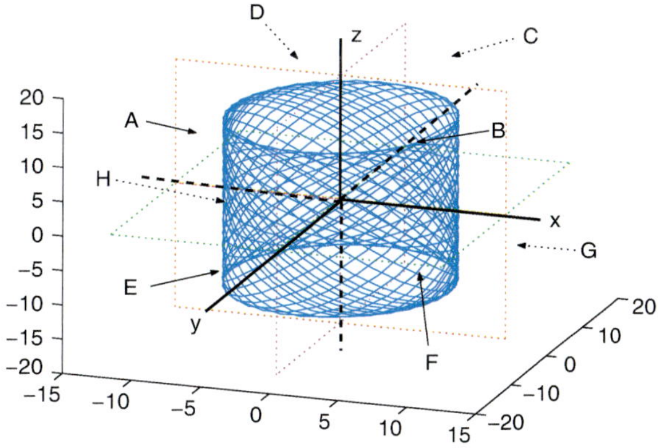


**Fig. 17** Exclusion zones in the effective phases plane (EPP) and Lissajous trajectory (*top*). Each labelled region represents different conditions on the Lissajous, as can be seen in the *bottom* picture

and are the ellipse like plots represented in Fig. 17. When the Lissajous trajectory, represented by a straight line of constant speed in the EPP, cuts one of these curves it means that the satellite is entering the exclusion zone. In this way we note that the computation of the time when the satellite will enter the exclusion zone reduces to compute the intersection of the straight line with the first exclusion zone it hits. Thus, time is just proportional to the distance to the intersection point.

For convenience, trajectories in the EPP will be reduced to $[0,2\pi] \times [0,2\pi]$ since, as we said, the values $\Phi=0$ and $\Phi = 2\pi$ are identified as well as $\Psi=0$ and $\Psi = 2\pi$ are. A typical LOEWE trajectory in the EPP looks like the one represented in Fig. 17 were the discontinuities correspond to manoeuvres. We also note that a point in the EPP gives as well information about the location of the satellite with respect to the libration point.

In Table 1 we represent the values of $\omega$ and $\nu$ that we have for the Earth+Moon-Sun case, as well as the usual radius of the exclusion zone. We note that the slope of a Lissajous trajectory in the EPP is $\frac{\nu}{\omega}$ which in any case is slightly less than one.

**Table 1** Values for the Earth+Moon-Sun case

| | $\omega$ | $\nu$ | usual $R$ (km) |
|---|---|---|---|
| $L_1$ | 2.08645356 | 2.01521066 | 90000 |
| $L_2$ | 2.05701420 | 1.98507486 | 14000 |

This means that the maximum time without eclipse is achieved by initial conditions near the lower tangential point to the exclusion zone.

In this discussion we will consider square Lissajous. This is, $A_y = A_z$ and this value will be denoted by $A$. In this case the border of the exclusion zones (21) is given by $\bar{k}^2 \sin^2 \Phi + \cos^2 \Psi = (R/A)^2$, meaning that the time without eclipse depends only on the relative size *R/A* of the exclusion zones. This is, missions with different amplitudes and different exclusion zone radius have identical representations in the EPP if the relative size *R/A* is the same. This fact allows us to compute the maximum time for a mission without hitting an exclusion zone as a function only of *R/A* which will be considered in percentage. It is just a matter of computing the the intersection time of the orbits starting at the lower tangential conditions. In all the usual cases (*R/A* less than 30%), the results are about 6 years.

### The Tangent to Tangent Cycle

As a result of our previous discussion we use the lower tangential points or, more specifically, lower tangential trajectories in the EPP, as target points of the strategies. The two strategies we discussed in the previous section can be considered: the *xy*-strategy given by equations (17) and (18) and the *z*-strategy given by equations (16) and (19). The basic idea comes from the fact that in the EPP the Lissajous trajectory looks like a set of parallel lines and that (18), (19) represent jumps in the $\Phi$ and $\Psi$ directions respectively. Then, essentially, for the *xy*-strategy we just have to measure the horizontal distance from the collision phases to the nearest lower tangential trajectory, whereas for the *z*-strategy, the vertical distance.

In Fig. 18 we represent the way these two strategies work. See [13] for other special symmetries of the *xy* and *z*-manoeuvres when performed in certain locations.

The LOEWE strategy consists in making the satellite to enter and follow a tangent to tangent cycle. Let us assume that the satellite is injected into the Lissajous trajectory at the point $(\Phi_i,\Psi_i)$ of the EPP. Take this point as time $t = 0$. As time increases, the satellite approaches an exclusion zone which would be intersected at time $t_c$. Prior to the collision, a time $t_m$ is selected in order to apply an *xy*-manoeuvre (i.e. a jump $\Delta\ \Phi$) or a *z*-manoeuvre (i.e. a jump $\Delta\ \Psi$) in such a way that the point $(\Phi_m,\Psi_m)$ at time $t_m$ in the EPP jumps to a new one $(\Phi_{tg},\Psi_{tg})$ (equal to $(\Phi_m + \Delta\Phi,\Psi_m)$ or to $(\Phi_m,\Psi_m + \Delta\Psi)$) which goes into a lower tangential trajectory to the exclusion zone as it is represented in Fig. 18. Then, the tangent to tangent cycle starts in the natural way by renaming $(\Phi_{tg},\Psi_{tg})$ to $(\Phi_i,\Psi_i)$ and reseting time to zero. Once in the cycle, after each manoeuvre the satellite will be about 6 years free of eclipse.

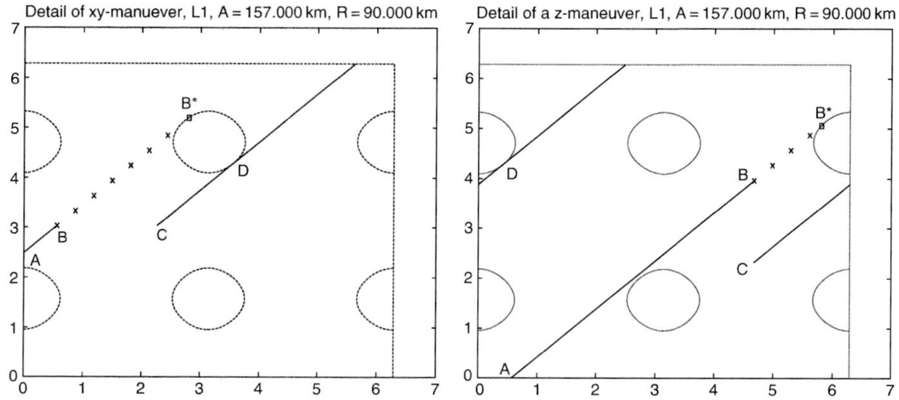


**Fig. 18** Detail of $xy$ (*left*) and $z$ (*right*) manoeuvres for the LOEWE strategy. Let us assume that the Lissajous trajectory starts at A. In both cases the manoeuvre is computed to reach a lower tangential trajectory to an exclusion zone before a collision would happen in B*. The result of the manoeuvre is to jump from B to C in the EPP. Note that in these cases the jump of the $xy$-manoeuvre is symmetrical with respect to $\Phi = \beta$ and the one for the $z$-manoeuvre is symmetrical with respect to $\Psi = \pi$. Lower tangency to the exclusion zone occurs at point D

There is a technical remark to be made about closing a cycle. We define a cycle as the part of trajectory in the EPP comprised between two lower tangencies and having a manoeuvre in between. However, several slightly different patterns in the $xyz$ representation of the trajectory can appear If two different manoeuvre points can be observed in the $xy$ projection, we say that the cycle is two-sided. Otherwise, only one manoeuvre point, one-sided (see Fig. 19). For both cases two cycles are needed to repeat the same starting position and velocity (and so the same pattern in the $xyz$-coordinates), but since both manoeuvres of these cycles have the same magnitude and the time span between them is the same, it is not necessary to make such distinctions for the total amount of delta-v expended and for the results we present. Only to say that the fact that for a given Lissajous, its LOEWE strategy is associated to a one sided or to a two-sided cycles, depends exclusively on the size of *R/A* i.e. the size of the exclusion zone in the EPP, since the Lissajous trajectories in the EPP are always seen as parallel lines separated by a constant distance. Figure 19 gives some examples

## Summary of Results

All Lissajous orbits considered have amplitude $A = A_y = A_z$ ("square Lissajous"). Qualitatively, the results depend only on the relative amplitude R/A. However, in terms of real cost, they also depend on A. The main result is that the $xy$-strategy is proportionally a little cheaper for orbits of relative size R/A less than 60%. When R/A is bigger than 60% then the $z$-strategy is cheaper than $xy$-strategy. Concerning the actual costs for the LOEWE strategies, these can be seen in Table 2 for the $L_1$ case and in Table 3 for the $L_2$ case.

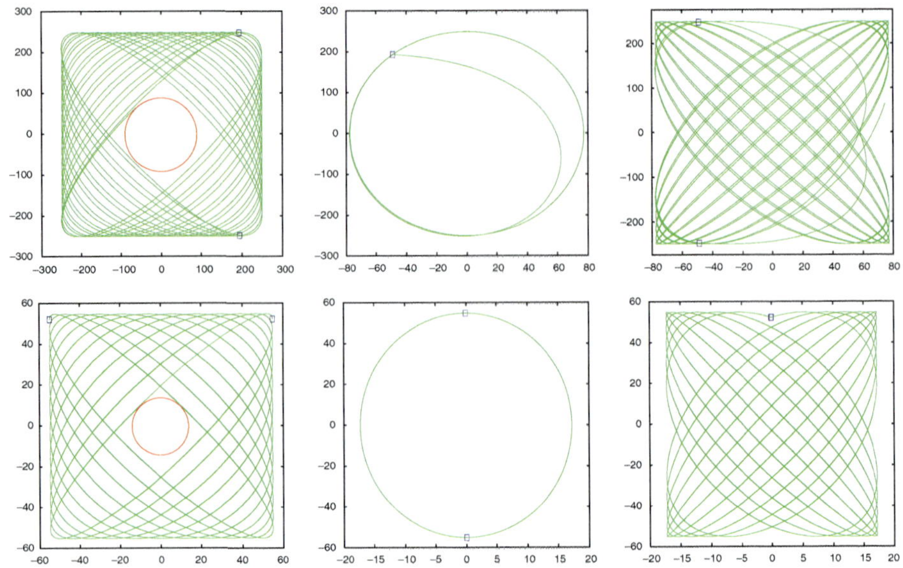

**Fig. 19** First row, example of one-sided cycle. Trajectory around $L_1$ with $A = 250000$ km, $R = 90000$ km ($xy$-manoeuvres). Second row, example of two-sided cycle. Trajectory about $L_2$ with with $A = 55000$ km, R = 14000 km. ($z$-manoeuvres). The manoeuvres are marked with a small box. ¿From *left* to *right* displaying the $yz$, $xy$ and $xz$ projections

## 5.5.2 Rendez-vous in the Libration Zone

Spatial rendez-vous consists of making two satellites meet at a particular point in an orbit. As explained above, the non-escape manoeuvres can be used to change either both the amplitudes and phases or only the phases. These two different cases can be useful for space rendez-vous, as sometimes the satellites for any reason may lay in the same orbit, but in different phases, and in some other cases, they can be in different orbits.

The methodology we present here works when the satellites are already in the Lissajous orbit and whichever their relative positions. Consequently, it can be useful essentially in two different scenarios: as the main rendez-vous strategy that the mission designer plans, or as a contingency plan because of a failure or of a lack of precision of other planned rendez-vous strategies.

Let us assume we have inserted a pair of satellites in a Lissajous orbit. The position of each one of the satellites is defined by the hyperbolic coefficients $A_1$ and $A_2$ (equal zero); the central part amplitudes $A_3$ and $A_4$ (in this example we consider the same for both of the satellites) and the phases. If the satellites are on the same Lissajous, the 4 amplitudes are the same, being $A_1 = A_2 = 0$, and so their positions only differ in the phases (in plane and/or out of plane). Our goal is to make them meet at some time $t_r$ in the future, using impulsive manoeuvres in the non-escape direction with an affordable cost.

**Table 2** Summary of results for avoiding the exclusion zone about $L_1$ using LOEWE strategies with $xy$ or $z$ manoeuvres. ACY $= K_* \times A$, is the average cost per year. TWE is the time without eclipse once a manoeuvre is performed and, MC $= M_* \times A$, is the cost of each manoeuvre in the cycle. In the formulae, the amplitude $A$ is considered to be in thousands of km

| R/A (%) | | ACY $= K_* \cdot A$ (cm/s) | | TWE (years) | | MC $= M_* \cdot A$ (cm/s) | |
|---|---|---|---|---|---|---|---|
| From | To | $K_{xy}$ | $K_z$ | $xy$ | $z$ | $M_{xy}$ | $M_z$ |
| 77.25 | 80.35 | 39.571 | 38.063 | 1.737 | 1.917 | 68.727 | 72.972 |
| 73.95 | 77.20 | 33.785 | 32.955 | 1.985 | 2.157 | 67.059 | 71.077 |
| 70.66 | 73.94 | 29.191 | 28.784 | 2.233 | 2.396 | 65.184 | 68.978 |
| 67.32 | 70.65 | 25.435 | 25.295 | 2.481 | 2.636 | 63.108 | 66.680 |
| 63.92 | 67.31 | 22.291 | 22.322 | 2.729 | 2.876 | 60.838 | 64.191 |
| 60.50 | 63.92 | 19.608 | 19.746 | 2.977 | 3.115 | 58.379 | 61.517 |
| 57.00 | 60.45 | 17.282 | 17.486 | 3.225 | 3.355 | 55.741 | 58.666 |
| 53.46 | 56.99 | 15.238 | 15.480 | 3.473 | 3.595 | 52.932 | 55.647 |
| 49.90 | 53.45 | 13.425 | 13.684 | 3.722 | 3.834 | 49.958 | 52.467 |
| 46.30 | 49.89 | 11.797 | 12.061 | 3.970 | 4.074 | 46.831 | 49.136 |
| 42.65 | 46.28 | 10.327 | 10.586 | 4.218 | 4.313 | 43.560 | 45.665 |
| 39.00 | 42.64 | 8.991 | 9.238 | 4.466 | 4.553 | 40.154 | 42.061 |
| 35.30 | 38.96 | 7.769 | 7.999 | 4.714 | 4.793 | 36.624 | 38.337 |
| 31.54 | 35.25 | 6.646 | 6.856 | 4.962 | 5.032 | 32.982 | 34.502 |
| 27.80 | 31.53 | 5.611 | 5.798 | 5.210 | 5.272 | 29.238 | 30.569 |
| 24.05 | 27.75 | 4.654 | 4.816 | 5.458 | 5.512 | 25.403 | 26.547 |
| 20.20 | 24.00 | 3.766 | 3.903 | 5.706 | 5.751 | 21.491 | 22.450 |
| 16.41 | 20.15 | 2.941 | 3.052 | 5.955 | 5.991 | 17.512 | 18.287 |
| 12.60 | 16.40 | 2.173 | 2.258 | 6.202 | 6.231 | 13.479 | 14.072 |
| 8.80 | 12.55 | 1.458 | 1.517 | 6.451 | 6.470 | 9.405 | 9.816 |
| 5.00 | 8.77 | 0.791 | 0.824 | 6.699 | 6.710 | 5.301 | 5.533 |
| 2.80 | 4.93 | 0.170 | 0.177 | 6.947 | 6.949 | 2.363 | 1.233 |

An immediate solution to the problem of making two satellites meet is letting one of them follow its way unperturbed along the Lissajous, and planning manoeuvres on the other one.

According to (18, 19), there is only one possible jump in each direction ($xy$ or $z$) at each moment of time. And the other way round, once the jump has been fixed, the time cannot be chosen. In our case, the jump in the phases is clearly determined: we want to jump from one trajectory to the other one.

Let $(\phi_i^1, \psi_i^1)$ and $(\phi_i^2, \psi_i^2)$ be the effective phases that determine the trajectory of satellite 1 and 2 respectively. The manoeuvres will be performed on satellite 1 (note that this choice is not relevant). Therefore, the phases determining the trajectory of satellite one after the manoeuvre have to coincide with the phases that define the trajectory of satellite 2, $(\phi_i^2, \psi_i^2)$. Assuming that the manoeuvre in $xy$ is performed at time $t_m^{xy}$ and the manoeuvre in $z$ at a different time $t_m^z$. The equations to be solved are,

$$\phi_i^2 = \phi_f^1 = -\phi_i^1 - 2\omega t_m^{xy} + 2\beta \ (\mathrm{mod} 2\pi), \qquad \psi_i^2 = \psi_f^1 = -\psi_i^1 - 2\nu t_m^z \ (\mathrm{mod} 2\pi).$$

**Table 3** Summary of results for avoiding the exclusion zone about $L_2$ using LOEWE strategies with $xy$ or $z$ manoeuvres. Same comments of Table 2 apply here

| R/A (%) | | ACY $= K_* \cdot A$ (cm/s) | | TWE (years) | | MC $= M_* \cdot A$ (cm/s) | |
|---|---|---|---|---|---|---|---|
| ¿From | To | $K_{xy}$ | $K_z$ | $xy$ | $z$ | $M_{xy}$ | $M_z$ |
| 48.53 | 52.18 | 12.688 | 12.966 | 3.778 | 3.889 | 47.939 | 50.428 |
| 44.81 | 48.52 | 11.099 | 11.376 | 4.030 | 4.132 | 44.732 | 47.009 |
| 41.10 | 44.80 | 9.664 | 9.931 | 4.282 | 4.375 | 41.380 | 43.449 |
| 37.30 | 41.05 | 8.358 | 8.609 | 4.534 | 4.618 | 37.894 | 39.758 |
| 33.50 | 37.28 | 7.164 | 7.394 | 4.786 | 4.861 | 34.286 | 35.947 |
| 29.65 | 33.45 | 6.067 | 6.274 | 5.037 | 5.104 | 30.566 | 32.027 |
| 25.80 | 29.64 | 5.056 | 5.260 | 5.289 | 5.347 | 26.747 | 28.021 |
| 22.00 | 25.75 | 4.122 | 4.276 | 5.541 | 5.591 | 22.842 | 23.905 |
| 18.02 | 21.90 | 3.256 | 3.380 | 5.793 | 5.834 | 18.862 | 19.717 |
| 14.15 | 18.00 | 2.452 | 2.551 | 6.045 | 6.077 | 14.821 | 15.504 |
| 10.20 | 14.10 | 1.704 | 1.776 | 6.297 | 6.320 | 10.733 | 11.224 |
| 6.30 | 10.15 | 1.009 | 1.053 | 6.549 | 6.563 | 6.610 | 6.910 |
| 2.80 | 6.26 | 0.362 | 0.379 | 6.801 | 6.806 | 2.465 | 2.577 |

So,

$$t_m^{xy} = \frac{\beta}{\omega} - \frac{\phi_i^2 + \phi_i^1}{2\omega} + \frac{k\pi}{\omega}, \qquad t_m^z = -\frac{\psi_i^2 + \psi_i^1}{2\nu} + \frac{\bar{k}\pi}{\nu}, \tag{22}$$

where $k,\bar{k} \in \mathbb{Z}$ can be used to adjust the times to mission requirements (i.e. positive, close to each other. . .).

The cost of the manoeuvres of making one satellite follow the other is proportional to the initial differences in phases. The closer to 0 or to $2\pi$ that the difference $|\phi_i^1 - \phi_i^2|$ is, the cheaper the manoeuvres in $xy$. For the $z$-manoeuvres, the same with $\psi$.

Particularly, the cost of the $xy$-manoeuvres can be measured by the size of $\alpha$,

$$\alpha_{xy}\left(\phi_i^1,\phi_i^2\right) = 2A_x \left|\sin\left(\phi_i^1 + \omega t_m^{xy} - \beta\right)\right| = 2A_x \left|\sin\left(\frac{\phi_i^1 - \phi_i^2}{2}\right)\right|. \tag{23}$$

The cost of the $xy$ manoeuvres is,

$$\text{cost}_{xy} = \alpha_{xy}\left(\phi_i^1,\phi_i^2\right)\sqrt{\frac{d_2^2 + \bar{k}^2 d_1^2}{c^2 + \bar{k}^2}},$$

and the cost of the $z$-manoeuvres is,

$$\alpha_z\left(\psi_i^1,\psi_i^2\right) = 2A_z \left|\sin\left(\psi_i^1 + \nu t_m^z\right)\right| = 2A_z \left|\sin\left(\frac{\psi_i^1 - \psi_i^2}{2}\right)\right|. \tag{24}$$

So both $\alpha_{xy}$ and $\alpha_z$ are maximum when the differences between the initial phases are around $\pi$.

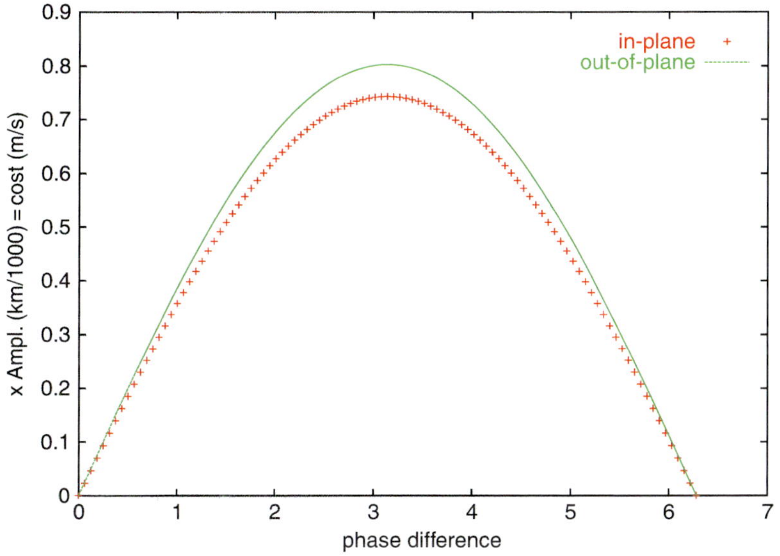


**Fig. 20** Cost of the persecution manoeuvres depending on the difference between the initial phases (Lissajous around $L_2$)

The cost coefficients for $xy$ and $z$ manoeuvres are represented in Fig. 20. These coefficients are never bigger than 0.8 for the chasing manoeuvres. This means that for an amplitude of 250000–300000 km, the maximum manoeuvres would be around 200 m/s, which is a considerable $\Delta$ v, but still affordable. However, the costs of the rendez-vous when the initial differences in phases lay in the range [–30,30] degrees is less than 20 m/s for Lissajous orbits in the Earth-Sun system of usable size.

In terms of time, no more than 50 days will be necessary in almost any case between one manoeuvre and the other one. The worst case is when the differences in phases are around $2\pi$ (despite being cheap in $\Delta v$). The general case, with initial differences in phases of less than $\pi$ corresponds to a waiting time between manoeuvre of 5–20 days.

The time of the $xy$-manoeuvres depends on $\frac{1}{2}(\phi_1 + \phi_2)$ (respectively in $z$, the time depends on $\frac{1}{2}(\psi_1 + \psi_2)$) as shown in equation (22). Therefore, a natural way of representing the time between manoeuvres is as a function of this semi-summations. For the $xy$ manoeuvres, there is another parameter playing a role in the computation of the time, which is $\beta$. Figure 21 shows the representation of the time between manoeuvres depending on the initial phases of both satellites. Values on the $x$-axis correspond to $\phi_1 + \phi_2$, while the values on the $y$-axis are $\psi_1 + \psi_2$. We can see how both directions show $2\pi$ periodicity, justified by equation (22), (in $x$ the base interval is $[2(\pi + \beta), 2(2\pi + \beta)]$, due to the aforementioned role of $\beta$).

We have seen that the EPP provide again nice ways to study rendez-vous in the libration zone. More details about this type of rendez-vous as well other strategies

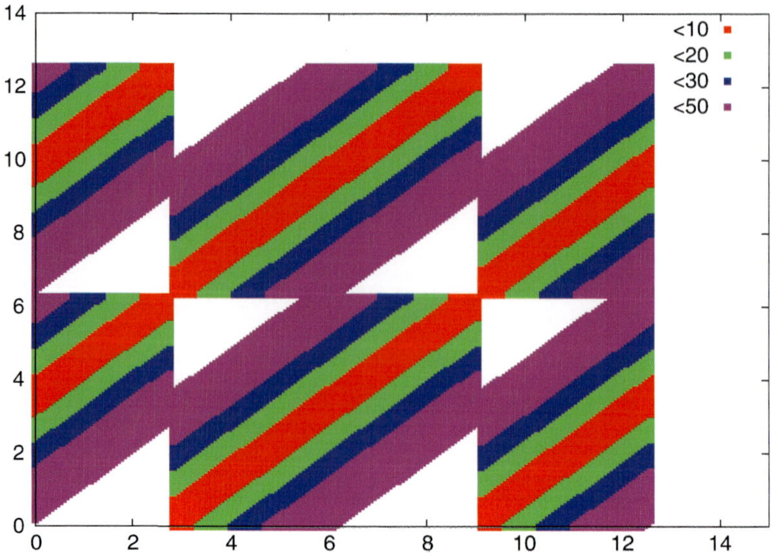


**Fig. 21** Time between manoeuvres (in days) depending on $\phi_1 + \phi_2$ ($x$-axis) and $\psi_1 + \psi_2$ ($y$-axis)

involving intermediate meeting trajectories also combined with amplitude changes can be found in [19].

## *5.6 Transfers Between Earth-Moon and Sun-Earth Lissajous Orbits*

In this section we consider a way to compute connecting trajectories between Earth-Moon and Sun-Earth $L_2$ Lissajous type orbits. Here we only summarise some results with a procedure that starts considering coupling two CRTBP. The interested reader will found more details, moreover the refinement of the trajectories in real ephemeris tackled in such a way that the cost of the coupling is iteratively reduced in [20] and [16].

A convenient way of tackling the four body problem Sun-Earth-Moon-Spacecraft is to decouple it in two different restricted three body problems: the Sun-Earth+Moon CRTBP (SE) and the Earth-Moon one (EM). Furthermore, we use the SE problem as general reference frame. Therefore, the natural thing to do is to deploy the Earth-Moon CRTBP inside the Sun-Earth+Moon model. The small primary of the SE problem represents the Earth and the Moon together in the point $(1 - \mu_{SE}, 0, 0)$. In fact, this point is the barycentre of the EM system. Then, this barycentre, which in the EM adapted coordinate frame acts as the origin of coordinates, circles the Sun at a constant angular rate on the ecliptic plane, completing a revolution every year. The Earth and the Moon, in turn, are assumed to circle around their common centre of mass, in a plane with fixed small inclination with respect to

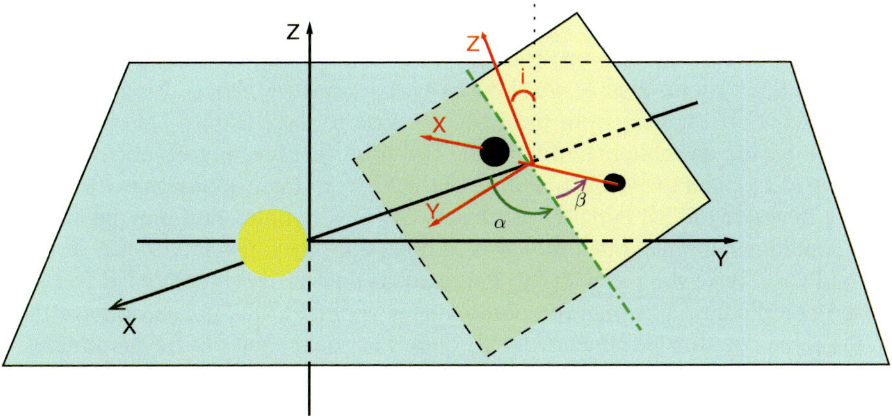


**Fig. 22** Coupling of the RTBPs. The figure shows the three angles which relate the SE coordinate frame with the EM one: $i$ (inclination), $\alpha$ and $\beta$

the ecliptic. Finally, the position of the Moon with respect to the Sun-Earth axis is described by two angles (see Fig. 22):

- $\alpha$: the angle from the axis joining the Sun and the Earth-Moon barycentre to the line of nodes of the Moon orbit, measured on the ecliptic plane.
- $\beta$: the angle from the line of nodes to the position of the Moon, measured on the plane of motion of the Moon around the Earth, or the plane of relative motion of the Earth and the Moon (mean longitude of the Moon).

Basically, we want to use the unstable manifold of a Lissajous orbit belonging to one of the models and integrate it until it meets the stable manifold of a Lissajous orbit of the other model. A Poincaré section is used as the meeting point. In particular, we choose the section,

$$\mathcal{S} = \{X = (x,y,z,\dot{x},\dot{y},\dot{z}) \mid x = -1 + \mu\},$$

(in SE coordinates). This plane forms a fixed angle of 90 degrees with the SE $X$-axis, and contains the Earth-Moon barycentre, i.e. the EM origin of coordinates. Note that when looked from the EM system, $\mathcal{S}$ is not fixed with respect to the $X$-axis.

Once $\mathcal{S}$ has been chosen, initial conditions on the stable (respectively unstable) invariant manifold of a particular Lissajous in the Sun-Earth+Moon system (respectively Earth-Moon) can be integrated backwards (respectively forwards) in time until they intersect $\mathcal{S}$. As the CRTBP model is autonomous (that is, time is not explicit in the equations of motion), we can set t = 0 on section $\mathcal{S}$. Actually, it is at this initial epoch when the phases $\alpha$ and $\beta$ depicted in Fig. 22 describe the relative position of the RTBPs. For times $t \neq 0$, the relative phases can be obtained using their rotation rates (which are assumed to be constant and have been computed by using the periods of revolution of the Earth and the Moon, 365.25 and 27.32 days, respectively).

Intersections on $\mathcal{S}$

Let us choose a Lissajous orbit around $L_2$ in the SE system, and another one around $L_2$ in the EM system. That is, we choose $(A_x,A_z)_{SE}$ and $(A'_x,A'_z)_{EM}$. Now, fix a sense of motion. For instance, from the EM Lissajous to the SE one. This means that we will use the unstable manifold of the EM orbit (because it goes away from the vicinity of $L_2$), and the stable manifold of the SE orbit (as it approaches it).

We integrate the SE initial conditions backwards in time, until they intersect $\mathcal{S}$. EM conditions are integrated forwards in time, until they also intersect $\mathcal{S}$. For this second case, once the cut with the Poincaré section has been computed in Earth-Moon coordinates, a change of coordinates is applied to the points, to have them in the same coordinate frame as the SE ones. For some values of the amplitudes in both sides, no intersecting region exists on the section. For some other values, we get an overlapping like the one depicted in Fig. 23. Bear in mind that connecting trajectories have to be continuous in position. Differences in velocity, on the contrary, can always be adjusted by using and adequate $\Delta$ v. On the Poincaré section $x$ is fixed to $-1+\mu_{SE}$. Therefore, we have to look for points on this section such that $(y,z)_{EM} = (y,z)_{SE}$ in order to obtain trajectories with common position coordinates.

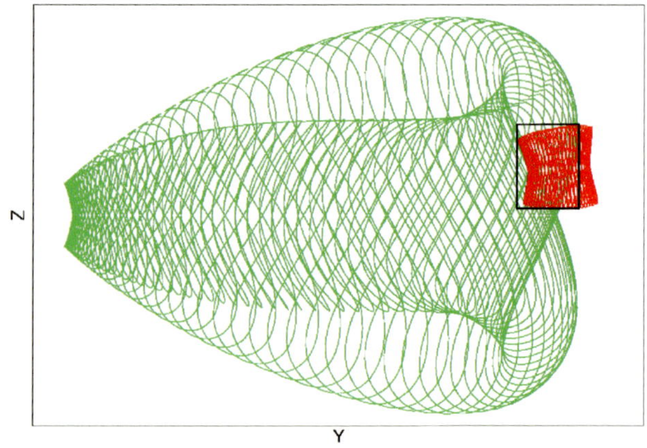


**Fig. 23** *yz* projection of the stable manifold of a Lissajous in the SE system (*green*) and the unstable manifold of a Lissajous in the EM system (*red*), at the crossing with the Poincaré section. The intersecting region in position coordinates is contained in the *black* square

Once an overlapping in the *yz* projection of the Poincaré section intersection has been detected between manifolds coming from the SE and the EM side, we can start computing connecting trajectories. For instance, pick a point $(y,z)$ in the overlapping region. There exist initial conditions on the SE and the EM Lissajous that result in this $(y,z)$ once integrated to the section. We want to find the exact phases that lead to $(y,z)$, taking as a seed a couple of phases such that the point on the section associated with $(\phi,\psi)$ is known and close to $(y,z)$ (the same for the EM case) by using, for instance, a Newton method on the Poincaré maps of both problems. Note that only the phases play a role in the Newton method, whereas all the amplitudes remain fixed.

Therefore, a connecting trajectory from a EM Lissajous orbit to a SE Lissajous orbit, crossing the Poincaré section at $t = 0$ through a particular point $(y,z)$ is represented by two couples of phases: $(\phi,\psi)_{SE}$ and $(\phi',\psi')_{EM}$. The complete trajectory is then obtained by integrating the initial condition represented by the EM phases forwards in time (respectively, the one represented by the SE phases backwards in time) until they meet at $(-1+\mu_{SE},y,z)$. At this point a manoeuvre is necessary, as the intersection occurs in position but not necessarily in velocity coordinates.

Moreover, a Lissajous orbit is a torus-like object, and its hyperbolic manifolds present several foldings when their projection in position coordinates on $\mathcal{S}$ is studied. This means that there may exist more than one couple $(\phi,\psi)$ that lead to a given $(y,z)$ on the section (in fact, up to 4 different couples of phases from each of the problems can be associated to the same $(y,z)$ on the section, leading to a final maximum number of possible connections equal to 16). This fact has to be taken into account when computing and storing the connecting trajectories. See for instance Fig. 24, where the starting phases from the EM side have been classified in two groups: the ones that lead to two points on the Poincaré section with common position coordinates and different velocity coordinates and the ones leading to four points with the same characteristics.

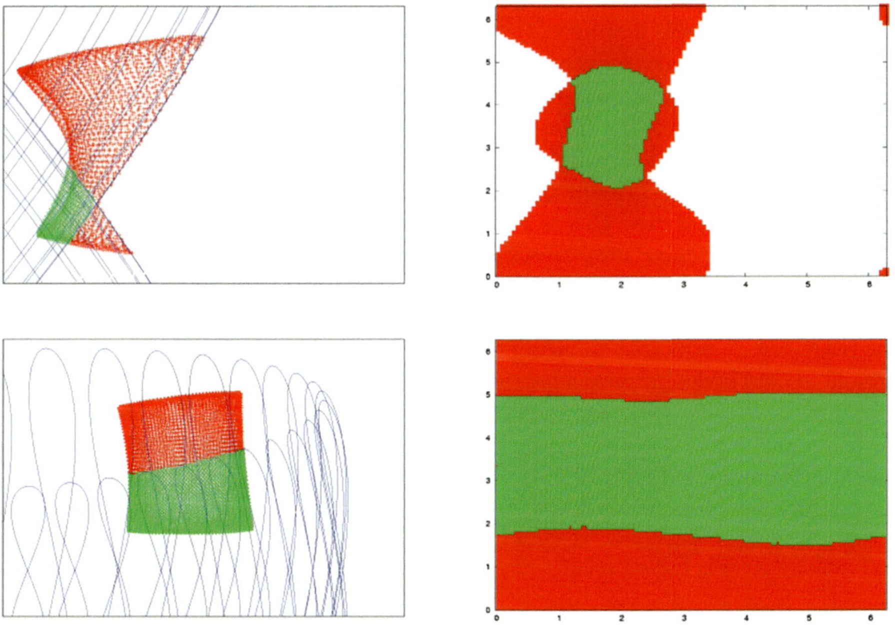

**Fig. 24** (*left*) $yz$ projection of the cuts of the manifolds on the Poincaré section (*blue* SE). Depending on how the manifolds fold on this projection, there exist 2 or 4 connecting trajectories through the same point. Starting phases on the EM manifold are classified on the right depending on the number of trajectories that they produce (2: *red*, 4:*green*)

**Table 4** Amplitudes of the Lissajous orbits around $L_2$ in the SE and EM problems which have been explored in the search for asymptotic connecting trajectories

| | SE (km) | EM (km) |
|---|---|---|
| $A_x$ | 80000–250000 | 1500–10000 |
| $A_z$ | 15000–550000 | 5000–35000 |

A wide range of in plane and out of plane amplitudes has been explored both for the SE and the EM side (see Table 4) and their intersections computed using the method explained in the previous section. The results can be found in [16] and it is not easy to draw general conclusions, due to the complexity of the problem and the great amount of parameters involved.

In this summary, the Poincaré section is fixed to the plane with fixed $x$ coordinate equal to $-1 + \mu_{SE}$. As for the sense of the motion, we choose to go from the EM to the SE libration regions (note, however, that the methodology that we develop is also usable for the opposite sense of motion). Concerning the relative initial configuration, only phases $\alpha$ and $\beta$ such that $\alpha + \beta \in (80,120)$ lead to intersections between the manifolds on $\mathcal{S}$.

Furthermore, the $\Delta vs$ which are necessary on the Poincaré section are usually between 100 and 400 m/s. manoeuvres of hundreds of meters per second are quite big, as the manifolds of the uncoupled RTBPs are not always close to each other. Despite this fact, note that no effort has been devoted to optimising them, as our final goal is to refine the trajectories to real ephemeris and it is in this more realistic model where the costs will be minimised.

Finally, Lissajous orbits from the SE side with big $A_x$ amplitudes (bigger than 100000 km) and small $A_z$ amplitudes (less than 150000 km) lead to much better results than other types of motions, with coupling manoeuvres ranging from slightly less than 100 to 200 m/s. This is a natural consequence of the fact that even moderately big out of plane oscillations in the lunar libration regions seem tiny when transformed to SE coordinates, and therefore, they cannot be expected to meet the manifold of a SE Lissajous orbit with big $A_z$ in a cheap way. On the other hand, the existence of zero cost connecting trajectories between planar Lyapunov orbits around SE and EM $L_2$ libration points was also been previously proved [18]. Among Lissajous orbits, the ones that are more similar to Lyapunovs are, the ones with big $A_x$ and small $A_z$. Consequently, the present results which point towards the aforementioned kind of Lissajous orbits as the most adequate for lunar to solar transfer, are consistent with previous results (Fig. 25).

Refinement to JPL Ephemeris

To start with, one can think of trying to refine the whole connecting trajectories using a multiple shooting method, after properly setting the initial epoch. However, this will rarely provide satisfying results, as the parallel shooting cannot generally couple the two different RTPBs with no additional help. That is to say that the connecting trajectory obtained from coupling two RTBPs may be too far from a real

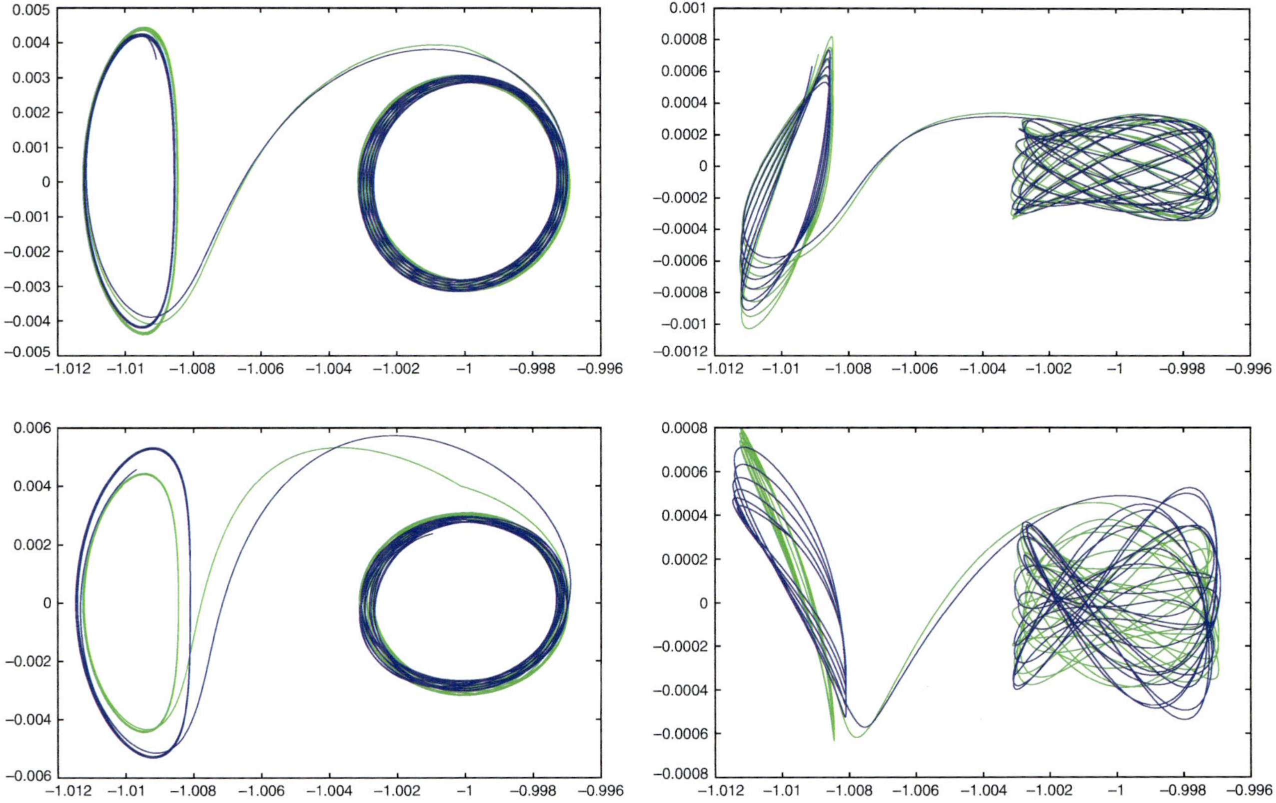


**Fig. 25** Trajectories joining a Lissajous around $L_2$ in the EM system and a Lissajous around the $L_2$ point of the SE system. Pictures on the *left* show $xy$ projections while pictures on the *right* show $xz$ projections. Trajectories in *green* correspond to JPL ephemeris trajectories at the first refinement from the coupled RTBP's ($\Delta v$ is noticeable in the figures), while trajectories in *blue* are zero cost JPL connections. The example in the first row is labelled as 15 in Table 5, and corresponds to the case where the final refined trajectory does not differ from the original seed. On the contrary, the second row corresponds to projections of trajectory 14, whose 0-cost refinement leads to deviations from the original path

**Table 5** Cost and transfer duration for several example connecting trajectories found in the coupled RTBPs model. The amplitudes of the Lissajous around $L_2$ in each of the problems, as well as the coupling angles at $t = 0$ are shown. Note that an infinite number of connecting trajectories exists between two given Lissajous which have an overlapping in $yz$ coordinates, and the table only shows the cost of one of them (close to the minimum cost one)

| | SE | Amplitudes (km) | EM | Amplitudes (km) | Phases | (deg) | Coupled | RTBPs |
|---|---|---|---|---|---|---|---|---|
| | | | | | | | Cost | Duration |
| | $A_x$ | $A_z$ | $A_x$ | $A_z$ | $\alpha$ | $\beta$ | (m/s) | (days) |
| 1 | 100072.0 | 320000 | 6560.7 | 20000 | 50 | 40 | 419.19 | 166.05 |
| 2 | 103200.0 | 330000 | 5576.6 | 17000 | 45 | 50 | 394.49 | 166.84 |
| 3 | 106327.0 | 340000 | 7544.8 | 23000 | 80 | 10 | 377.00 | 167.27 |
| 4 | 112581.4 | 360000 | 6560.7 | 20000 | 50 | 50 | 347.84 | 170.33 |
| 5 | 115708.7 | 370000 | 6560.7 | 20000 | 35 | 60 | 303.18 | 169.51 |
| 6 | 121963.2 | 390000 | 5904.6 | 18000 | 65 | 30 | 295.20 | 169.48 |
| 7 | 125090.3 | 400000 | 6232.6 | 19000 | 55 | 45 | 284.60 | 170.71 |
| 8 | 134472.2 | 430000 | 4592.5 | 14000 | 30 | 75 | 287.24 | 171.96 |
| 9 | 137600.0 | 440000 | 5576.0 | 17000 | 65 | 40 | 268.67 | 174.98 |
| 10 | 137600.0 | 440000 | 7216.7 | 22000 | 80 | 20 | 249.16 | 169.51 |
| 11 | 244088.3 | 75382.4 | 5248.5 | 16000 | 15 | 105 | 163.08 | 183.92 |
| 12 | 244088.3 | 75382.4 | 4592.5 | 14000 | 20 | 100 | 167.55 | 204.89 |
| 13 | 244088.3 | 75382.4 | 6560.7 | 20000 | 0 | 120 | 155.71 | 201.98 |
| 14 | 212964.4 | 75382.4 | 5248.5 | 16000 | 10 | 100 | 209.59 | 179.18 |
| 15 | 212964.4 | 75382.4 | 2952.3 | 9000 | 60 | 45 | 243.53 | 169.33 |
| 16 | 194619.3 | 90458.9 | 4920.5 | 15000 | 25 | 95 | 102.99 | 177.71 |
| 17 | 177341.7 | 60305.9 | 4920.5 | 15000 | 35 | 70 | 173.87 | 172.57 |
| 18 | 175638.0 | 90458.9 | 6560.7 | 20000 | 30 | 85 | 72.23 | 173.94 |
| 19 | 177341.7 | 60305.9 | 6888.7 | 21000 | 75 | 35 | 52.74 | 169.98 |
| 20 | 157075.8 | 45229.4 | 6888.7 | 21000 | 5 | 100 | 87.69 | 169.02 |

JPL ephemeris trajectory at some points for the parallel shooting to smooth it at the first try.

On the contrary, each one of the legs (SE and EM) can be easily refined from CRTBP to JPL coordinates, if these transformations are done separately. The problem is, naturally, that the point corresponding to the initial epoch, which in CRTBP coordinates was the same on the section for both branches, is changed during the refinement. Consequently, some kind of forcing in the conditions corresponding to the point on the section, in order to obtain connecting trajectories which are continuous in position in JPL coordinates, has to be applied. A good way to proceed is to use a forcing that consists of refining the EM leg with no restrictions, and then using the final $(x,y,z)$ coordinates that have been obtained as the position coordinates of the first point of the SE leg. In other words, we force the initial point of the SE leg to be the final one of the EM leg, which we have obtained by freely refining the part of the connecting trajectory coming from the lunar region. This method is called the *section-forced refinement to JPL coordinates* [20, 16].

The section-forced refinement to JPL coordinates can be applied to any of the connecting trajectories that we obtain by coupling the two RTBPs. As a result, we get trajectories in JPL coordinates, which are continuous in position but need a $\Delta v$

**Table 6** Manoeuvres at the coupling point which are necessary when the coupled CRTBP connecting trajectories of Table 6 have been refined to JPL ephemeris using the section forced refinement, for several epochs of section crossing

| | | JPL cost (m/s) | |
|---|---|---|---|
| | CRTBP cost (m/s) | 28-Dec-2009 | 24-Aug-2015 |
| 1 | 419.19 | 516.55 | 523.34 |
| 2 | 394.49 | 491.29 | 496.22 |
| 3 | 377.00 | 484.13 | 491.20 |
| 4 | 347.84 | 430.68 | 432.47 |
| 5 | 303.18 | 362.52 | 362.80 |
| 6 | 295.20 | 376.61 | 374.95 |
| 7 | 284.60 | 351.09 | 348.88 |
| 8 | 287.24 | 323.10 | 316.74 |
| 9 | 268.67 | 332.53 | 327.02 |
| 10 | 249.16 | 315.88 | 310.41 |
| 11 | 163.08 | 214.65 | 227.30 |
| 12 | 167.55 | 219.24 | 232.40 |
| 13 | 155.71 | 173.06 | 187.83 |
| 14 | 209.59 | 255.45 | 271.59 |
| 15 | 243.53 | 170.58 | 155.16 |
| 16 | 102.99 | 128.47 | 137.01 |
| 17 | 173.87 | 187.99 | 199.09 |
| 18 | 72.23 | 87.46 | 95.88 |
| 19 | 52.74 | 79.43 | 80.92 |
| 20 | 87.69 | 128.81 | 144.96 |

in the coupling point. Moreover, the cost of the manoeuvre in JPL coordinates is of the same order of magnitude as the original one in the coupled RTBPs, as observed in Table 6 (the characteristics of the Lissajous orbits joined by these connecting trajectories are contained in Table 5).

Moreover, the connecting trajectories in JPL coordinates which are obtained from the coupled RTBPs ones by using the section forced refinement have to be regarded just as a good initial seed for obtaining cheaper realistic trajectories and not as final usable trajectories themselves. For instance, trajectory optimisation procedures could be applied to them and low cost transfers are bound to be obtained. No optimiser has been used in the current work. On the contrary, a search for free connecting trajectories in JPL coordinates has been successfully performed by modifying the multiple shooting algorithm [20, 16].

# 6 Station Keeping

## *6.1 The Target Mode Approach and the Floquet Mode Approach*

The problem of controlling a spacecraft moving near an inherently unstable libration point orbit is of current interest. In the late 1960's, Farquhar [24, 15] suggested several station–keeping strategies for nearly–periodic solutions near the collinear points. Later, in 1974, a station–keeping method for spacecraft moving on halo orbits in the vicinity of the Earth–Moon translunar libration point ($L_2$) was, published by Breakwell, Kamel, and Ratner [11]. These studies assumed that the control could be modelled as continuous. In contrast, specific mission requirements influenced the station–keeping strategy for the first libration point mission. Launched in 1978, the International Sun–Earth Explorer–3 (ISEE–3) spacecraft remained in a near–halo orbit associated with the interior libration point ($L_1$,) of the Sun–Earth/Moon barycentre system for approximately three and one half years [26]. Impulsive manoeuvres at discrete time intervals (up to 90 days) were successfully implemented as a means of trajectory control. Since that time, more detailed investigations have resulted in various station–keeping strategies, including the two identified here as the Target Point and Floquet Mode approaches.

The Target Point method (as presented by Howell and Pernicka [42], Howell and Gordon [43], and Keeter [47], which is based in Breackwell's ideas) computes correction manoeuvres by minimising a weighted cost function. The cost function is defined in terms of a corrective manoeuvre as well as position and velocity deviations from a nominal orbit at a number of specified future times $t_i$. The non–final state vectors at each time $t_i$ are denoted as "target points." The target points are selected along the trajectory at discrete time intervals that are downstream of the manoeuvre. In contrast, the Floquet Mode approach, as developed by Simó et al. [58, 59], incorporates invariant manifold theory and Floquet modes to compute the manoeuvres. Floquet modes associated with the monodromy matrix are used to determine the unstable component corresponding to the local error vector. The manoeuvre is then computed in such a way that the dominant unstable component of the error is eradicated. It is noted that both approaches have been demonstrated in a complex model such as the Earth–Moon system.

### Target Point Approach

The goal of the Target Point station–keeping algorithm is to compute and implement manoeuvres to maintain a spacecraft "close" to the nominal orbit, i.e., within a region that is locally approximated in terms of some specified radius centred about the reference path. To accomplish this task, a control procedure is derived from minimisation of a cost function. The cost function, $J$, is defined by weighting both the control energy required to implement a station–keeping manoeuvre, $\Delta v$, and a series of predicted deviations of the six–dimensional state from the nominal orbit at specified future times. The cost function includes several sub-matrices from the state transition matrix. For notational ease, the state transition matrix is partitioned into four $3 \times 3$ sub-matrices as

$$\Phi(t_k,t_0) = \begin{bmatrix} A_{k0} & B_{k0} \\ C_{k0} & D_{k0} \end{bmatrix}.$$

The controller, in this formulation, computes a $\Delta v$ in order to change the deviation of the spacecraft from the nominal path at some set of future times. The cost function to be minimised is written in general as

$$J = \Delta v^T Q \Delta v + p_1^T R p_1 + v_1^T R_\upsilon v_1 + p_2^T S p_2 + v_2^T S_\upsilon v_2 + p_3^T T p_3 + v_3^T T_\upsilon v_3, \quad (26)$$

where superscript $T$ denotes transpose. The variables in the cost function include the corrective manoeuvre, $\Delta v$ at some time $t_c$, and $p_1, p_2$ and $p_3$ that are defined as $3 \times 1$ column vectors representing linear approximations of the expected deviations of the actual spacecraft trajectory from the nominal path (if no corrective action is taken) at specified future times $t_1$, $t_2$ and $t_3$, respectively. Likewise, the $3 \times 1$ vectors $v_1$, $v_2$ and $v_3$ represent deviations of the spacecraft velocity at the corresponding $t_i$. The future times at which predictions of the position and velocity state of the vehicle are compared to the nominal path are denoted as target points. They are represented as $\Delta t_i$ such that $t_i = t_0 + \Delta t_i$. The choice of identifying *three* future target points is arbitrary.

In equation (26), $Q$, $R$, $S$, $T$, $R_\upsilon$, $S_\upsilon$, and $T_\upsilon$, are $3 \times 3$ weighting matrices. The weighting matrix $Q$ is symmetric positive definite; the other weighting matrices are positive semi–definite. The weighting matrices are generally treated as constants that must be specified as inputs. Selection of the appropriate weighting matrix elements is a trial and error process that has proven to be time–consuming. A methodology has been developed that automatically selects and updates the weighting matrices for each manoeuvre. This "time-varying" weighting matrix algorithm is based solely on empirical observations.

Determination of the $\Delta v$ corresponding to the relative minimum of this cost function allows a linear equation for the optimal control input, i.e.,

$$\Delta v^* = -A^{-1} \times (B + C), \quad (27)$$

with

$$A = Q + B_{10}^T R B_{10} + B_{20}^T S B_{20} + B_{30}^T T B_{30} + D_{10}^T R_\nu D_{10} + D_{20}^T S_\nu D_{20} + D_{30}^T T_\nu D_{30},$$
$$B = \left(B_{10}^T R B_{10} + B_{20}^T S B_{20} + B_{30}^T T B_{30} + D_{10}^T R_\nu D_{10} + D_{20}^T S_\nu D_{20} + D_{30}^T T_\nu D_{30}\right) v_0,$$
$$C = \left(B_{10}^T R A_{10} + B_{20}^T S A_{20} + B_{30}^T T A_{30} + D_{10}^T R_\nu C_{10} + D_{20}^T S_\nu C_{20} + D_{30}^T T_\nu C_{30}\right) p_0,$$

where $v_0$, is the residual velocity ($3 \times 1$ vector) and $p_0$ is the residual position ($3 \times 1$ vector) relative to the nominal path at the time $t_0$. The performance of the modified Target Point algorithm is not yet truly "optimal," though it has been proved to successfully control the spacecraft at reasonable costs. This accomplishment alone provides the user with a quick and efficient way to obtain reasonable station–keeping results. Given some procedure to select the weighting matrices, the manoeuvre is computed from equation (27). The corrective manoeuvre ($\Delta v^*$) is a function of spacecraft drift (in both position and velocity with respect to the nominal orbit), the state transition matrix elements associated with the nominal orbit, and the weighting matrices. It is assumed here that there is no delay in implementation of the manoeuvre; the corrective manoeuvre occurs at the time $t_0$, defined as the current time. Note that this general method could certainly accommodate inclusion of additional target points. Although the nominal orbit that is under consideration here is quasi-periodic, the methodology does not rely on periodicity; it should be applicable to any type of motion in this regime.

In this application, three additional constraints are specified in the station–keeping procedure to restrict manoeuvre implementation. First, the time elapsed between successive manoeuvres must be greater than or equal to a specified minimum time interval, $t_{min}$. This constraint may be regulated by the orbit determination process, scientific payload requirements, and/or mission operations. Time intervals of 1–3 days are considered in the Earth-Moon system. The second constraint is a scalar distance ($p_{min}$) and specifies a minimum deviation from the nominal path (an isochronous correspondence) that must be exceeded prior to manoeuvre execution. For distances less than $p_{min}$ manoeuvre computations do not occur. Third, in the station–keeping simulation, the magnitude of position deviations are compared between successive tracking intervals. If the magnitude is decreasing, a manoeuvre is not calculated. For a corrective manoeuvre to be computed, all three criteria must be satisfied simultaneously.

After a manoeuvre is calculated by the algorithm, an additional constraint is specified on the minimum allowable manoeuvre magnitude, $\Delta v_{min}$. If the magnitude of the calculated — $\Delta v$ is less than $\Delta v_{min}$ then the recommended manoeuvre is cancelled. This constraint is useful in avoiding “small” manoeuvres that are approximately the same order of magnitude as the manoeuvre errors. It also serves to model actual hardware limitations.

### Floquet Mode Approach

An alternative strategy for station–keeping is the Floquet Mode approach, a method that is significantly different from the Target Point approach. It can be easily formulated in the circular restricted three–body problem. In this context, the nominal halo orbit is periodic. The variational equations for motion in the vicinity of the nominal trajectory are linear with periodic coefficients. Thus, in general, both qualitative and quantitative information can be obtained about the behaviour of the nonlinear system from the monodromy matrix, $M$, which is defined as the state transition matrix (STM) after one revolution along the full halo orbit.

The knowledge of the dynamics of the flow around a halo orbit, or any solution close to it, allows possibilities other than the station–keeping procedure described here, such as the computation of transfer orbits both between halo orbits and from the Earth to a halo orbit [30, 32]. The behaviour of the solutions in a neighbourhood of the halo orbits is determined by the eigenvalues, $\lambda_i$, $i = 1,\ldots,6$ and eigenvectors $e_i$, $i = 1,\ldots,6$ of $M$. Gathering the eigenvalues by pairs, their geometrical meaning is the following:

a) The first pair $(\lambda_1, \lambda_2)$ with $\lambda_1 \cdot \lambda_2 = 1$ and $\lambda_1 \approx 1500$, is associated with the unstable character of the small and medium size halo orbits. The eigenvector, $e_1(t_0)$, associated with the largest eigenvalue, $\lambda_1$, defines the most expanding direction, related to the unstable nature of the halo orbit. The image under the variational flow of the initial vector $e_1(t_0)$, together with the vector which is tangent to the orbit, defines the linear approximation of the unstable manifold of the orbit. In a similar way, $e_2(t_0)$ can be used to compute the linear approximation of the stable manifold.

b) The second pair $(\lambda_3, \lambda_4) = (1, 1)$ is associated with neutral variables (i.e., unstable modes). However, there is only one eigenvector with eigenvalue equal to one. This vector, $e_3(t_0)$ is the tangent vector to the orbit. The other eigenvalue, $\lambda_4 = 1$, is associated with variations of the energy (or the period) of the orbit through the family of halo orbits. Along the orbit, the vectors $e_3$ and $e_4$ span an invariant plane under the flow.

c) The third couple, $(\lambda_5, \lambda_6)$, is formed by two complex conjugated eigenvectors of modulus one. The restriction of the flow to the corresponding two–dimensional invariant subspace, is essentially a rotation. This behaviour is related to the existence of quasi-periodic halo orbits around the halo orbit (see [34]).

When considering dynamical models of motion different from the restricted three–body problem, halo orbits are no longer periodic, and the monodromy matrix is not defined. Nevertheless, for quasi-periodic motions close to the halo orbit (and also for the Lissajous orbits around the equilibrium point) the unstable and stable manifolds subsist. The neutral behaviour can be slightly modified including some instability which, from a practical point of view, is negligible when compared with the one associated with $\lambda_1$.

Instead of the vectors $e_i(t)$ it is convenient to use the Floquet modes $\overline{e}_i(t)$ which, for the periodic case, are defined as six periodic vectors from which the $e_i(t)$ can be easily recovered (see [62]). For instance $\overline{e}_1(t)$ is defined as $e_1(t)\cdot\exp[-(t/T)\log\lambda_1]$, where $T$ is the period of the halo orbit. The control algorithm is developed to use this information for station–keeping purposes. The emphasis is placed on formulating a controller that will effectively eliminate the unstable component of the error vector, $\delta(t) = (\delta x,\delta y,\delta z,\delta\dot{x},\delta\dot{y},\delta\dot{z})$ defined as the difference between the actual coordinates obtained by tracking and the nominal ones computed isochronously on the reference orbit. At any epoch, $t$, $\delta$ can be expressed in terms of the Floquet modes

$$\delta(t) = \sum_{i=1}^{6} \alpha_i \overline{e}_i(t). \tag{28}$$

The objective of the controller is to add a manoeuvre such that the magnitude of the component of the error vector in the unstable direction, $\alpha_1$, is reduced to zero. The five remaining components do not produce large departures from the reference orbit. In contrast, the component of the error vector along the unstable mode increases by a factor of $\lambda_1$ in each revolution.

Denoting the impulsive manoeuvre as $\Delta = (0,0,0,\Delta_x,\Delta_y,\Delta_z)^T$, cancelling the unitary unstable Floquet mode requires

$$\frac{\overline{e}_1}{\|\overline{e}_1\|} + (0,0,0,\Delta_x,\Delta_y,\Delta_z)^T = \sum_{i=2}^{6} c_i \overline{e}_i(t). \tag{29}$$

From these equations $\Delta_x$, $\Delta_y$, $\Delta_z$ can be obtained as a function of $c_5$ and $c_6$. These free parameters are determined by either imposing a constraint on the available directions of the control or minimising a suitable norm of $\Delta$.

For practical implementation it is useful to compute the so–called projection factor along the unstable direction. It is defined as the vector $\pi$ such that $\delta \cdot \pi = \alpha_1$. Note that for the computation of $\pi$ only the Floquet modes are required, so it can be computed and stored together with the nominal orbit. To annihilate the unstable projection, $\alpha_1$, with a manoeuvre, $\Delta v = (0,0,0,\Delta_x,\Delta_y,\Delta_z)^T$, we ask $(\delta + \Delta v) \cdot \pi = 0$. In this way,

$$\Delta_x \pi_4 + \Delta_y \pi_5 + \Delta_z \pi_6 + \alpha_1 = 0, \tag{30}$$

is obtained, where $\pi_4$, $\pi_5$ and $\pi_6$ are the last three components of $\pi$. Choosing a two axis controller, with $\Delta_z = 0$, and minimising the Euclidean norm of $\Delta v$, the following expressions for $\Delta_x$ and $\Delta_y$ are obtained,

$$\Delta_x = -\frac{\alpha_1 \pi_4}{\pi_4^2 + \pi_5^2}, \quad \Delta_y = -\frac{\alpha_1 \pi_5}{\pi_4^2 + \pi_5^2}. \tag{31}$$

In a similar way, a one or three axis controller can be formulated.

Once the magnitude of the manoeuvre is known, the determination of the epoch at which it must be applied is an important issue to be addressed. The study of this question requires the introduction of the gain function, $g(t) = ||\Delta||^{-1}$, where $\Delta$ is the unitary impulsive manoeuvre. It measures the efficiency of the control manoeuvre along the orbit to cancel the unitary unstable component. This component is obtained using the projection factors and the error vector. As the projection factor changes along the orbit, the same error vector has different unstable components. Then, it is natural to consider a delay in the manoeuvre until reaching a better epoch with less cost. So, the function to be studied is

$$R(t) = \frac{\exp\left(t \log\left(\frac{\lambda_1}{T}\right)\right)}{g(t)}. \tag{32}$$

However, as shown in [58], this function is always increasing. Therefore, it is never good to wait for a manoeuvre except for operational reasons.

As it has been said, when the station keeping has to span for a long time, the satellite can tend to deviate far away from the nominal orbit. This could happen since the cancellation of the unstable component does not take care of the neutral components which might grow up to the limit of loosing controllability. In order to prevent large deviations of the satellite from the nominal orbit, it is advisable to perform manoeuvres of insertion into the stable manifold. The main idea of the strategy is to put the satellite in a such state that approaches the nominal orbit asymptotically in the future. This strategy is, in principle, much cheaper than to target to the nominal orbit itself since the latter case can be considered, from an implementation point of view, as a sub-case of targeting to the stable manifold. Moreover, even when the controllability using only unstable component cancellation manoeuvres (UCCM) is assured, it can be advantageous to perform an insertion into the stable manifold since the control effect of this manoeuvres usually persist for a longer time span than UCCM. Moreover, subsequent UCCM would be cheaper due to the fact that the satellite is closer to the nominal orbit, and consequently the projection of the deviation in the unstable component is smaller.

Although the idea is simple, the implementation is not so easy since, in the first place, the target state in the stable manifold can not be accomplished with a single manoeuvre as it happens with UCCM and secondly, the actual state of the satellite is known but affected by tracking errors. Moreover the manoeuvres to be done will be noised by some errors too. We refer to [39] for the details of the implementation.

As a final remark, several constraints that impact the manoeuvres must be specified in the procedure. The most relevant are the time interval between two consecutive tracking epochs (tracking interval), the minimum time interval between manoeuvres, and the minimum value of $\alpha_1$, that can not be considered due solely to tracking errors.

Special emphasis must be placed on the evolution of $\alpha_1$. With no tracking errors, the evolution of this parameter is exponential in time (see Figs. 26 and 27). When adding tracking errors, and in order to prevent a useless manoeuvre, this value must be greater than the minimum. So, the minimum value must be selected as a function of the orbit determination accuracy. On the other hand, the value of $\alpha_1$ should not be too large because this increases the value of the manoeuvre in an exponential way. Thus, a maximum value is chosen such that if $\alpha_1$, is greater than the maximum, a control manoeuvre will be executed to cancel the unstable component. When $\alpha_1$, is between minimum and maximum values, the error can be due to small oscillations around the nominal orbit. In this case, a manoeuvre is executed only if the error has been growing at an exponential rate in the previous time steps and the time span since the last manoeuvre agrees with the selected one. Also, if the magnitude of the calculated $\Delta v$ is less than $\Delta v_{min}$, then the recommended manoeuvre is cancelled. Once these parameters have been fixed, there are no more free variables allowing any further minimisation.

## *6.2 Numerical Results*

In Figs. 26 and 27 we show some results of simulations done for a halo orbit around $L_2$ in the Earth-Moon system. We display the evolution of the unstable component and the deviations from the nominal trajectory, in position and velocity in different situations. In Fig. 26 the manoeuvres are done without any error, while in Fig. 27 they are performed with errors. In each Figure we show the results of the station keeping strategy with and without stable manifold insertion manoeuvres. From them, the exponential grow of $\alpha_1$ as well as the role of the insertion manifold manoeuvres, become clear.

Finally, in Fig. 28 the averaged $\Delta$ v used for the station keeping is displayed. As in the previous case, the simulations correspond to a halo orbit around $L_2$ in the

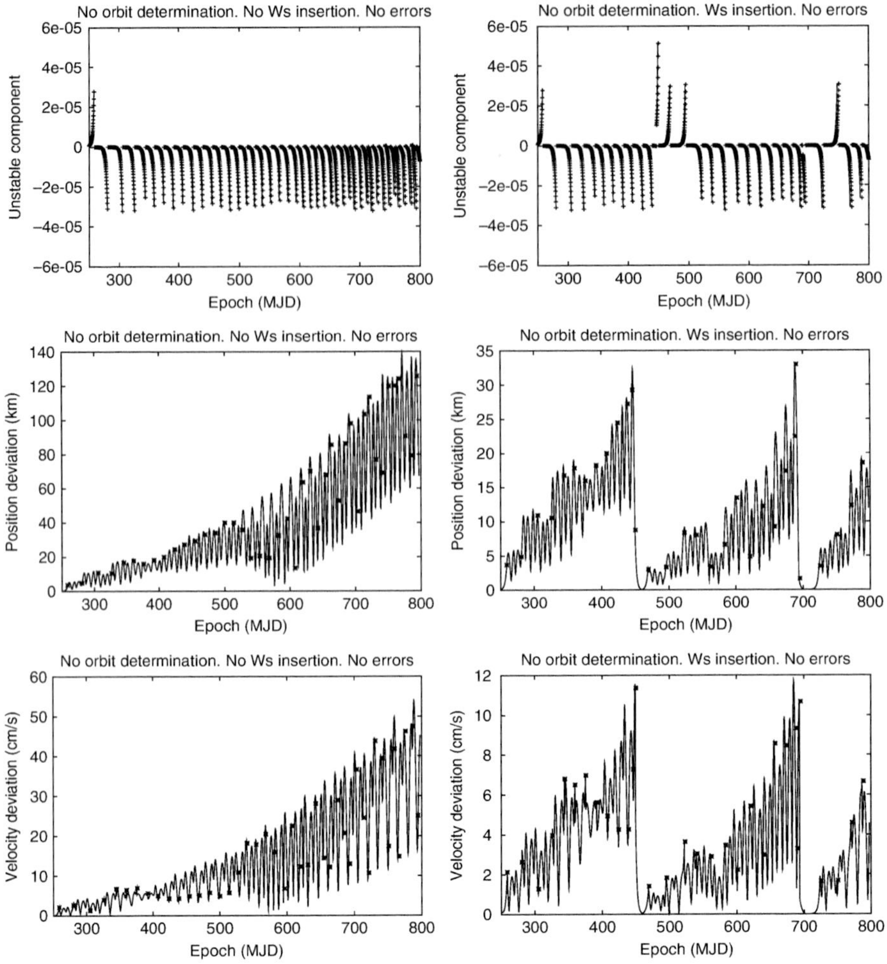


**Fig. 26** From *top* to *bottom* evolution with time of the unstable component, position deviations with respect to the nominal trajectory and velocity deviations. In the all figures no orbit determination has been performed because the simulations have been done with no errors for the tracking and the execution of the manoeuvres. There is only an error at the initial insertion epoch. In the *left* hand side figures there are no manoeuvres for the insertion in the stable manifold while in the *right* hand side yes. These manoeuvres can be clearly seen because after its execution the distance to the nominal orbit goes to zero both in position and in velocity. The discontinuities that appear in these two figures are associated to the execution of the manoeuvres. The points marked with a cross are those at which the tracking has been performed and then ones marked with a star are those at which a manoeuvre has been executed

Earth-Moon system. If no insertion manifold manoeuvres are done, there appears an exponential grow of the $\Delta$ v. On the contrary, if these manoeuvres are executed the station keeping cost per year remains constant. From this Figure it is also clear that a good orbit determination procedure can be useful to reduce the total $\Delta$ v.

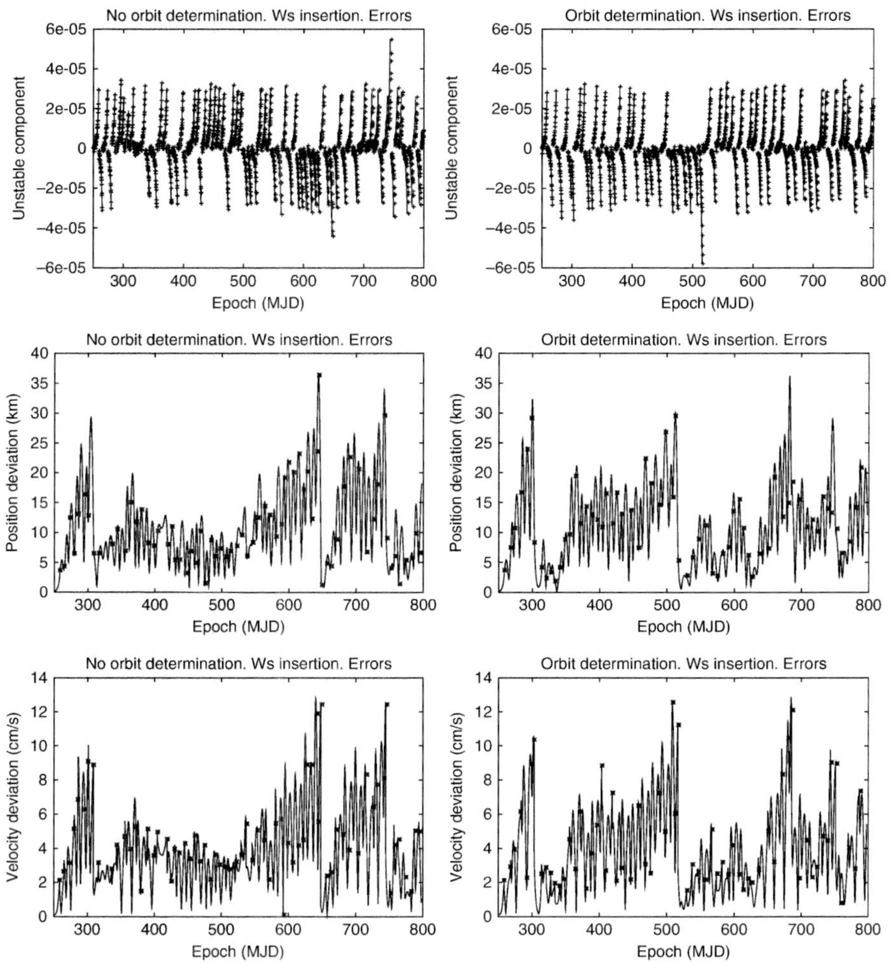


**Fig. 27** From *top* to *bottom* evolution with time of the unstable component, position deviations with respect to the nominal trajectory and velocity deviations. In the *left* figures no orbit determination if performed while in the *right* hand side ones yes. In all the figures the manoeuvres and the tracking are performed with errors. There is also an error at the initial insertion epoch. There are manoeuvres for the insertion in the stable manifold that can be clearly seen at the moments at which the distance to the nominal orbit decreases to very small values (which is not equal to zero because there is an error added to the manoeuvres). The points marked with a *cross* are those at which the tracking has been performed and then ones marked with *a star* are those at which a manoeuvre has been executed

## 7 Low Energy Transfers and the Weak Stability Boundary

According to Simó [60]: "It seams feasible to produce accurate and enough complete descriptions of the dynamics on the centre manifolds of the collinear libration points as well as large parts of the corresponding stable and unstable manifolds. Having these concepts in hand, the design of space missions *à la carte*, involving

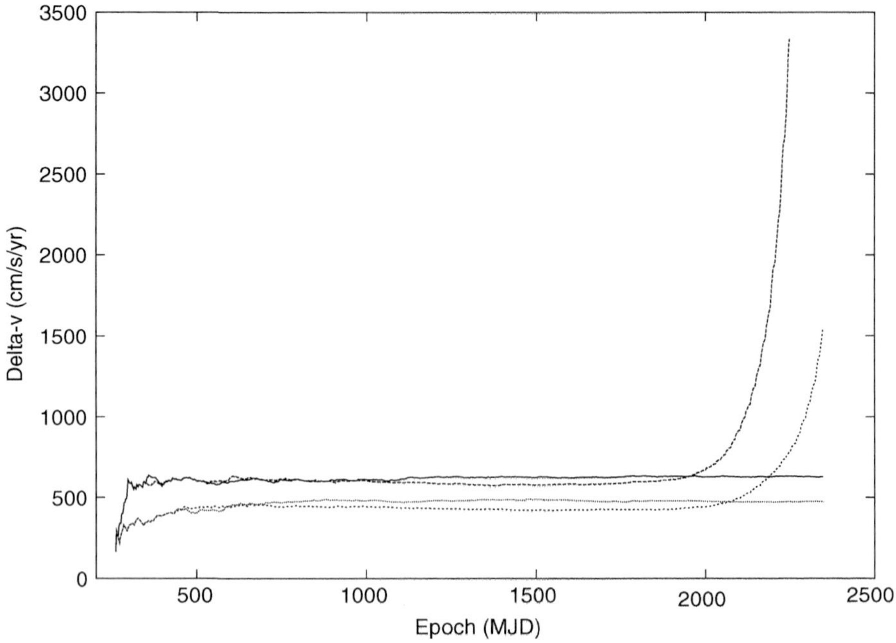


**Fig. 28** Averaged Δ v used for the station keeping in cm/s/year in different situations. The two curves with an exponential grow of the Δ v correspond to simulations with no insertions in the stable manifold. For the *upper* curve there was no orbit determination. The other two curves, for which there seems to be a finite limit for the Δ v, we have used insertion manoeuvres in the stable manifold. The one with lower cost uses orbit determination and the other no

the vicinities of these points, could be done in an automatic way". Although not all the theoretical and practical questions underlying the above idea and required for its implementation have been solved, some progress has been done and will be summarised in this section.

The invariant manifold structures associated to the collinear libration points, provide not only the framework for the computation of complex spacecraft mission trajectories, but also can be used to understand the geometrical mechanisms of the material transport in the solar system. This approach has been used recently for the design of low energy transfers from the Earth to the Moon [50] and for a "Petit Grand Tour" of the Moons of Jupiter [48, 36]. It has also been used to explain the behaviour of some captured Jupiter comets, see [46, 49].

The weak stability boundary, although it is not a clearly defined concept, provides also some kind of low energy transfers, as will be explained in what follows.

## *7.1 The Weak Stability Boundary*

The weak stability boundary (WSB) is a concept mainly developed after [3] and the rescue of the Hiten spacecraft [4]. The transfer trajectory of this spacecraft from the Earth to the Moon first visited a neighbourhood of the collinear libration point $L_1$ of the Earth-Sun system. Afterwards, it went to the vicinity of the $L_2$ point of the

Earth-Moon system and finally reached the Moon. This kind of transfer trajectories require a large transfer time (between 60 and 100 days) but a small $\Delta v$ (they can save up to 150 m/s with respect to a Hohmann transfer) since they eliminate the hyperbolic excess velocity at lunar periapsis upon arrival and use the dynamics of the problem in a more natural. For this reason, they are also called low energy transfers.

The WSB (also named fuzzy boundary) has not got a precise definition (at least in the mathematical sense of the word). Some definitions that can be found in the literature are: "a generalisation of the Lagrange points and a complicated region surrounding the Moon", "a region in phase space supporting a special type of chaotic motion for special choices of elliptic initial conditions with respect $m_2$",.... According to Belbruno [9], the WSB can also be defined algorithmically in the following way:

In the framework of the CRTBP, consider a radial segment, $l$, departing from the small primary $m_2$. Take trajectories for the infinitesimal body, $m$, starting on $l$ which satisfy

1. The particle $m$ starts its motion at the periapsis of the oscullating ellipse.
2. The initial velocity vector of the trajectory is perpendicular to the line $l$.
3. The initial two-body Kepler energy of $m$ with respect to $m_2$ is negative.
4. The eccentricity of the initial two-body Keplerian motion is held fixed along $l$.

According to [8], the motion of a particle with the above initial conditions is *stable* about $m_2$ if: after leaving $l$ it makes a full cycle about $m_2$ without going around $m_1$ and returns to $l$ at a point with negative Kepler energy with respect to $m_2$. Of course, the motion will be *unstable* if the above condition is not fulfilled. (Fig. 29)

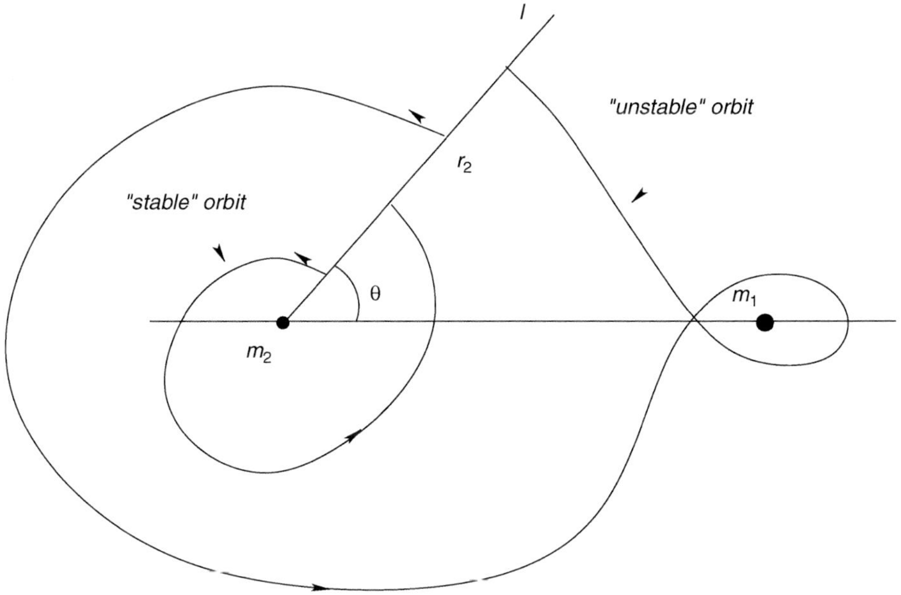


**Fig. 29** Qualitative representation of stable and unstable orbits about $m_2$

Belbruno claims (without giving any proof or numerical evidence) that as the initial conditions vary along $l$, there is a finite distance $r^*(\theta,e)$, depending on the polar angle $\theta$ (which $l$ makes with the $x$-axis) and the eccentricity $e$ of the initial osculating ellipse, such that

- If $r_2 < r^*$, the motion is stable.
- If $r_2 > r^*$, the motion is unstable.

Furthermore, $r^*(\theta,e)$ is a smooth function of $\theta$ and $e$ which defines the WSB

$$\mathcal{W} = \left\{ r^*(\theta,e) \mid \theta \in [0,2\pi], e \in [0,1] \right\}$$

The above definition has several weak points:

1. The requirements 1. and 2. on the initial conditions fix the modulus of the velocity and its direction, but not the sense. So, there are two different orbits with the initial conditions specified in Belbruno's definition, which can have different stability behaviour.
2. It is not true that for fixed values of $\theta$ and $e$ there is a finite distance $r^*(\theta,e)$ defining the boundary of the stable and unstable orbits. Figure 30 shows this fact clearly. In this Figure we have represented for a fixed value of $e = 0.70$ and for $\theta \in [0,2\pi]$, the points associated to initial conditions producing an stable orbit, using the two possible senses of the velocity at the perigee. As it has been said, the two plots do not agree since two orbits with the same initial position and

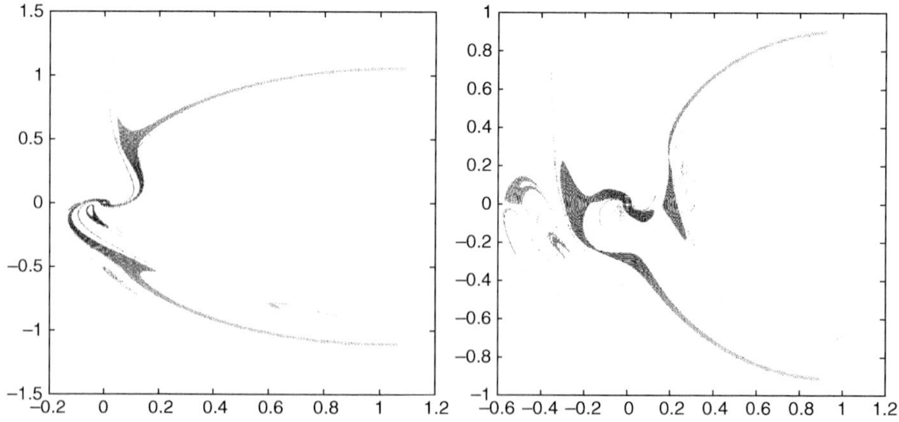


**Fig. 30** Stable initial positions $(x,y)$ about $m_2$ (at the origin) for $\theta \in [0,2\pi]$ according to Belbruno's definition. $\mu = 0.012$ (Earth-Moon mass ratio) and $e = 0.70$. The two plots correspond to the two possible senses of the velocity at the perigee of the osculating ellipse

opposite velocities may have different stability properties. If we draw any line through the origin (which is the point where we have set $m_2$), there are several transitions from stability to instability. In fact, for a fixed value of $\theta$, the set of stable points recalls a Cantor set.

The suitable setting for the WSB is the restricted four body problem, as follows from the kind of trajectories associated to the low energy transfers and as it is also established in [5]. This paper has an heuristic explanation, using invariant manifolds and Hill's regions, about how a ballistic capture by the Moon can take place but it has not any numerical computation supporting the claims.

## 7.2 Numerical Determination of WSB and Low Energy Transfers

There is only a few number of papers dealing with the numerical computation of WSB, now with a different definition from the algorithmic one given in the preceding section. They are, in fact, procedures to determine transfer trajectories with low $\Delta v$ requirements of two different kinds:

1. Direct numerical search (see [10,7,6] and [14]).
2. Computations using invariant manifolds (see [50]).

The direct numerical search is usually done directly in accurate models of motion. The method developed in [7] is a forward targeting algorithm that uses as initial orbit a Keplerian ellipse with an Earth apoapsis of 1.5 million kilometres. The one developed in [6, 10 and 14] is backward/forward integration with three steps:

1. First a backward integration from the final injection conditions up to the vicinity of the $L_1$ libration point of the Earth-Sun system (at 1.5 million kilometres from the Earth). This vicinity is what is defined, in an ambiguous way, as the WSB.
2. Second, a forward propagation from the Earth departure to the WSB region. An optimal intermediate lunar gravity assist may be incorporated in this trajectory.
3. A matching of the forward/backward trajectories using a constrained parameter optimisation algorithm.

Using this methodology, a large number of trajectories can be obtained for certain Sun-Earth-Moon configurations, as is shown in the above mentioned papers.

The computations using invariant manifolds, which uses the natural dynamics of the problem, proceeds along the following steps:

1. The first step is to decouple the restricted four body problem (spacecraft-Moon-Earth-Sun) in two coupled three body systems: spacecraft-Moon-Earth and spacecraft-Earth-Sun.

2. Within each three body system they transfer from the vicinity of the Earth into the region where the invariant manifold structure of the Earth-Sun libration points interacts with the invariant manifold structure of the Moon-Earth libration points. The region of intersection is computed using a Poincaré section (along a line of constant $x$-position passing through the Earth) which helps to glue the Sun-Earth Lagrange point portion of the trajectory with the lunar ballistic capture portion.
3. The Earth to Moon trajectory is integrated in the bi-circular four body model where both the Moon and the Earth are assumed to move in circular orbits about the Earth and the Sun respectively. Finally, the bi-circular solution is differentially corrected to a fully integrated trajectory with the JPL ephemerides.

The second item of this procedure must be done as follows: first a suitable Sun–Earth $L_2$ periodic orbit is computed as well as their stable and unstable manifolds. Some orbits on the stable manifold come close to the Earth and, at the same time, points close to the unstable manifold propagated backwards in time come close to the stable manifold. So, with an small $\Delta v$, is possible to go from the Earth to the unstable manifold of this periodic orbit. At the same time, when we consider the $L_2$ point of the Earth–Moon system, is has periodic orbits whose stable manifold "intersect" the unstable manifold that we have reached departing from the Earth and are temporary captured by the Moon. With a second small $\Delta v$ we can force the intersection to behave as a true one.

This second procedure, which is also known as the Shoot the Moon method, gives the good approach for the computation, in a systematic way, of all the possible low energy transfer orbits. The term "low energy" applied to these trajectories is due to the small manoeuvres that must be done.

Using the same ideas of the Shoot the Moon procedure, the "Petit Grand Tour" of the moons of Jupiter can be designed. In a first step, the Jovian–moons n-body system is decoupled into several three–body systems. The tour starts close to the $L_2$ point of an outer moon (for instance Ganymede). Thanks to an heteroclinic connection between periodic orbits around $L_1$ and $L_2$, we can go from the vicinity of $L_2$ to the vicinity of $L_1$ and, in between, perform one o several loops around Ganymede. Now, we can look for "intersections" between the unstable manifold of the periodic orbit around the $L_1$ point and the stable manifold of some p.o. around the $L_2$ point of some inner moon (for instance Europa). By the same considerations, we can turn around Europa and leave its influence through the $L_1$ point. Once the orbits have been obtained, they are refined to a more realistic model easily.

## *7.3 Solar System Low Energy Transfers and Astronomical Applications*

As Lo and Ross [52] suggested, the exploration of the phase space structure as revealed by the homoclinic/heteroclinic structures and their association with mean motion resonances, may provide deeper conceptual insight into the evolution and structure of the asteroid belt (interior to Jupiter) and the Kuiper belt (exterior to Neptune), plus the transport between these two belts and the terrestrial planet region.

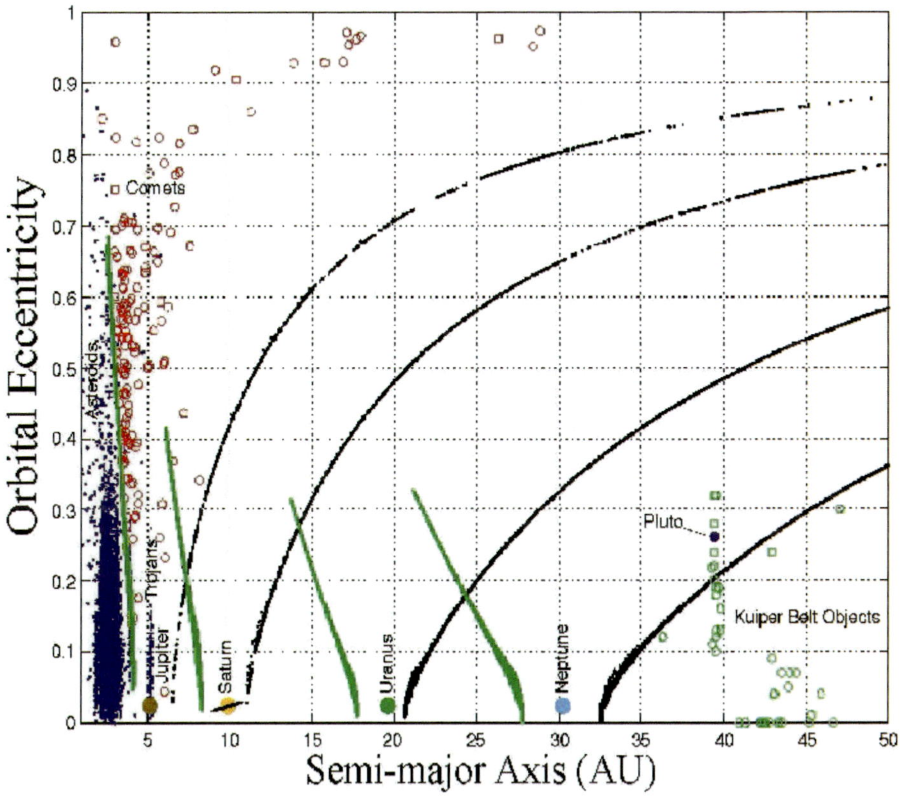


**Fig. 31** Dynamical channels in the solar system. Local semi-major axis versus orbital eccentricity. The $L_1$ (*grey*) and $L_2$ (*black*) manifolds for each of the giant outer planets is displayed. Notice the intersections between manifolds of adjacent planets, which leads to chaotic transport. Also shown are the asteroids (*dots*), comets (*circles*) and Kuiper Belt objects (*lighter circles*). (see [52])

Potential Earth–impacting asteroids may use the dynamical channels as a pathway to Earth from nearby heliocentric orbits in resonance with the Earth (Fig. 31). This phenomena has been observed recently in the impact of comet Shoemaker–Levy 9 with Jupiter, which was in 2:3 resonance with Jupiter just before impact. Also, the behaviour of comet Oterma that switches from a complicated trajectory outside the orbit of Jupiter to one lying within, can be explained with this kind of ideas. To make the transition, the comet passed through a bottleneck near two of Jupiter's libration points–where objects maintain a fixed distance relative to the planet and the Sun.

Numerical simulations of the orbital evolution of asteroidal dust particles show that the earth is embedded in a circumsolar ring of asteroidal dust known as the zodiacal dust cloud. Both simulations and observations reveal that the zodiacal dust cloud has structure. When viewed in the Sun–Earth rotating frame, there are several high density clumps which are mostly evenly distributed throughout the Earth's orbit. The dust particles are believed to spiral towards the Sun from the asteroid belt, becoming trapped temporarily in exterior mean motion resonances with the Earth. it is suspected that the gross morphology of the ring is given by a simpler CRTBP model involving the homoclinic and heteroclinic structures associated with the libration points.

A complete exploration of all the possible connections between the manifolds related to the libration points of the Solar System bodies should be of interest, not only for spacecraft mission design, but also to understand better the dynamics of the solar system.

## 8 Rescue Trajectories from the Moon's Surface

We look for rescue orbits that departing from the Moon reach a halo or a Lissajous orbit around the collinear libration points $L_1$ and $L_2$ of the Earth–Moon system. This will be done exploring the behaviour of the stable manifold of this kind of orbits (more details can be found in [1]). We recall that a stable manifold is composed by orbits which approach the periodic/quasi–periodic orbit forwards in time, so if they depart from the surface of the Moon is because they have reached it backwards in time after starting close to the reference periodic/quasi–periodic orbit.

Recently, Baoyin and McInnes [2] studied the same problem, in the framework of the Planar Circular Restricted Three–Body Problem and using the planar Lyapunov periodic orbits around the $L_1$ and the $L_2$ point instead of the three–dimensional halo and Lissajous orbits. We remark that the planar Lyapunov orbits correspond to Lissajous orbits with the vertical amplitude, $\alpha_4$, equal to zero. They focused their attention to primaries surface coverage, initial and arrival flight path angles, transfer time and initial velocity. Concerning the transfers between planar Lyapunov periodic orbits and the Moon, they found out that a whole surface coverage can be obtained only in specific energy levels. In particular, among the periodic orbits with this property, the smallest one around $L_1$ is characterised by a value of the Jacobi constant $C = 3.12185282430647$ and it provides a transfer of about 11 days; the smallest one around $L_2$ is characterised by $C = 3.09762627487867$ and it provides a transfer of about 14 days. However, from the contents of the paper it is not clear how long the departure from the libration point orbit takes to the spacecraft, that is, the time necessary to be at a significant distance (for instance, 100 km) from the nominal periodic orbit.

### *8.1 Stable Manifolds Associated with Halo Orbits*

The first explorations that we have done consist in the numerical globalisation of the stable manifold associated with orbits of the halo families around $L_1$ and $L_2$, until they reach the Moon's surface (considered as a sphere). Not all the orbits of a stable manifold can get to the Moon: it depends on the parameter along the orbit (phase) associated with the trajectory of the invariant manifold.

To check if a trajectory reaches the Moon's surface, at each integration step we compute

$$S = r_2^2 - r_M^2, \tag{33}$$

where $r_M$ is the radius of the Moon ($r_M = 1737.53\,km$). If $S$ changes sign, then we compute the arrival point at the surface of the Moon by means of the Newton's Method until $|S| \leq 10^{-12}$; this is, denoting by $\mathbf{X}^{(i)}$ the $i$th–iteration of the procedure, we compute the sequence of points

$$\mathbf{X}^{(i+1)} = \mathbf{X}^{(i)} - \frac{S}{\nabla S}, \qquad \text{where} \qquad \nabla S = 2(x - \mu + 1)\dot{x} + 2y\dot{y} + 2z\dot{z}, \tag{34}$$

until the stopping condition is fulfilled.

Once an orbit reaches the Moon's surface, we compute the latitude $\varphi$ and the longitude $\lambda$ corresponding to the arrival point, namely,

$$\varphi = \tan^{-1}\left(\frac{z}{\sqrt{(x - \mu + 1)^2 + y^2}}\right), \qquad \lambda = \tan^{-1}\left(\frac{y}{x - \mu + 1}\right). \tag{35}$$

In our exploration ($\varphi = 0°, \lambda = 0°$) corresponds to the Moon's point which is closest to the Earth, that is, we have considered negligible the inclination of the lunar equatorial plane with respect to the plane of the Earth–Moon orbit. Actually, this inclination is approximately of $6°41'$ (Bussey and Spudis, [12]).

In addition, we calculate the physical velocity of arrival, the physical transfer time and the arrival angle $\vartheta$, defined as the angle between the velocity vector and the Moon's surface normal vector, that is,

$$\cos\vartheta = \frac{v \cdot \nabla S}{\|v\|\;\|\nabla S\|}. \tag{36}$$

Of course, not all the orbits of a certain manifold can reach the surface of the Moon. To prevent for long time integrations we have set some controls. If after 10 adimensional time units (about 43.5 days) the function $S$ has not changed sign, we move to the next point of the manifold to be explored. Increasing the final time we get some more collision orbits but, from a qualitative point of view, the results are almost identical.

The second control takes into account how many times the orbit has gone close to the Moon without getting to it. We are interested in almost direct transfers and we do not see operational advantages in trajectories winding around the Moon indefinitely. For this purpose we compute the number of minima of the $r_2$ function along the orbits. This is, how many times the function

$$\dot{r}_2 = \frac{(x - \mu + 1)\dot{x} + y\dot{y} + z\dot{z}}{r_2} \tag{37}$$

changes sign and simultaneously

$$\ddot{r}_2 = \frac{\dot{x}^2 + \dot{y}^2 + \dot{z}^2 + (x - \mu + 1)\ddot{x} + y\ddot{y} + z\ddot{z} - \dot{r}_2^2}{r_2} > 0. \qquad (38)$$

If we get more than 5 minima, then we discard such trajectory and we proceed to explore the next point of the manifold. Indeed, we have seen that in this case the orbit remains revolving around the Moon up to a considerably large time.

Just for illustration purposes, in Fig. 32 we display three trajectories of the stable manifold of the halo orbit around $L_1$ with $z$–amplitude $\alpha_4 = 0.2$ normalised units ($\approx 11000$ km). The one on the left reaches the Moon directly, the one in the middle performing one loop around the Moon and the one on the right performing four.

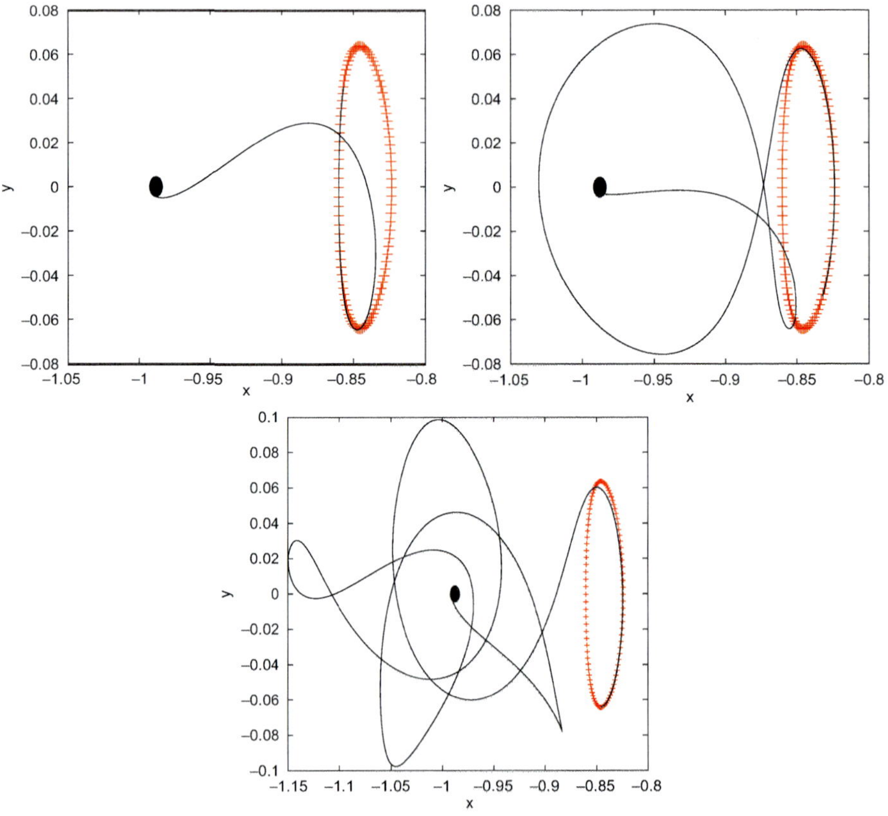


**Fig. 32** $(x,y)$ projection of three trajectories of the stable manifold associated with the halo orbit around the $L_1$ point of the Earth–Moon system with $\alpha_4 = 0.2$ normalised units ($\approx 11000$ km). From left to right, the trajectory reaches the Moon without minima of the $r_2$ function, after one close encounter and after four loops around the Moon

We remark that we discard the minima associated with the loops exhibited by the trajectories before leaving the neighbourhood of the periodic orbit.

### 8.1.1 Numerical Results

The above procedure has been applied using as reference orbits those of the halo families around the $L_1$ and the $L_2$ point of the Earth–Moon system. Since we have used Lindstedt–Poincaré series expansions for both the periodic orbits and the stable manifolds, we have considered $z$–amplitudes within the convergence domain of the semi–analytical approximation (see Gómez et al. [35]). This means that for the $L_1$ point, the $z$–amplitude $\alpha_4$ takes values up to 0.45 normalised units, that is, approximatively 23000 km (or, equivalently, the Jacobi constant ranges from 3.14100 to 3.18633). For the halo orbits around $L_2$, we allow the $z$–amplitude to vary up to 0.35 normalised units, that is, approximatively 28000 km or $C \in [3.14254, 3.16410]$. The exploration have been carried out varying the $z$–amplitude at step equal to $10^{-3}$ in both cases.

In Figs. 33 and 34, we display the accessible regions on the Moon associated with the halo orbits of class I around the $L_1$ and $L_2$ equilibrium points. The regions are reached by orbits of the stable manifolds that, starting at the halo orbits (and integrating backwards in time), reach the surface of the Moon after having 0 and 3 or less close approaches to the Moon (minima of the $r_2$ function).

The accessible regions associated with the halo orbits of class II having the same number of minima, are the symmetric, with respect to the $\{\varphi = 0\}$ axis, to the ones obtained for class I; this is due to the symmetry of the orbits with respect to the $\{z = 0\}$ plane. In all the cases, the patterns defining the accessible regions displayed in the figures are composed by different curves, each one associated with a halo orbit of a certain amplitude. The curves at the centre of each pattern, that is, the shortest ones, are associated with the halo orbits with the smallest amplitudes. For the $L_1$ case, the curves covering almost all the longitude's range are those with $\alpha_4 \geq 0.43$ normalised units ($\approx$ 22000 km).

As is clearly seen from the plots corresponding to 0 minima, the regions on the surface of the Moon from which we can reach a halo orbit without performing any loop to the Moon in between, cover (approximatively) only one half of the total surface. As we increase the number of allowed loops, the area of the region increases and for three o more loops the surface of the Moon is completely covered.

These results show that we cannot reach a halo orbit, either around $L_1$ or $L_2$, with a rescue orbit starting at an arbitrary point of the surface of the Moon if it does not perform at least one loop around the Moon. As we increase the number of loops, the area of the rescue region on the surface of the Moon increases. In fact, if we allow at least 3 minima, one can reach the halo families departing from any point of the surface of the Moon.

However, we remark that the points of allowed departure are not uniformly distributed on the Moon's surface. In other words, there exist regions where we have more chances to take off joining the invariant manifold associated with a given halo

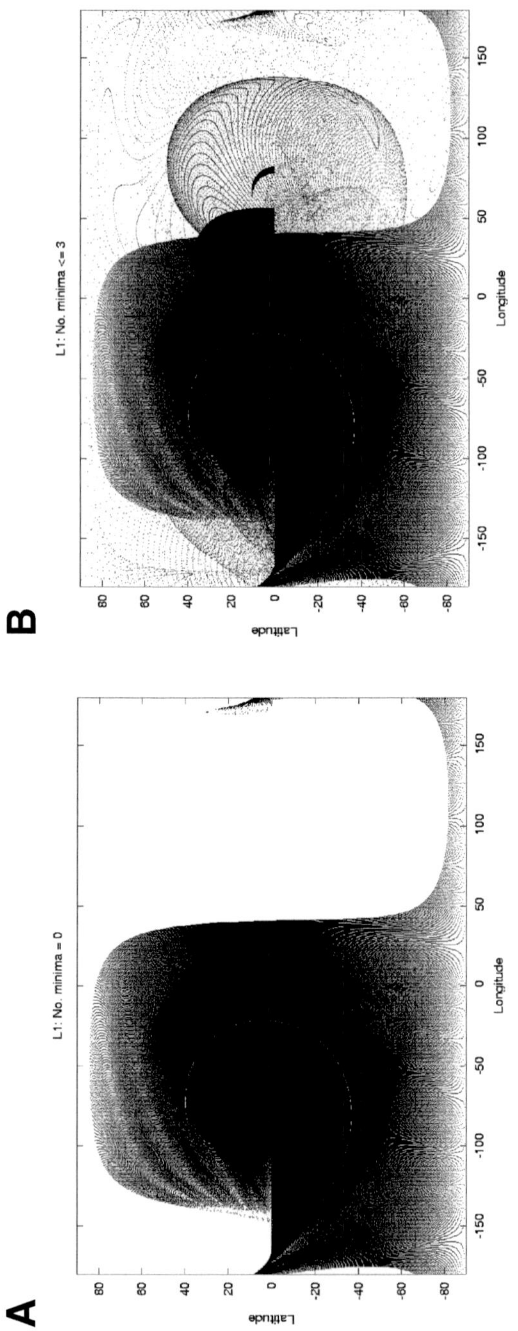


**Fig. 33** Longitude and latitude of the accessible zones of the Moon associated with the halo orbits of class I around the $L_1$ libration point. The two plots correspond to rescue orbits with 0 loops around the Moon before reaching the halo orbit (*left*) and with three or less loops (*right*). The range of amplitudes explored goes from $\alpha_4 = 0.01$ normalised units ($\approx$ 500 km) to $\alpha_4 = 0.45$ normalised units ($\approx$ 23000 km)

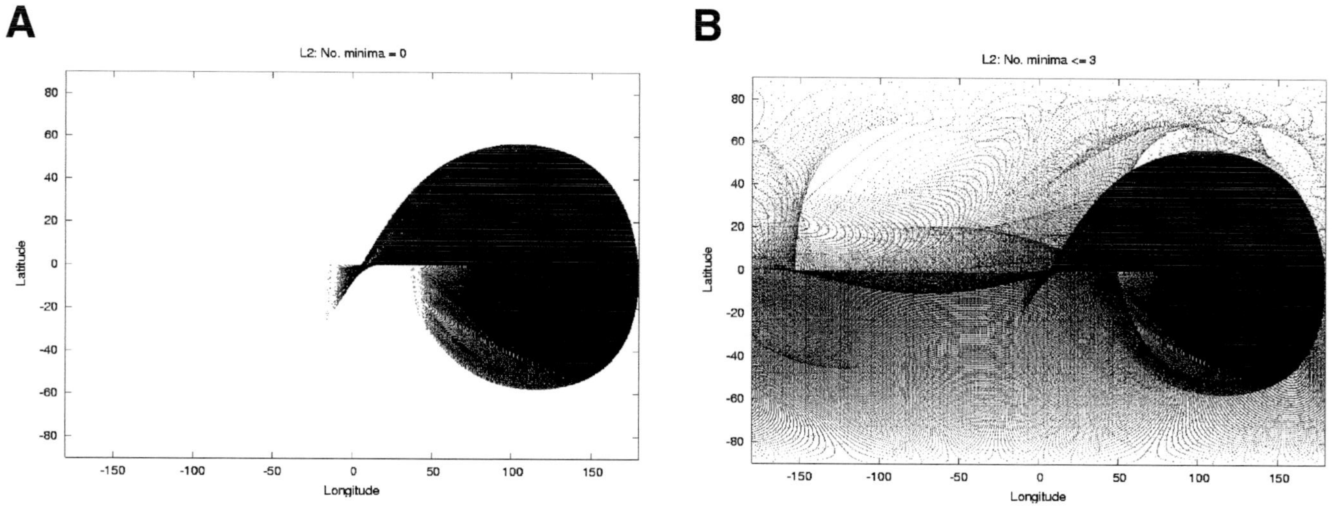


**Fig. 34** Longitude and latitude of the accessible zones of the Moon associated with the halo orbits of class I around the $L_2$ libration point. The two plots correspond to rescue orbits with 0 loops around the Moon before reaching the halo orbit (*left*) and with three or less loops (*right*). The range of amplitudes explored goes from $\alpha_4 = 0.01$ normalised units ($\approx 1000$ km) to $\alpha_4 = 0.35$ normalised units ($\approx 28000$ km)

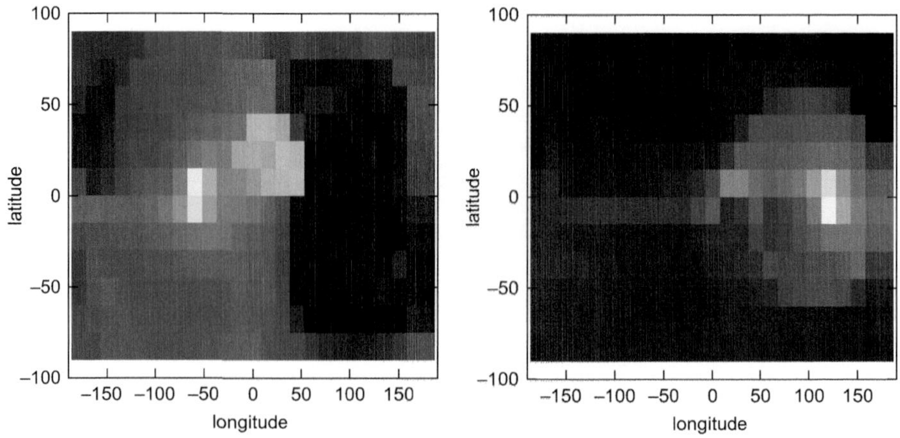


**Fig. 35** Density of opportunities of departure from the Moon's surface per unit of length of the arrival orbit and per unit of area element. Lighter colour corresponds to greater chance. On the *left*, the $L_1$ case; on the *right*, the $L_2$ one

orbit. This is illustrated in Fig. 35, where lighter colour corresponds to greater probability of departure. For this representation, we considered the Moon's surface as a rectangle of dimensions $[-180°,180°] \times [-90°,90°]$ in terms of longitude and latitude and we discretise it in small squares of $15° = 180°/12$ of side. For each nominal halo orbit, independently of the number of minima needed, we compute the number of departures found in each of these squares and we weight this value taking into account the length of the arrival periodic orbit. Finally, the sum of the values obtained is divided by the total number of departures and by the area[1] of the spherical square considered. Concerning the modulus of the velocity at the departure from the surface of the Moon, we note that in all the cases it is almost equal to the escape velocity of the Moon (about 2.4 km/s). This result follows from the analysis of the Jacobi integral and taking into account that all the orbits of the stable manifold of a certain halo orbit have the same value of the Jacobi constant as the halo orbit itself.

If $|v|$ denotes the modulus of the velocity, the Jacobi integral can be written as

$$|v|^2 = 2\Omega - C = (x^2 + y^2) + 2\left(\frac{1-\mu}{r_1} + \frac{\mu}{r_2}\right) + (1-\mu)\,\mu - C. \tag{39}$$

When we move on the surface of the Moon, $r_2 = r_M$ is constant and, since $x$ varies within $[\mu - 1 - r_M, \mu - 1 + r_M]$, the value of $r_1 = \sqrt{r_2^2 - 1 + 2\mu - 2x} = \sqrt{r_M^2 - 1 + 2\mu - 2x}$ goes from 0.995492 to 1.004508 adimensional units. If we

[1]We recall that, assuming the radius of the sphere equal to $r_M$, the area of the spherical rectangle with $\lambda \in [\lambda_1,\lambda_2]$ and $\varphi \in [\varphi_1,\varphi_2]$ is given by $S = r_M^2(\lambda_2 - \lambda_1)(\sin\varphi_2 - \sin\varphi_1)$.

compute the extrema of 2 $\Omega$ we get that it varies within the interval [8.3537898, 8.3538706] which means that the variations of the modulus of the velocity, for a fixed value of $C$, are of the order of $10^{-5}$ adimensional velocity units or, equivalently, of $10^{-2}$ m/s. As a consequence, the main variations on the modulus of the departure velocity are due to the variations of the Jacobi constant along the families of halo orbits. In Fig. 36, we display how this modulus varies, as a function of the Jacobi constant, for the two families of halo orbits considered. As it can be seen from the plots, the maximum variations along the two families are of the order of 10 m/s.

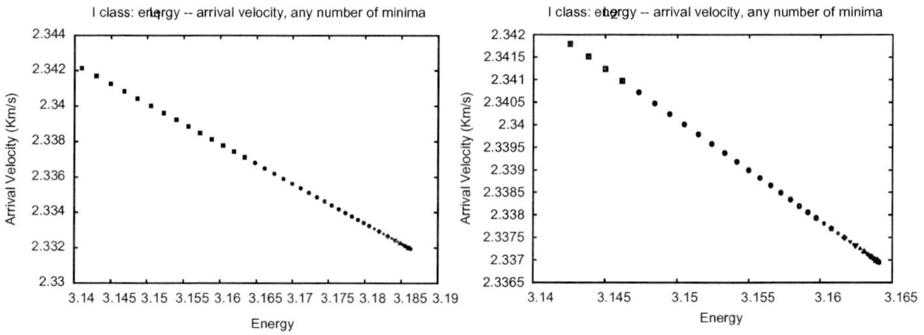


**Fig. 36** Modulus of the departure velocity (km/s) as a function of the Jacobi constant of the target halo orbit. On the *left*, the plot corresponds to the results obtained for the halo orbits around $L_1$; on the *right*, for the halo orbits around $L_2$. We remark that the value of the departure velocity depends only on the value of the Jacobi constant of the target halo orbit

As mentioned before, we have also computed the departure angle from the surface of the Moon. Because of the definition we assumed, it takes values between 0 (orthogonal departure) and 90° (tangent departure). The results are shown in Fig. 37 for the halo families around $L_1$ and $L_2$. For these figures we have normalised the parameter along the orbit to the interval $[0,2\pi]$. If we consider rescue trajectories without any loop around the Moon and any halo orbit as target orbit, we find rescue trajectories with departure angles taking any value in [0°,90°]. On the other hand, in Fig. 38, we display the regions characterised by almost orthogonal departure, that is, from which we can leave with an angle less than 10°.

There is also a very large range of phases (more than one half of the full range) which are not reached by the rescue orbits. Increasing the number of loops the range of phases clearly increases and there always seem to be more rescue orbits with a large value for the departure angle than with a small one.

Concerning the transfer time, we first recall that a stable manifold is established on asymptotic trajectories. This means that to go from the Moon to a nominal libration point orbit on such trajectories would take, in principle, an infinite time. Concretely, we consider the interval of time elapsed going from the initial conditions on the stable manifold, that is, about 70–90 km from the given halo orbit, to the Moon's surface. In this way, both in the $L_1$ and in the $L_2$ case, direct orbits need approximately 10 days, while the non–direct ones need about 10 more days

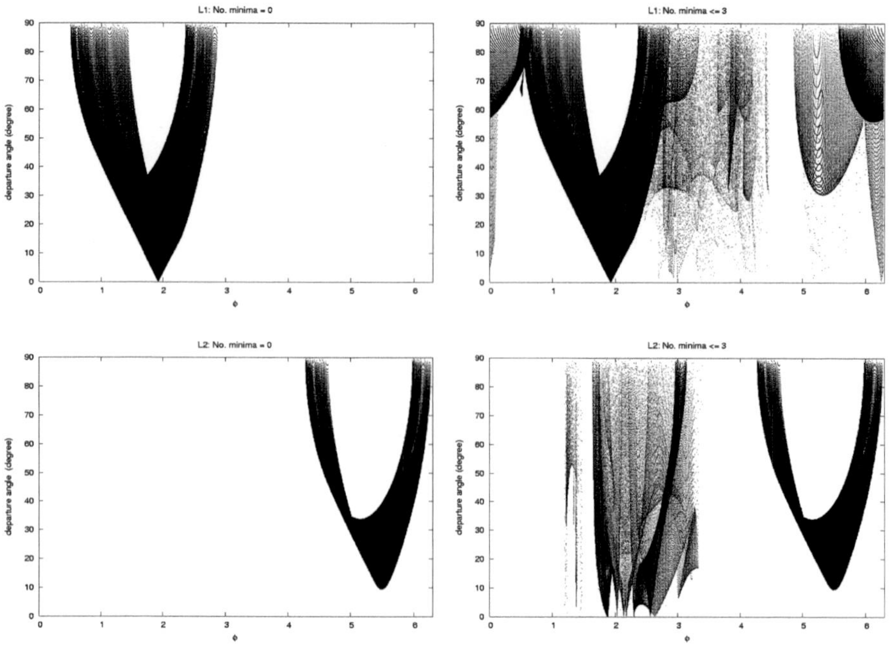


**Fig. 37** Departure angle from the Moon ($y$–axis, in degrees) versus the arrival phase ($x$–axis, in radians) to the periodic orbits of the halo family around $L_1$ (*top*) and $L_2$ (*bottom*). The two plots for each case correspond to 0 loops around the Moon before reaching the halo orbit (*left*) and 3 or less loops (*right*)

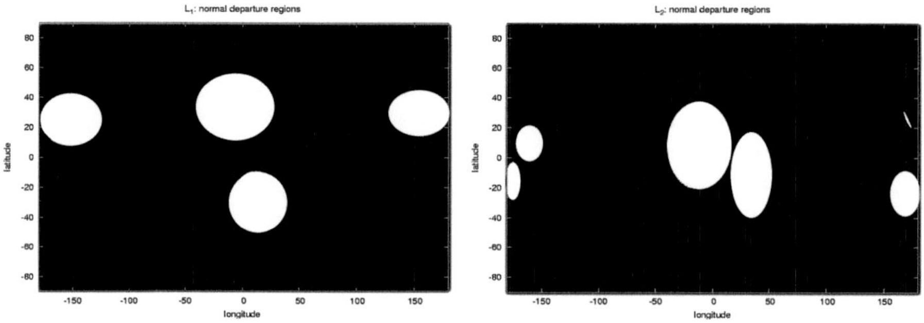


**Fig. 38** Regions of normal departure (the *white* ones) from the Moon to the halo family around $L_1$ (*left*) and $L_2$ (*right*). The two plots correspond to trajectories performing up to 3 loops around the Moon before reaching the halo orbit

for each further loop. However, we notice that, in all the cases considered, approximately 5 days, that is, half of the transfer time for direct trajectories, are devoted to approach asymptotically the nominal periodic orbit. When the distance to the halo orbit is less than 1000 *km* the velocity on the trajectory is less than 0.25 km/s.

## 8.2 Stable Manifolds Associated with Lissajous Orbits

To complete the analysis of rescue orbits from the surface of the Moon to a certain libration point orbit, we have proceeded to the numerical globalisation of the stable manifold associated with Lissajous orbits around $L_1$ and $L_2$. Looking to their intersections with the surface of the Moon we are able to determine the basic characteristics of the rescue transfer orbits. As in the previous case, we integrate backwards in time the trajectories on the stable manifold starting close to the Lissajous orbit. If they reach the Moon, then we compute the longitude and the latitude of the intersection point, the velocity and the angle of arrival and the time of flight.

Since we have approximated the stable manifold of the Lissajous orbits by an order 25 Lindstedt–Poincaré method, the values of the amplitudes to be considered are limited by the convergence domain of the expansions. In Fig.39, we show the results obtained by comparing the semi–analytical approach with the numerical one: the white region corresponds to the convergence domain of the series expansion (see Masdemont, [55]).

We stress that the main difference between this exploration and the one corresponding to the halo orbits is the amount of data to treat and the computational time. The reason is clear: in the halo case, for each equilibrium point and for a fixed energy level we had only two symmetric periodic orbits, now we have an infinite number of Lissajous orbits characterised by an in–plane and an out–of–plane amplitude, $\alpha_3$ and $\alpha_4$. In addition, for each pair of amplitudes we have two phases to explore, in order to reach all the points of the orbit. To afford the consequent computer effort, we have restricted the exploration to square Lissajous orbits ($\alpha_3 = \alpha_4$).

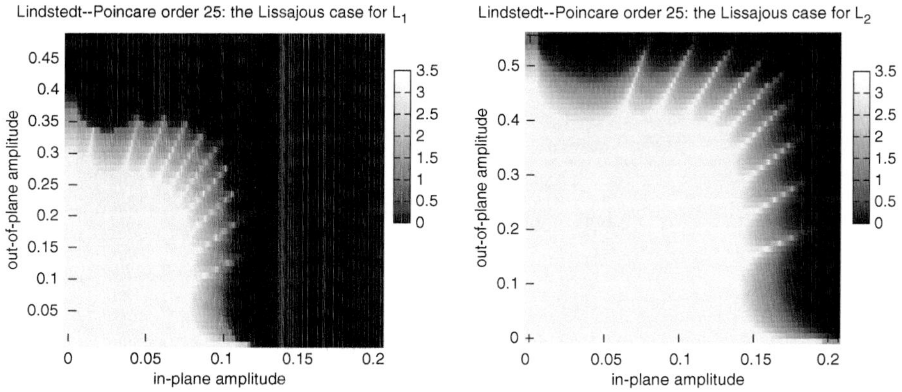


**Fig. 39** Test on the order 25 Lindstedt–Poincaré series expansion in the Lissajous case. Setting $\phi_1 = \phi_2 = 0$, we compare the values of position and velocity offered by the Lindstedt–Poincaré series and a Runge–Kutta–Fehlberg method of order 7–8. The colour box indicates the time of agreement (in terms of adimensional units) between the semi–analytical series expansion and the numerical integration. On the *left*, the $L_1$ case; on the *right*, the $L_2$ one. Darker the colour of the points worse the reliability of the expansion

### 8.2.1 Numerical Results

For halo orbits, the hyperbolic manifolds are two–dimensional, so if they intersect the Moon's surface, the intersection is a curve. Those associated with Lissajous orbits are three–dimensional, so their intersections with the surface of the Moon, in principle, will be two–dimensional regions. For a given Lissajous orbit, that is, for fixed values of the amplitudes $\alpha_3$ and $\alpha_4$, its stable manifold can be parametrised by the two phases $\phi_1$, $\phi_2$ and the time. Assume that this stable manifold reaches the Moon. If we fix one of the phases, say $\phi_1$, and allow the other one to vary within $[0,2\pi]$, the intersection with the surface of the Moon is a curve, which is closed if any value of $\phi_2$ gives rise to an orbit which intersects the Moon's surface. As we change the value of $\phi_1$, the intersection curve changes both in shape and in position in the $(\lambda,\varphi)$ plane. Of course, it might happen that for certain intervals of values of any of the phases, the intersection disappears. The envelope of all the intersection curves defines the boundary, on the surface of the Moon, of the two–dimensional intersecting region associated with the stable manifold of the Lissajous orbit.

In the left plot of Fig. 40, we show the intersection curves of the stable manifold associated with the square Lissajous orbit of amplitude equal to 0.09 ($\approx$ 6000 km) around the $L_1$ point. In this case, we allow $\phi_1$ and $\phi_2$ to vary in a continuous way in $[0,2\pi]$, although only values of $\phi_1 \in (0.75,2.1)$ rad give rise to orbits that reach the Moon without performing any loop around it. On the right, we display a detail of the two–dimensional region found, that is, just some curves corresponding to toy–values of $\phi_1$.

For the two kinds of Lissajous orbits that we have considered as target orbits, we have found that most of the rescue orbits reach the Lissajous orbit in a direct way if the Lissajous is around $L_1$ and needs at least two loops around the Moon in the $L_2$ case. Moreover, we can get to the neighbourhood of $L_1$, with any number of minima, only departing from almost half of an equatorial strip of 40° in terms of latitude. In the $L_2$ case, the accessible region on the Moon is quite larger, though we do not have full coverage. This is shown in Fig. 41. Certainly, the reason of the

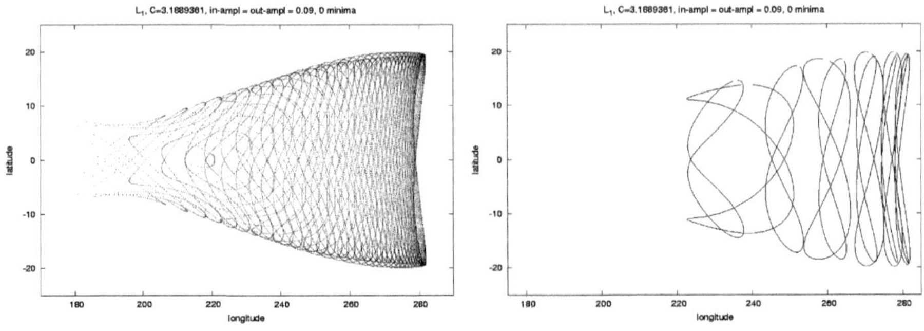


**Fig. 40** Departure region from the surface of the Moon associated with the square Lissajous orbit of amplitude equal to 0.09 ($\approx$ 6000 km) around the $L_1$ obtained from direct (without loops around the Moon) rescue orbits. See explanation in the text

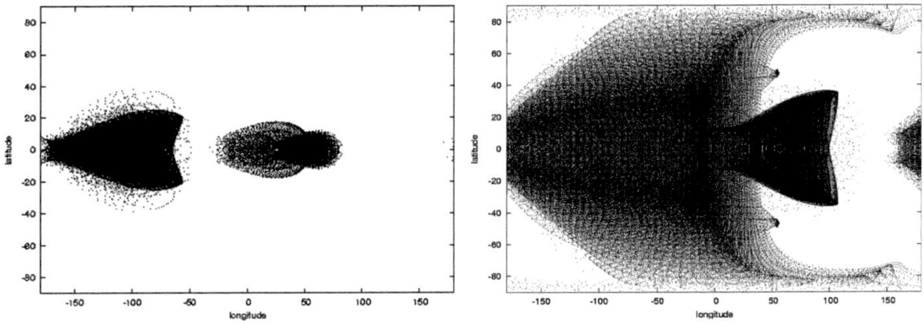


**Fig. 41** Regions of allowed departure from the surface of the Moon, considering as targets square Lissajous orbits and trajectories performing up to 5 loops around the Moon. On the *left*, the $L_1$ case; on the *right*, the $L_2$ one

two different behaviour resides on the wider range of amplitudes explored in the second case.

In Figs. 42 and 43, we display the allowed departure regions for direct trajectories and for those characterised by 2 minima, in the $L_1$ and in the $L_2$ case. Apart from

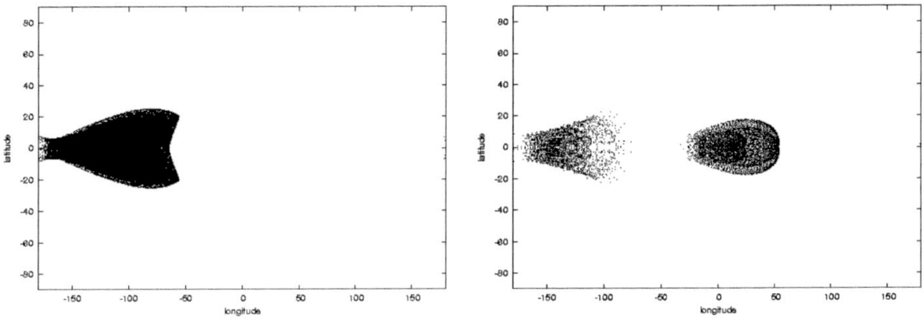


**Fig. 42** Regions of allowed departure from the surface of the Moon, considering as targets square Lissajous orbits around $L_1$. On the *left*, direct rescue; on the *right*, rescue characterised by two loops around the Moon

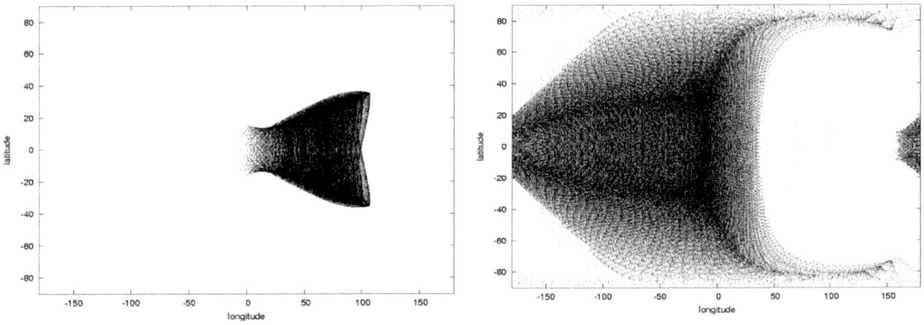


**Fig. 43** Regions of allowed departure from the surface of the Moon, considering as targets square Lissajous orbits around $L_2$. On the *left*, direct rescue; on the *right*, rescue characterised by two loops around the Moon

the forbidden region, in the $L_2$ case we have found a greater density of opportunities on the near-side of the Moon at equatorial latitudes.

As a further consideration, our simulation has revealed that the stable manifolds intersect the Moon only if the amplitudes $\alpha_3 = \alpha_4$ are big enough, this is: $\alpha_3 = \alpha_4 > 0.064$ ($\approx$ 4000 km) for $L_1$ and $\alpha_3 = \alpha_4 > 0.075$ ($\approx$ 5000 km) for $L_2$.

Concerning the phases, in Fig. 44 we display the range of phases $\phi_1$ and $\phi_2$ which generate direct rescue trajectories for two square Lissajous orbits around each libration point. For the quasi–periodic orbits we explored, greater the amplitude of the Lissajous greater the interval of values $\phi_1$ that give rise to rescue orbits.

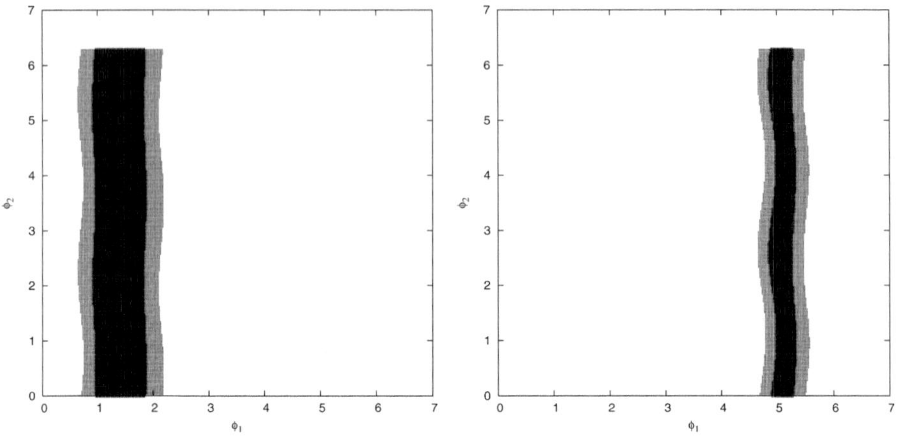


**Fig. 44** The range of phases $\phi_1$ and $\phi_2$ associated with direct lunar departing trajectories for different square Lissajous orbits. On the *left*, the $L_1$ case for amplitudes equal to 0.075 ($\approx$ 5000 km) and 0.09 ($\approx$ 6000 km); on the *right*, the $L_2$ case for amplitudes equal to 0.141 ($\approx$ 10000 km) and 0.149 ($\approx$ 11000 km) . We note that greater the amplitude greater the interval of phases

By the same reasons described in the halo case, the departure velocity is almost the same for all the transfer orbits associated with a given Lissajous orbit and depends only on the value of the Jacobi constant. The behaviour of the transfer time is quite analogous to that found for the halo orbits. The departure angle for direct rescue trajectories is represented in Fig. 45 as a function of the amplitudes $\alpha_3 = \alpha_4$ and one phase $\phi_1$. We do not have almost orthogonal departure in the $L_1$ case, while for $L_2$ the departure region characterised by an angle less than 10° is shown in Fig. 46.

## *8.3 Summary of Results*

We have seen that the dynamics corresponding to the stable manifold associated with the central manifold of the collinear points $L_1$ and $L_2$ allows the design of rescue orbits from the surface of the Moon to a libration point orbit. We have considered as nominal arrival orbits square Lissajous orbits and halo orbits of different class and size, taking advantage of various numerical tools.

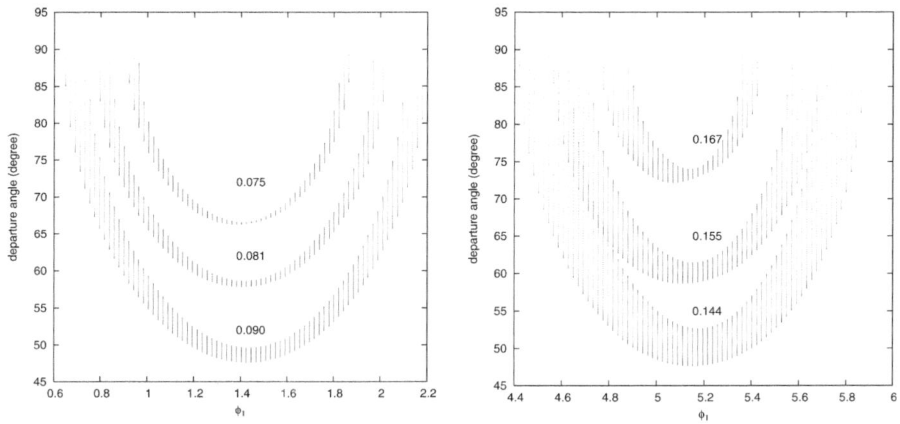


**Fig. 45** Departure angle for direct rescue trajectories in degrees as a function of the phase $\phi_1$ and of the amplitude of square Lissajous orbits of different size. On the *left*, the $L_1$ case; on the *right*, the $L_2$ case

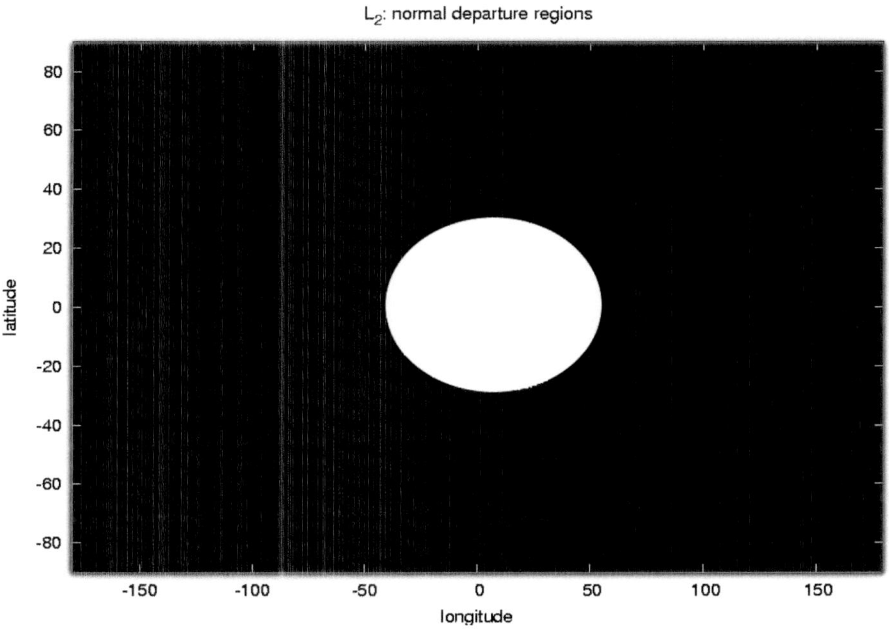


**Fig. 46** Region of normal departure (the *white* one) from the Moon to square Lissajous orbits around $L_2$. The plot corresponds to trajectories performing up to 5 loops around the Moon before reaching the quasi–periodic orbit. In the $L_1$ case, we have not found rescue orbits characterised by a departure angle less than 10°

The results offer some distinctions depending on the arrival orbit and on the type of trajectory. In particular, we have shown that, in case of direct rescue orbits, there are large regions on the Moon's surface from which the departure is not possible, independently of the nominal libration point orbit selected. For non–direct orbits, which are characterised by a longer transfer time, rescue can take place from much larger regions. In all the cases, the departure velocities are similar.

Concerning the departure angle, we can say that there exist precise correlations between its value and the type of trajectory and between its value and the arrival orbit (in terms of size and phase/phases). Moreover, we have identified the regions where the departure is almost orthogonal.

# 9 Spacecraft Missions to LPO's

## *9.1 Introduction*

The use of LPO, and more generally the use of Lagrangian Point Dynamics is very important for space agencies for scientific and exploration mission design:

- Orbits around Libration Points are extensively used today for scientific missions (mainly solar observatories at $L_1$ and astronomy at $L_2$ of the Earth-Sun system).
- Future lunar and planetary missions can *save energy* by using trajectories flying the Weak Stability Boundary (WSB) regions.
- Scientific and exploration missions to the Moon and planets may obtain an incredible *flexibility* in mission design by using this type of trajectories.
- After a system failure or under-performance, the use of Lagrangian Point Dynamics may be essential for *Mission Recovery*.
- Additionally, libration points can be used for *communications* (relay to the far side of the Moon) and *navigation*.

## *9.2 Missions to LPO*

The following tables illustrate some of the past, on-going and future spacecraft missions to fly Libration Point Orbits (some of those missions have been only designed and later on postponed due to budgetary problems).

| | | |
|---|---|---|
| Mission | ISEE-3 | 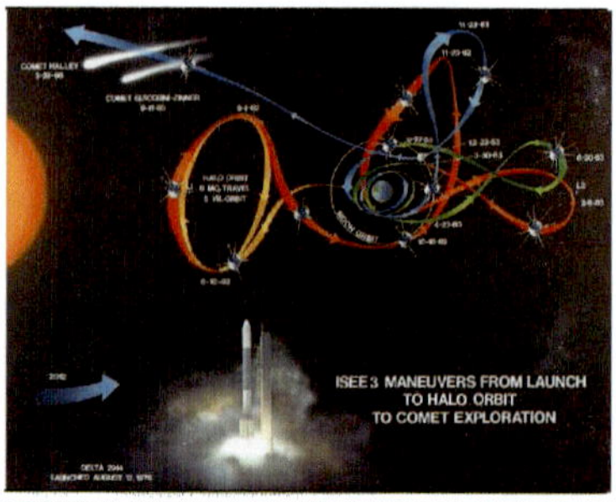  |
| Agency | NASA | |
| Target | $L_1$ Earth-Sun | |
| Launch | 1978 | |
| Objective | Solar wind, cosmic rays | |

| | |
|---|---|
| Mission | WIND |
| Agency | NASA |
| Target | $L_1$ Earth-Sun |
| Launch | 1994 |
| Objective | Solar wind, magnetosphere |

| | |
|---|---|
| Mission | SOHO |
| Agency | NASA–ESA |
| Target | $L_1$ Earth-Sun (Halo orbit) |
| Launch | 1996 |
| Objective | Solar observatory |

| | |
|---|---|
| Mission | ACE |
| Agency | NASA |
| Target | $L_1$ Earth-Sun |
| Launch | 1997 |
| Objective | Solar wind, particles |

| | |
|---|---|
| Mission | MAP |
| Agency | NASA |
| Target | $L_2$ Earth-Sun |
| Launch | 2001 |
| Objective | Background cosmic radiation |

| | |
|---|---|
| Mission | GENESIS |
| Agency | NASA |
| Target | L1 – $L_2$ Earth-Sun |
| Launch | 2001 |
| Objective | Solar wind composition |

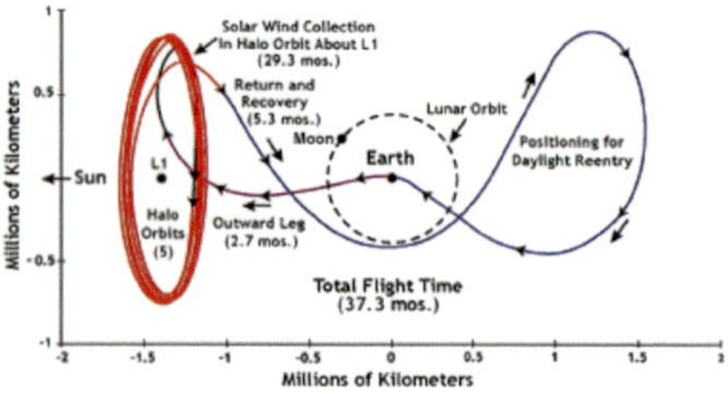


| | |
|---|---|
| Mission | WSO |
| Agency | International |
| Target | $L_2$ Earth-Sun |
| Launch | 2009 |
| Objective | Ultraviolet astronomy |

| | |
|---|---|
| Mission | Herschel |
| Agency | ESA |
| Target | $L_2$ Earth-Sun (large) |
| Launch | 2008 (Ariane 5) |
| Objective | Infrared astronomy |

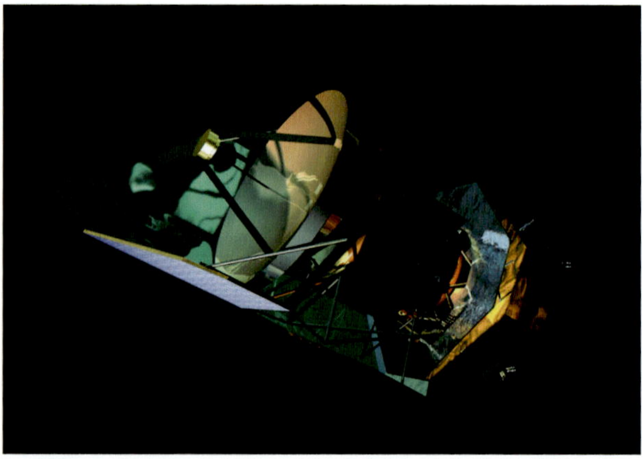

| | |
|---|---|
| Mission | Plank |
| Agency | ESA |
| Target | $L_2$ Earth-Sun (small Lissa.) |
| Launch | 2008 (Ariane 5) |
| Objective | Cosmic microwave backgnd. |

| | |
|---|---|
| Mission | Triana |
| Agency | NASA |
| Target | $L_1$ Earth-Sun |
| Launch | 2008 |
| Objective | Earth observation |

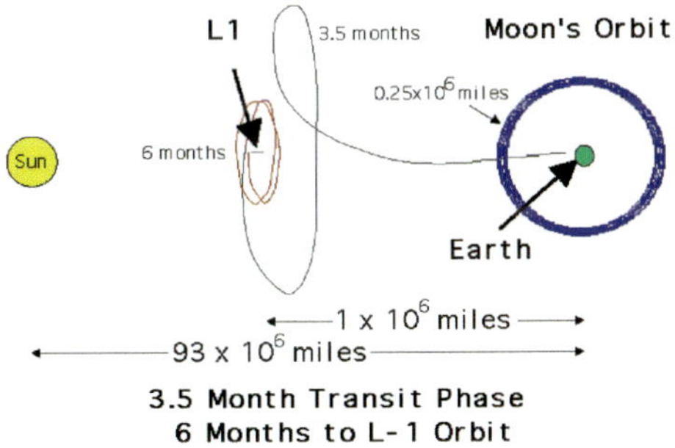

| | | |
|---|---|---|
| Mission | GAIA |  |
| Agency | ESA | |
| Target | $L_2$ Earth-Sun (small Lissa.) | |
| Launch | 2011 (Soyuz–Fregat) | |
| Objective | Astrometry | |

| | | |
|---|---|---|
| Mission | NGST/JWST |  |
| Agency | NASA | |
| Target | $L_2$ Earth-Sun | |
| Launch | 2013 | |
| Objective | Space telescope | |

| | |
|---|---|
| Mission | Constellation X |
| Agency | NASA |
| Target | $L_2$ Earth-Sun |
| Launch | 2013 |
| Objective | X–ray astronomy |

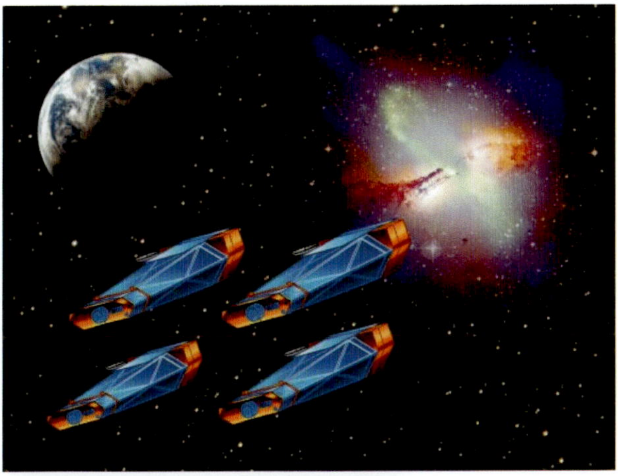

| | |
|---|---|
| Mission | DARWIN |
| Agency | ESA |
| Target | $L_2$ Earth-Sun |
| Launch | 2014 |
| Objective | Planetary systems |

| | |
|---|---|
| Mission | TPF |
| Agency | NASA |
| Target | $L_2$ Earth-Sun |
| Launch | 2015 |
| Objective | Planetary systems |

| | | |
|---|---|---|
| Mission | SAFIR |  |
| Agency | NASA | |
| Target | $L_2$ Earth-Sun | |
| Launch | 2015 | |
| Objective | Infrared telescope | |

| | | |
|---|---|---|
| Mission | LISA Pathfinder | 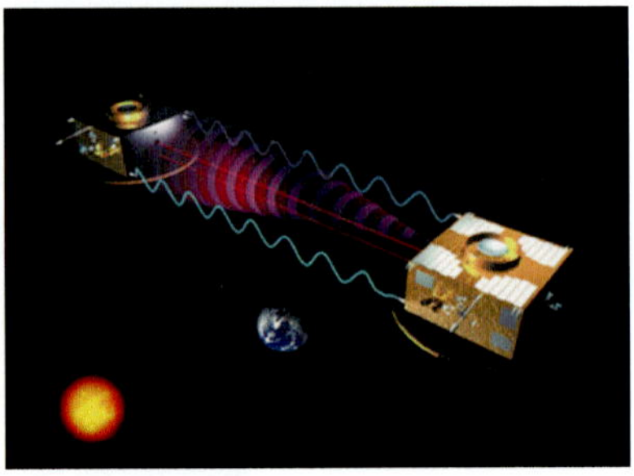 |
| Agency | ESA | |
| Target | $L_1$ Earth-Sun | |
| Launch | 2010 (VEGA) | |
| Objective | Technology demonstrator | |

| | | |
|---|---|---|
| Mission | Eddington |  |
| Agency | ESA | |
| Target | $L_2$ Earth-Sun | |
| Launch | pending | |
| Objective | Astronomy | |

## 9.3 Mission Design Problems

The mission design of satellite projects flying orbits around Libration Point includes the following aspects:

- Definition of *nominal trajectory*: the first step is the selection of the environment (the two- body system: Earth-Sun, Earth-Moon), the libration point (collinear L1, L2, L3 or triangular points L4 or L5) and the type of trajectory (Halo, Lissajous,...)
- *Transfer trajectories* to Libration Point Orbits from initial launch conditions or parking orbits.
- *Launch window* calculations taking into account the main mission constraints imposed for scientific or technical reasons.
- *Navigation* of transfer and nominal trajectories: calculation of the required propellant to correct launch injection dispersion, orbit determination errors and manoeuvres mechanisation errors.
- *Orbit maintenance*: strategies to keep the spacecraft flying around the selected nominal path.
- *Formation flying* techniques: new astronomy missions to LPO imposes the formation flying of several probes to implement interferometric techniques, the design of the formation architecture, the tight control and the collision avoidance techniques must be defined.
- *Eclipse avoidance*: most of the missions flying LPO orbits must avoid the eclipses in order to continue nominal operations.
- *Transfer between libration point orbits*: in some cases there is a need to transfer the probe from one initial LPO orbit to another larger or smaller amplitude trajectory.

## 9.4 LPO in Lunar and Exploration Missions

The use of Libration Point Dynamics or trajectories through the Fuzzy Boundary (or Weak Stability Boundary WSB) regions of planets and moons is very important to *save energy* and to obtain *mission design flexibility* for scientific and exploration projects.

Application of fuzzy boundary region (WSB) transfers to lunar missions have been studied by many authors, and provides an important energy saving factor with respect to classical Hohmann type transfers. It is possible to perform a ballistic capture in a Moon elliptic orbit from a WSB trajectory, with an important saving at the lunar injection manoeuvre (up to 40% in missions like *Lunarsat*). The penalisation of this energy saving lunar transfer method is an additional mission duration of about 90 days.

The design of interplanetary transfers from the Earth to the planets can be optimised from an energy point of view by incorporating SLS (Single Lunar Swingby), DLS (Double Lunar Swingby) or TLS (Triple Lunar Swingby) at the departure from the Earth sphere of influence. Those techniques incorporate trajectory legs through the WSB region and could save abut 150 kg of propellant to missions like Mars Express, with the penalisation of a larger transfer duration and more complex operations by performing several lunar swingbyes (Fig. 47).

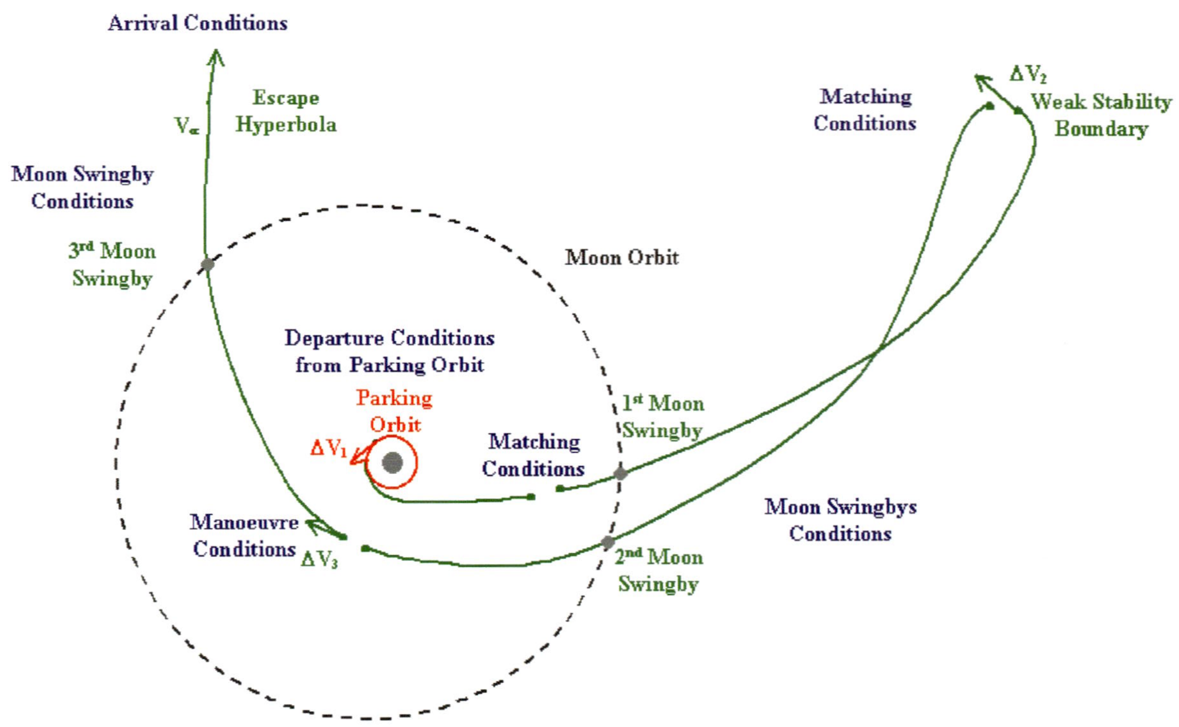


**Fig. 47** Example of TLS Earth departure technique

The use of Libration Point Dynamics methods has been also applied to the inner planet capture of some ESA missions like:

- Bepi Colombo at Mercury.
- Venus Express at Venus.
- Mars Express at Mars.

The main finding of those analysis is that the energy saving is very low, but the mission design is highly *flexible* compared with classical orbit capture techniques. In particular, the use of classical (based on patched conics analysis) capture imposes a given argument of pericenter and right ascension of ascending node of the resulting planetary orbit, while the use of LPO techniques give practically a full freedom to select above parameters, with a small penalisation in mission duration. From a scientific point of view, the capacity to choose freely the orbital plane orientation gives an extraordinary increase in the final outcome of the mission.

A similar conclusion can be obtained for the application of LPO techniques to the outer planet capture (Jupiter, Saturn, Uranus, Neptune). However, if a tour of giant planet natural moons (Jupiter tour) is designed, the use of LPO techniques gives again an important energy saving factor in addition to the high flexibility. The first analysis of giant planet moon tour was done by Koon and Lo, subsequent work was performed at DEIMOS under ESA contract, finding energy optimum ways for the moon tour trajectory design (Figs. 48 and 49).

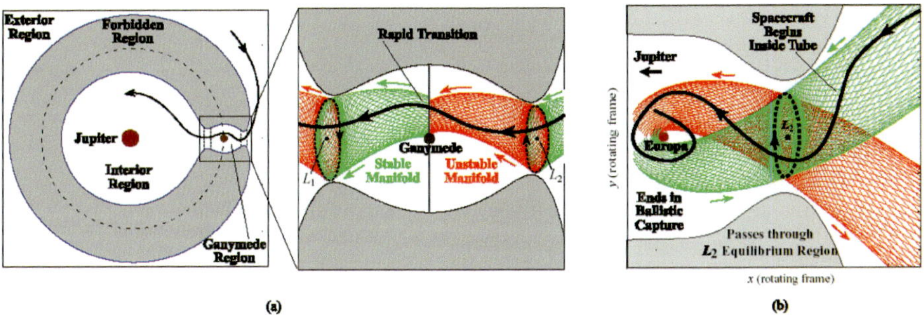


**Fig. 48** Moon Tour design by Koon and Lo

## *9.5 LPO in Mission Recovery*

Science and exploration missions require challenging spacecraft engineering, and it is possible that some system under-performance or system failures occur.

If an important propulsion system failure occur, the use of Fuzzy Boundary Region trajectories use to be an important asset for the mission analyst to redesign the mission and achieve its final scientific goals.

| Natural Moon | SMMA (km) | WSB min (km) | WSB max (km) |
|---|---|---|---|
| Io | 421600 | 314899 | 578619 |
| Europa | 670900 | 527741 | 866993 |
| Ganymede | 1070000 | 756974 | 1561307 |
| Callisto | 1810000 | 1380032 | 2633379 |

**Fig. 49** Jupiter Moon Tour design by DEIMOS

**Table 7** Transfer cost for Jupiter Moon Tour design by DEIMOS

| Transfer | Hohmann transfer $\Delta v$ (km/s) | Best comb. | Koon's results | Current study |
|---|---|---|---|---|
| Amalthea–Io | 9.739 | 6.860 | – | – |
| Io–Europa | 3.546 | 1.211 | – | – |
| Europa–Ganymede | 2.823 | 0.794 | 1.214 | 1.117 |
| Ganymede–Calixto | 2.473 | 0.560 | – | 1.099 |

The first example was the Japanese mission MUSES. MUSES A and B were launched by ISAS in January 1990. They were put in a high elliptic orbit. MUSES B had some technical problems while MUSES A, renamed Hiten, was proposed then for lunar orbit. Due to its propellant constraints, a classical transfer was not possible. Dr. E. Belbruno and J. K. Miller proposed a trajectory design by using Weak Stability Boundary techniques. This was the first example of WSB.

The second example was again a Japanese mission. Nozomi was a Japanese Mars mission launched on a M-V-3 vehicle the 4th of July 1998. The original mission design was a Double Lunar Assist (DLS) escape trajectory, with Mars arrival on October 1999. A malfunction of a latching valve left the spacecraft in a wrong heliocentric trajectory. The mission was recovered by an additional 4 years heliocentric trajectory including two Earth swingbyes. Final arrival to March was January 2004 (Fig. 50).

In 1997, a Proton launched the AsiaSat 3 towards a GEO orbit. Unfortunately, it was left stranded in a lower orbit for the Proton's upper stage fired for only 1 second out of a scheduled 110-second burn. After the launch, the satellite was declared a total loss for its original purposes. The manufacturer, Hughes Global Services Inc., reached an agreement with the insurers to try a high-risk salvage

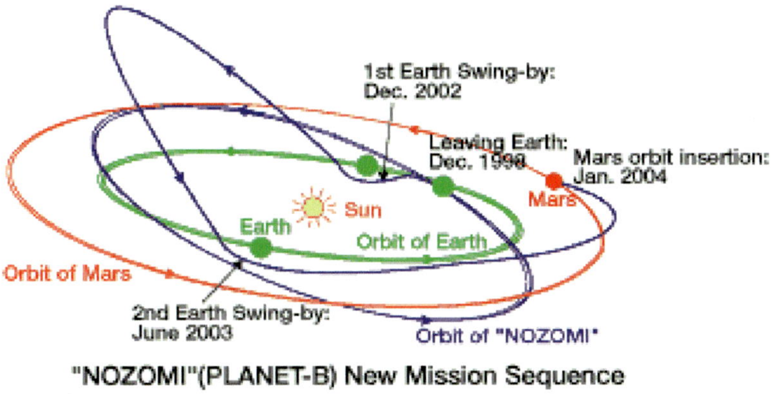


**Fig. 50** NOZOMI Mission recovery

mission of the spacecraft. In 1998, R. Ridenoure and Ed Belbruno, proposed a recovery solution using the satellite propellant and the Moon to head the satellite to its geosynchronous orbit via WSB transfer. They also ascertained that around 35% of the station-keeping propellant budget would remain when the satellite arrived to its final orbit. Contrary to that approach, Hughes chose a more conventional "free-return" trajectory, simply performing a swingby around the Moon. With that final solution, the s/c spent more fuel to reduce the inclination and circularise its orbit, leaving little, if any, left for station-keeping (Fig. 51).

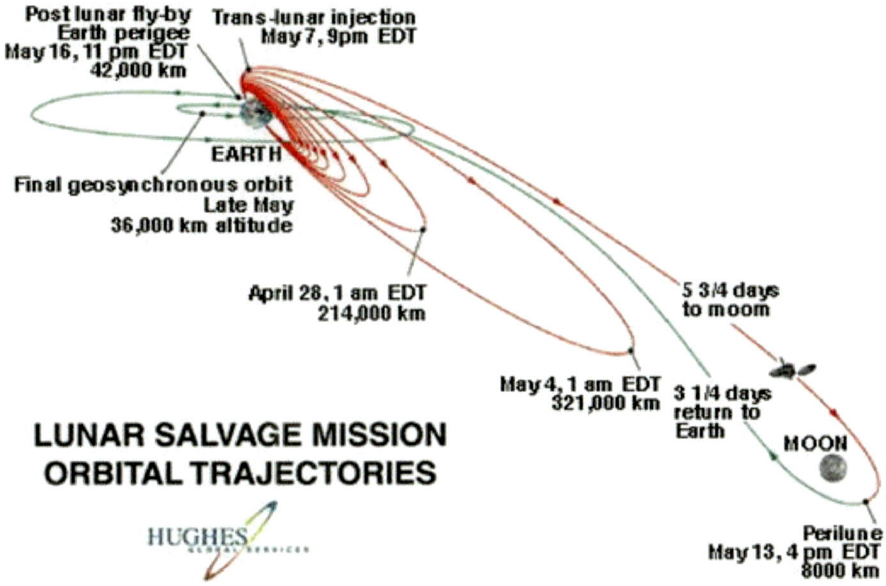


**Fig. 51** ASIASAT Mission recovery

### 9.6 Conclusions

The analysis and understanding of three- and four-body dynamics is very important for the mission design of scientific and exploration space missions for:

- Design of Libration Point Orbit projects (nominal and transfer trajectories, eclipse avoidance, formation flying, navigation, orbit maintenance, etc.).
- Energy saving for future exploration missions (WSB trajectories).
- Increase of flexibility in interplanetary mission design.
- Mission recovery in case of system failure.

An important background is available thanks to the fundamental work of several groups (JPL, ESA, University of Barcelona, Purdue, La Sapienza Rome, DEIMOS, etc.). However, the design of future scientific and exploration missions continuously requires the solution of new and each time more difficult problems related with the Lagrangian Point Dynamics.

## 10 Summary and Outlook

The invariant manifold structures of the collinear libration points for the restricted three-body problem give the framework for understanding some transport phenomena from a geometric point of view. In particular, the stable and unstable invariant manifold tubes associated to libration point orbits are the phase space conduits transporting material between primary bodies for separate three-body systems. The new trends in the assessment of mission design should include, among others, these methodologies, mainly derived from the application of results coming from Dynamical Systems Theory. The capabilities of modern computers make possible new tools that just some years ago could only be considered in a theoretical base.

New space missions are increasingly more complex, requiring new and unusual kinds of orbits to meet their scientific goals, orbits which are not easily found by the traditional conic approach. The delicate heteroclinic dynamics employed by the Genesis Discovery Mission illustrates the need for a new paradigm: study of $n$-body problems ($n \geq 3$) using dynamical systems theory as laid out by Poincaré. This significant increase of libration point missions with high degree of complexity, and the navigation in the Solar System using low energy transfers, claim for a design tool of easy use. The human presence in a space gateway station located at $L_1$ Earth-Moon system to service missions to the Moon and to the libration points of the Sun-Earth system, among other applications, will benefit from a design tool that, in an interactive way, could provide estimations and real trajectories to fullfill a mission on demand.

The dynamical structures of the three-body problem (such as stable and unstable manifolds, and bounding surfaces), reveal much about the morphology and transport of particles within the solar system, be they asteroids, dust grains, or spacecraft.

The cross-fertilisation between the study of the natural dynamics in the solar system and engineering applications has produced a number of new techniques for constructing spacecraft trajectories with desired behaviours, such as rapid transition between the interior and exterior Hill's regions, resonance hopping, and temporary capture. These dynamical structures, which exist for a range of energies, provide a framework for understanding the aforementioned dynamical phenomena from a geometric point of view. In particular, the stable and unstable invariant manifold tubes associated to periodic and quasi-periodic orbits around the libration points $L_1$ and $L_2$ are phase space structures that conduct particles to and from the smaller primary body (e.g., Jupiter in the Sun-Jupiter-comet three-body system), and between primary bodies for separate three-body systems (e.g., Saturn and Jupiter in the Sun-Saturn-comet and the Sun-Jupiter-comet three-body systems).

These invariant manifold tubes can be used to produce new systematic techniques for constructing spacecraft trajectories with interesting characteristics. These may include mission concepts such as a low energy transfer from the Earth to the Moon and a "Petit Grand Tour" of the moons of Jupiter.

One problem that appears in this approach is that we have to face with objects inside a high dimensional space, and so very difficult to be represented, wherein a four dimensional set of libration point orbits act as saddle points. The mathematical tools developed up to present, and essentially based in normal forms or in Lindstedt-Poincaré procedures, pretend, in some way, to decrease the degrees of freedom as well as to decouple the hyperbolic parts from the elliptical ones in order to simplify and study the problem from a mathematical point of view. They should be the base for the construction of new devices that permit the mission analysis in a systematic way and avoiding trial and error as much as possible. Let us enumerate briefly strong and weak points of these methodologies.

- **Normal Form Methods**
  - Provide a full description of the orbits in any selected energy level of interest.
  - Provide the uncoupling between central, stable and unstable manifolds.
  - The change of variables between the usual position and velocity and the normal form variables can be explicitly obtained in both ways.
  - The variables in the central part have not an special meaning. This is, in general they do not provide explicitly to which type of libration orbit refer.
  - The change of variables is implemented as a long series expansion and is relatively time consuming to evaluate.

- **Lindstedt-Poincaré Methods**
  - They provide full descriptions of levels of energy close to the one of the libration point.
  - The variables, amplitudes and phase have a very concrete and dynamical meaning. In fact they are the well known action-angle variables of the Hamiltonian systems.

- Because they are invariant tori, libration point orbits are selected fixing the amplitudes and varying the phases in a linear way. This is libration point orbits are seen as straight lines moving uniformly in time in suitable coordinates.
- The usual coordinates of position and velocity are obtained by a straightforward evaluation of the series.
- They provide only a partial description of levels of energy containing both Lissajous and halo orbits.
- The change of variables from the usual position and velocity to the amplitudes and phases is not direct and must be implemented through iterative numerical procedures.

The combination of the best properties of each methodology, together with an effective implementation of the numerical procedures, is the basis for the obtention of very efficient gadgets for the mission design in the Solar System using the libration points.

One of the best properties of the Poincaré sections of the normal forms about libration points is that they provide a full description of the orbits in a neighbourhood of them. Moreover, for levels of energy close to the libration point the invariant curves pointing to Lissajous trajectories distribute in a neat way about the fixed points corresponding to periodic orbits, using interpolation techniques and the normal form computations, it is not difficult to construct a database for any libration point in the solar system. A database can be used to generate, to display and to interact with trajectories in an immediate way.

For the solar system navigation (this includes transfer trajectories between planets or transfer trajectories between the moons of the planets, as well as transfers from or to stations in libration point regimes) it is very important to have a map of the main roads. Nevertheless, this is one of the parts that still needs more theoretical study. The building blocks (these are the orbits of the libration point regimes) are well known but, up to now, few effort has been put on the knowledge of their interconnections. Apart from some examples, the computation of transfer orbits relating libration point regimes of different primaries in a systematic way is still missing, although the numerical tools to analyse them are already developed.

As we already mentioned, the homoclinic and heteroclinic orbits are the ones that depart asymptotically from a libration point orbit and tend also asymptotically to the same one or to another one, inside the same level of energy. This is: they belong to both unstable and stable manifolds of libration point orbits, and provide natural transfers between orbits in the same level of energy at zero cost.

Asymptotic orbits can be considered in the same system (i.e. joining for instance libration point orbits from $L_1$ and $L_2$ in the Sun-Earth system or in the Earth-Moon system) or in different systems (i.e joining a libration point orbit in one system and with another libration point orbit of another system). The knowledge of these orbits can be very useful in the design of missions and may provide the backbone for other interesting orbits in the future. Up to now, and at least using coherent models, research has been focused on the computation of this type of orbits joining libration point orbits of the same system, and the tools to compute them, using either normal forms or Lindstedt methods. So it would be possible to have a database of these type of trajectories in a CRTBP framework and considering libration point orbits about $L_1$ and $L_2$.

In connection with interplanetary missions and satellite tours, there is some evidence that invariant manifolds associated with libration point orbits of different planets intersect in position. In this case, the techniques of mission design associated with libration point orbits of different planets could be used. The drawback is that, at least for the exterior planets, the time to obtain the encounter could be of the order of hundreds of years. In this case, small manoeuvres to obtain big deviations of the trajectory in hyperbolic regimes could be used to asses the mission design. Other possibilities should be the use of low trust propulsion, or the traditional transfer approach (conic orbits) in combination with the libration point regimes.

The exploration of the phase space structure as revealed by the homoclinic/heteroclinic structures and their association with mean motion resonances (mainly with Jupiter) , may provide deeper conceptual insight into the evolution and structure of the asteroid belt (interior to Jupiter) and the Kuiper belt (exterior to Neptune), plus the transport between these two belts and the terrestrial planet region.

It is estimated that about 1% of the asteroids are in a regime with potential close encounters with the Earth. Potential Earth–impacting asteroids may use the dynamical channels as a pathway to Earth from nearby heliocentric orbits in resonance with the Earth. This phenomena has been observed recently in the impact of comet Shoemaker–Levy 9 with Jupiter, which was in 2:3 resonance with Jupiter just before impact. Of course at some time, many of these asteroids or comets driven by the invariant manifolds of the Sun-Earth libration points will transit the bottleneck that the zero velocity curves draw about the Earth. This fact constitutes a serious hazard, and in fact, there is some evidence that the collision that caused the extinction of the dinosaurs might have been produced by an object in these energy levels.

Sky-watchers located at selected places, such as $L_2$ in the Sun-Earth libration point, could prevent from possible hazards. In case of detection, only small manoeuvres would be needed to deflect the dangerous body. At the same time, small deviations of the trajectories of some asteroids could be used to station them in selected orbits with the purpose of mining.

The idea of flying multiple spacecraft in formation with a mutual scientific goal is not new. However, attacking this problem in a demanding dynamical regime, as it is posed by the unique behaviour in the vicinity of the libration points, has recently become apparent. Only through a combination of the mathematical tools that have essentially affected a paradigm shift in three-body mission design (e.g., dynamical systems theory), creativity, and confidence in the existing tools has such a mission scenario been realized. One of the most exciting determinations in studies of this type is that many more valuable mission options are available.

The excellent observational properties of the $L_2$ point of the Earth–Sun system have lead to consider this location for missions requiring a multiple spacecraft in a controlled formation flying. Darwing, LISA and TPF are three of the more challenging examples of such missions.

The TPF Mission (Terrestrial Planet Finder) is one of the pieces of the NASA Origins Program. The goal of TPF is to identify terrestrial planets around stars nearby the Solar System. For this purpose, a space-based infrared interferometer with a baseline of approximately 100 m is required. To achieve such a large baseline, a distributed system of five spacecraft flying in formation is an efficient approach. Since the TPF instruments need a cold and stable environment, near Earth orbits are not suitable. Two potential orbits have been identified: a SIRTF-like heliocentric orbit and a libration orbit near the $L_2$ Lagrange point of the Earth-Sun system.

There are several advantages of using a libration orbit near $L_2$. Such orbits are easy and inexpensive to get to from Earth. Moreover, for missions with heat sensitive instruments (e.g. IR detectors), libration orbits provide a constant geometry for observation with half of the entire celestial sphere available at all times. The spacecraft geometry is nearly constant with the Sun, Earth, Moon always behind the spacecraft thereby providing a stable observation environment, making observation planning much simpler.

For any of the above mentioned missions, the first approach has to be the study of the problem in the simplest but meaningful situation. This means that simple idealistic models are used to get a first idea of the different magnitudes involved. Some of these models are:

- For the motion, of either a single spacecraft or a constellations, around the equilibrium points of the (Earth+Moon)-Sun system, the CRTBP is a good starting model. The effect of the remaining bodies of the solar system is small. This problem is rather well known and the dynamics around the equilibrium points is well understood.
- For motions around the equilibrium points of the Earth–Moon system the CRTBP is not a good model of motion. The effect of the Sun is large and the real dynamics around the equilibrium points seems to be, in some cases, far from the one corresponding to the RTBP model. More accurate four body problems should be introduced and studied including the Sun, the Earth and the Moon as primaries.

These four body models should be used to reinforce the study problems like

- The ones associated with "Le Petit Grand Tour"
- Low energy transfers to a ballistic capture at the Moon.
- Low energy transfers to mild unstable orbits around the triangular equilibrium points of the Earth-Moon system.

The better understanding of these problems should clarify the weak stability boundary concept. Some of the magnitudes that should be estimated with these simple models are:

- The nominal orbit to be used as a first guess for more refined models.
- The requirements for the station keeping within the desired accuracy.
- Simple transfer trajectories to change formation configurations.

Aside from the gravitational models, some effort should be done in the development of models including the solar radiation pressure. This can be useful for the station keeping strategies since, if the station keeping manoeuvres can be executed using solar radiation pressure, the contamination of the spacecraft due to the execution of an impulsive manoeuvre is avoided, and this enlarges the observation intervals of the spacecraft. In addition, the solar radiation pressure can play an important role in the formation flight problem, when the requirements for the mutual distances are very severe.

## References

1. E.M. Alessi, G. Gómez, and J.J. Masdemont: 'Leaving the Moon by Means of Invariant manifolds of Libration Point Orbits'. To appear in *Communications in Nonlinear Science and Numerical Simulations*, 2008.
2. H. Baoyin and C.R. McInnes: 'Trajectories to and from the Lagrange Points and the Primary Body Surfaces'. *Journal Guidance, Control and Dynamics*, 29:998–1003, 2006.
3. E. Belbruno: 'Lunar Capture Orbits, a Method for constructing Earth-Moon Trajectories and the Lunar GAS Mission'. *AIAA paper No. 87-1054*, 1987.
4. E. Belbruno and J.K. Miler: 'A Ballistic Lunar Capture Trajectory for the Japanese Spacecraft Hiten'. *JPL, IOM 312/90.4–1371*, 1990.
5. E. Belbruno: 'The Dynamical Mechanism of Ballistic Lunar Capture Transfers in the Four Body Problem from the Perspective of Invariant Manifolds and Hill's Regions'. *CRM, Preprint n. 270*, Barcelona, 1994.
6. E. Belbruno R. Humble, and J. Coil: 'Ballistic Capture Lunar Transfer Determination for the US Air Force Academy Blue Moon Mission'. *AIAA/AAS Astrodynamics Specialist Conference.* Paper No. 97-171, 1997.
7. E. Belbruno: 'Calculation of Weak Stability Boundary Ballistic Lunar Transfer Trajectories'. *AIAA/AAS Astrodynamics Specialist Conference.* Paper No. 2000-4142, 2000.
8. E. Belbruno: 'Analytic Estimation of Weak Stability Boundaries and Low Energy Transfers'. *Contemporary Mathematics.* 292 17–45, 2002.
9. E. Belbruno: *Capture Dynamics and Chaotic Motions in Celestial Mechanics.* Princeton University Press, 2004.
10. M. Belló-Mora, F. Graziani, P. Teofilatto, C. Circi, M. Porfilio, and M. Hechler: 'A Systematic Analysis on Weak Stability Boundary Transfers to the Moon'. *International Astronautical Federation*, 2000.
11. J.V. Breakwell, A.A. Kamel, and M.J. Ratner: 'Station-Keeping for a Translunar Communications Station'. *Celestial Mechanics*, 10(3):357–373, 1974.
12. B. Bussey and P. Spudis: *The Clementine Atlas of the Moon.* Cambridge University Press, 2004.
13. E. Canalias, J. Cobos, and J.J. Masdemont: 'Impulsive Transfers Between Lissajous Libration Point Orbits'. *Journal of the Astronautical Sciences*, 51: 361–390, 2003.
14. C. Circi and P. Teofilato: 'On the Dynamics of Weak Stability Boundary Lunar Transfers'. *Celestial Mechanics and Dynamical Astronomy.* 79: 41–72, 2001.
15. G. Colombo: 'The Stabilization of an Artificial Satellite at the Inferior Conjunction Point of the Earth–Moon System'. Technical Report 80, Smithsonian Astrophysical Observatory Special Report, 1961.
16. E. Canalias: 'Contributions to Libration Orbit Mission Design using Hyperbolic Invariant Manifolds' PhD thesis, Universitat Politècnica de Catalunya, Barcelona, Spain, 2007.
17. E. Canalias and J.J. Masdemont: 'Eclipse Avoidance for Lissajous Orbits Using Invariant Manifolds'. IAC–04–A.6.07, http://www.iac-paper.org, 2004.

18. E. Canalias and J.J. Masdemont: 'Lunar Space Station for Providing Services for Solar Libration Point Missions' IAC–05–C1.1.07, http://www.iac-paper.org, 2005.
19. E. Canalias and J.J. Masdemont: 'Rendez-vous in Lissajous Orbits using the Effective Phases Plane' IAC–06–C1.8.03, http://www.iac-paper.org, 2006.
20. E. Canalias and J.J. Masdemont: 'Computing Natural Transfers between Sun-Earth and Earth-Moon Lissajous Libration Point Orbits' *Acta Astronautica*, 63: 238–248, 2008.
21. P. Di Donato, J.J. Masdemont, P. Paglione, and A.F. Prado: 'Low Thrust Transfers from the Earth to Halo Orbits around the Libration Points of the Sun-Earth/Moon System' *Proceedings of COBEM 2007*, 19th International Congress of Mechanical Engineering, 2007.
22. P. Di Donato: 'Transfers from the Earth to Libration Point Orbits using Low Thrust with a Constrain in the Thrust Direction.' Master Thesis. Universitat Politècnica de Catalunya (ETSEIB), and Instituto Tecnológico de Aeronautica (eng aero). http://eprints.upc.es/pfc/handle/2099.1/2980, http://www.bd.bibl.ita.br/TGsDigitais, 2006.
23. N. Eismont, D. Dunham, J. Sho-Chiang, and R.W. Farquhar: 'Lunar Swingby as a Tool for Halo–Orbit Optimization in Relict–2 Project'. In *Third International Symposium on Spacecraft Flight Dynamics, ESA SP–326*. European Space Agency, Darmstadt, Germany, 1991.
24. R.W. Farquhar: 'Far Libration Point of Mercury'. *Astronautics & Aeronautics*, 5(8):4, 1967.
25. R.W. Farquhar. 'The Control and Use of Libration Point Satellites'. Technical Report TR R346, Stanford University Report SUDAAR–350 (1968). Reprinted as NASA, 1970.
26. R.W. Farquhar, D.P. Muhonen, C.R. Newman, and H.S. Heuberger: 'Trajectories and Orbital Maneuvers for the First Libration–Point Satellite'. *Journal of Guidance and Control*, 3 (6):549–554, 1980.
27. D. Folta and M. Beckman: 'Libration Orbit Mission Design: Applications of Numerical and Dynamical Methods'. In *Libration Point Orbits and Applications*, World Scientific, 2003.
28. G. Gómez, J. Llibre, and J. Masdemont: 'Homoclinic and Heteroclinic Solutions in the Restricted Three-Body Problem'. *Celestial Mechanics*, 44:239–259, 1988.
29. G. Gómez, À. Jorba, J.J. Masdemont, and C. Simó: 'A Dynamical Systems Approach for the Analysis of the SOHO Mission'. In *Third International Symposium on Spacecraft Flight Dynamics*, pages 449–454. European Space Agency, Darmstadt, Germany, October 1991.
30. G. Gómez, A. Jorba, J. Masdemont, and C. Simó: 'Study of the Transfer from the Earth to a Halo Orbit Around the Equilibrium Point $L_1$ '. *Celestial Mechanics*, 56 (4): 541–562, 1993.
31. G. Gómez, J.J. Masdemont, and C. Simó: 'Lissajous Orbits Around Halo Orbits'. *Advances in the Astronautical Sciences*, 95:117–134, 1997.
32. G. Gómez, A. Jorba, J. Masdemont, and C. Simó: 'Study of the Transfer Between Halo Orbits'. *Acta Astronautica*, 43 (9–10):493–520, 1998.
33. G. Gómez, J. Llibre, R. Martínez, and C. Simó: *Dynamics and Mission Design Near Libration Point Orbits – Volume 2: Fundamentals: The Case of Triangular Libration Points*. World Scientific, 2000.
34. G. Gómez, ÀA. Jorba, J.J. Masdemont, and C. Simó: *Dynamics and Mission Design Near Libration Point Orbits – Volume 4: Advanced Methods for Triangular Points*. World Scientific, 2000.
35. G. Gómez, ÀA. Jorba, J.J. Masdemont, and C. Simó: *Dynamics and Mission Design Near Libration Point Orbits – Volume 3: Advanced Methods for Collinear Points*. World Scientific, 2000.
36. G. Gómez, W.S. Koon, M.W. Lo, J.E. Marsden, J.J. Masdemont, and S.D. Ross: 'Invariant Manifolds, the Spatial Three-Body Problem and Space Mission Design'. *Advances in The Astronautical Sciences*, 109, 1:3–22, 2001.
37. G. Gómez, M. Marcote, and J.J. Masdemont: 'Trajectory Correction Manoeuvres in the Transfer to Libration Point Orbits'. In *Proceedings of the Conference Libration Point Orbits and Applications*. World Scientific, G. Gómez et al. Eds. pages 287–310, 2003.

38. G. Gómez, M. Marcote, and J.J. Masdemont: 'Trajectory Correction Manoeuvres in the Transfer to Libration Point Orbits'. *Acta Astronautica*, 56:652–669, 2005.
39. G. Gómez and J.J. Masdemont: 'Refinements of a Station–Keeping Strategy for Libration Point Orbits'. Preprint.
40. M. Hechler: 'SOHO Mission Analysis Transfer Trajectory'. Technical Report MAO Working Paper No. 202, ESA, 1984.
41. L.A. Hiday and K.C. Howell: 'Transfers Between Libration–Point Orbits in the Elliptic Restricted Problem'. In *AAS/AIAA Spaceflight Mechanics Conference, Paper AAS 92-126.*, 1992.
42. K.C. Howell and H.J. Pernicka: 'Stationkeeping Method for Libration Point Trajectories'. *Journal of Guidance, Control and Dynamics*, 16 (1):151–159, 1993.
43. K.C. Howell and S.C. Gordon: 'Orbit Determination Error Analysis and a Station–Keeping Strategy for Sun–Earth $L_1$ Libration Point Orbits'. *Journal of the Astronautical Sciences*, 42 (2):207–228, April–June 1994.
44. K.C. Howell and L.A. Hiday-Johnston: 'Time-Free Transfers Between Libration-Point Orbits in the Elliptic Restricted Problem'. *Acta Astronautica*, 32:245–254, 1994.
45. K.C. Howell and B.T. Barden: 'Brief Summary of Alternative Targeting Strategies for TCM1, TCM2 and TCM3'. Private communication, 1999.
46. K.C. Howell, B.G. Marchand, and M.W. Lo: 'Temporary Satellite Capture of Short–Period Jupiter Family Comets from the Perspective of Dynamical Systems'. In *AAS/AIAA Space Flight Mechanics Meeting, AAS Paper 00-155*, 2000.
47. T.M. Keeter: 'Station–Keeping Strategies for Libration Point Orbits: Target Point and Floquet Mode Approaches'. Master's thesis, School of Aeronautics and Astronautics, Purdue University, West Lafayette, Indiana, 1994.
48. W.S. Koon, M.W. Lo, J.E. Marsden, and S.D. Ross: 'Heteroclinic Connections Between Periodic Orbits and Resonance Transitions in Celestial Mechanics'. *Chaos*, 10 (2):427–469, 2000.
49. W.S. Koon, M.W. Lo, J.E. Marsden, and S.D. Ross: 'Resonance and capture of Jupiter comets'. *Celestial Mechanics and Dynamical Astronomy*, 81 (1):27–38, 2001.
50. W.S. Koon, M.W. Lo, J.E. Marsden, and S.D. Ross: 'Low Energy Transfer to the Moon'. *Celestial Mechanics and Dynamical Astronomy*. 81 (1):63–73,2001.
51. D.F. Lawden: *Optimal Trajectories for Space Navigation*. Butterworths & Co. Publishers, London, 1963.
52. M.W. Lo and S. Ross: 'SURFing the Solar Sytem: Invariant Manifolds and the Dynamics of the Solar System'. Technical Report 312/97, 2-4, JPL IOM, 1997.
53. M.W. Lo, B.G. Williams, W.E. Bollman, D. yHan, Y. Hahn, J.L. Bell, E.A. Hirst, R.A. Corwin, P.E. Hong, K.C. Howell, B.T. Barden, and R.S. Wilson: 'Genesis Mission Design'. In *AIAA Space Flight Mechanics, Paper No. AIAA 98–4468*, 1998.
54. J.J. Masdemont: *Estudi i Utilització de Varietats Invariants en Problemes de Mecànica Celeste'*. PhD thesis, Universitat Politècnica de Catalunya, Barcelona, Spain, 1991.
55. J.J. Masdemont: 'High Order Expansions of Invariant Manifolds of Libration Point Orbits with Applications to Mission Design'. *Dynamical Systems; an International Journal*, 20: 59–113, 2005.
56. D.L. Richardson: 'Analytical Construction of Periodic Orbits About the Collinear Points'. *Celestial Mechanics*, 22(3):241–253, 1980.
57. R. Servan, W.S. Koon, M.W. Lo, J.E. Marsden, L.R. Petzold, S.D. Ross, and R.S. Wilson: 'Halo Orbit Mission Correction Maneuvers Using Optimal Control'. 2000.
58. C. Simó, G. Gómez, J. Llibre, and R. Martínez: 'Station Keeping of a Quasiperiodic Halo Orbit Using Invariant Manifolds'. In *Second International Symposium on Spacecraft Flight Dynamics*, pages 65–70. European Space Agency, Darmstadt, Germany, October 1986.
59. C. Simó, G. Gómez, J. Llibre, R. Martínez, and R. Rodríquez: 'On the Optimal Station Keeping Control of Halo Orbits'. *Acta Astronautica*, 15 (6):391–397, 1987.

60. C. Simó: ‘Dynamical Systems Methods for Space Missions on a Vicinity of Collinear Libration Points’. In C. Simó, editor, *Hamiltonian Systems with Three or More Degrees of Freedom*, pages 223–241. Kluwer Academic Publishers, 1999.
61. V. Szebehely: *Theory of Orbits*. Academic Press, 1967.
62. W. Wiesel and W. Shelton: ‘Modal Control of an Unstable Periodic Orbit’. *Journal of the Astronautical Sciences*, 31 (1):63–76, 1983.
63. R.S. Wilson, K.C. Howell, and M.W. Lo: ‘Optimization of Insertion Cost Transfer Trajectories to Libration Point Orbits’. *Advances in the Astronautical Sciences*, 103:1569–1586, 2000.

# Chaos and Diffusion in Dynamical Systems Through Stable-Unstable Manifolds

**Massimiliano Guzzo**

**Abstract** The phase-space structure of conservative non-integrable dynamical systems is characterized by a mixture of stable invariant sets and unstable structures which possibly support diffusion. In these situation, many practical and theoretical questions are related to the problem of finding orbits which connect the neighbourhoods of two points $A$ and $B$ of the phase-space. Hyperbolic dynamics has provided in the last decades many tools to tackle the problem related to the existence and the properties of the so called stable and unstable manifolds, which provide natural paths for the diffusion of orbits in the phase-space. In this article we review some basic results of hyperbolic dynamics, through the analysis of the stable and unstable manifolds in basic mathematical models, such as the symplectic standard map, up to more complicate models related to the Arnold diffusion.

## 1 Introduction

Diffusion in conservative dynamical systems has been intensively studied in the last decades by means of analytical and numerical techniques. The problem is particularly complicate and interesting for non-integrable systems, because the phase-space is filled by a mixture of stable invariant sets and structures of peculiar topology which possibly support diffusion. The most famous example of non-integrable system is represented by the three body problem, which has motivated most of the researches done in this field. Many branches of dynamical systems theory, such as the KAM theory, hyperbolic theory, numerical investigations of dynamical systems with dynamical indicators, have been developed and tested on gravitational problems, such as the classical three body problem and the stability problems in our Solar System, up to the more recent problems related to space flight dynamics. But, why practical problems, such as those related to space flight dynamics, should

M. Guzzo (✉)
Dipartimento di Matematica Pura ed Applicata, Università degli Studi di Padova, Via Trieste, 63-35121, Padova, US
e-mail: guzzo@math.unipd.it

E. Perozzi, S. Ferraz-Mello (eds.), *Space Manifold Dynamics*,
DOI 10.1007/978-1-4419-0348-8_2, 

be concerned with these dynamical systems theories? A theoretical formulation of a practical problem could be the following one: given two points $A$ and $B$ of the phase-space, one would find an orbit $x(t)$ with initial condition $x(0)$ in a small neighbourhood of $A$ and $x(T)$ in a small neighbourhood of $B$ at some time $T$. In the hypothesis that such an orbit is found, it is usually affected by chaos, i.e. very small changes in the initial condition $x(0)$ can result in big changes in the complete orbit $x(t)$. In addition, in presence of chaos, the orbits display topological and statistical complexities, and the transfer in the phase-space is often called (chaotic) diffusion. Numerical integrations of chaotic diffusion are intrinsically affected by large numerical errors, which become relevant on a finite (and usually short) time called Lyapunov time. The analysis of the orbits on times of the order of the Lyapunov times requires a detailed dynamical analysis of the phase-space structure of the systems. In the last decades many numerical indicators, based essentially on the Lyapunov characteristic exponents (such as the so called Fast Lyapunov Indicators, introduced in [4]) or on the Fourier analysis (such as frequency analysis [14, 15]) have been successfully used to detect the phase-space structure of non-integrable systems (see, for example, [5]).

The peculiarity of chaos related to the existence of many different orbits with very close initial conditions on the one hand is clearly a problem for numerical integrations, on the other hand it provides the opportunity of constructing orbits with behave, in some sense, as one wishes, with very small corrections to a reference orbit. The branch of dynamical systems which studies the behavior of chaotic orbits under small corrections is known as shadowing theory, and it is motivated by the Anosov–Bowen theorem.

The problems which I have discussed above are related to the existence of peculiar structures in the phase-space called stable and unstable manifolds. Stable and unstable manifolds are defined (see Section 2) as the sets of points whose orbits are asymptotic in the past or in the future respectively to an hyperbolic invariant structure of the phase-space, such as an hyperbolic fixed point, periodic orbit or invariant torus. Specifically, having in mind the problem of finding transfer orbits between points $A$ and $B$ of the phase-space, when the points $A$ and $B$ belong to different hyperbolic invariant sets which we denote by $\mathcal{A}$, $\mathcal{B}$ and the unstable manifold $W^u$ of $\mathcal{A}$ intersects the stable manifold $W^s$ of $\mathcal{B}$, there exists the possibility of such transfer orbits between a neighbourhood of $\mathcal{A}$ and a neighbourhood of $\mathcal{B}$. If the intersection of these manifolds is transverse, the possibility of transfer orbits is coupled to complicate dynamics, usually called "chaotic" dynamics. In this paper I describe the structure of stable/unstable manifolds for a class of dynamical systems which have been often used as the prototype of conservative systems, that is the symplectic maps. In Section 2 we analyze the structure of these manifolds in one of the simplest examples in which the structure is already complex, i.e. the homoclinic tangle of hyperbolic saddle points of the standard map and of more general two dimensional dynamical systems. In Section 3 we describe how diffusion can be supported by the heteroclinic tangle related to different periodic orbits. In Section 4 we describe some higher dimensional examples related to Arnold diffusion.

## 2 The Homoclinic Tangle of Hyperbolic Saddle Points

The paradigm of chaotic dynamics in conservative systems is represented by the so-called standard map $(I,\varphi) \mapsto (I',\varphi')$ defined by:

$$\begin{aligned}\varphi' &= \varphi + I \\ I' &= I + \varepsilon \sin(\varphi + 1),\end{aligned} \tag{1}$$

where $I \in \mathbb{R}$ is an action variable, $\varphi \in \mathbb{S}^1$ is an angle and $\varepsilon \in \mathbb{R}$ is a parameter. For $\varepsilon = 0$ the standard map is integrable, i.e. the action $I$ is constant while the angle $\varphi$ rotates with angular velocity $I$.

Instead, for $\varepsilon \neq 0$ the phase-portraits of the map (represented in Fig. 1) show the existence of invariant curves, as well as of two dimensional regions where motions seem to be spread, and certainly not organized in invariant curves. This peculiar structure of the phase-plane qualifies the standard map as a non integrable system.

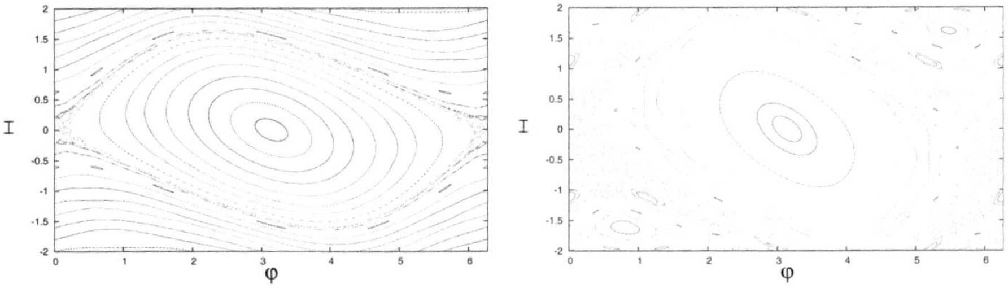


**Fig. 1** Phase portraits of the standard map for $\varepsilon = 0.6$ (*left* panel) and $\varepsilon = 1$ (*right* panel). In both cases the phase-plane contains a mixture of invariant curves and two dimensional regions where motions seem to spread uniformly. This peculiar structure of the phase-plane qualifies the standard map as a non integrable system

We remark also the presence of two fixed points $(I,\varphi) = (0,0),(0,\pm\pi)$, and (0,0) is an hyperbolic saddle point.

Two dimensional symplectic maps similar to the standard map can be obtained as Poincaré sections of higher dimensional continuous dynamical systems. An interesting example is provided by a simplified model of the spin-orbit rotations of oblate satellites, which is described in Fig. 2.

Because the fixed point $x_* = (I_*,\varphi_*) = (0,0)$ is a saddle point, one can apply the so called stable (unstable) manifold theorem (see [13]) to prove that the sets:

$$W^s = \{x\colon \lim_{t\to\infty} \psi^t(x) = x_*\},\ W^u = \{x\colon \lim_{t\to\infty} \psi^{-t}(x) = x_*\} \tag{2}$$

are smooth curves, locally tangent in $x_*$ to the eigenvectors of the Jacobian matrix of $\psi$: $\frac{\partial \psi}{\partial x}(x_*)$. The manifold $W^s$ is called stable manifold of $x_*$, $W^u$ is called unstable manifold of $x_*$.

In every two-dimensional dynamical system defined by a smooth map $\psi\colon M \to M$ ($M$ denotes the two-dimensional phase-space) with a saddle fixed point $x_*$ the

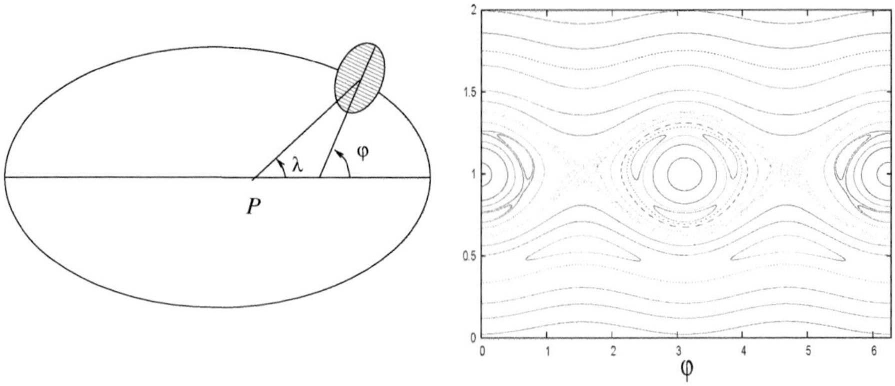


**Fig. 2** *Left* panel: a tri-axial satellite, whose center of mass moves on a Keplerian orbit, is constrained to rotate in the plane $x$,$y$ around an axis of inertia. The equation of motion for the libration angle $\varphi$ is: $\ddot{\varphi} = -\frac{3}{2}\Omega^2\left(\frac{a}{|r|}\right)^3\frac{I_2-I_1}{I_3}\sin(2(\varphi-\lambda))$, where $\lambda$,$a$,$r$ denote the true longitude, the semi-major axis and the distance from the center of mass to the central body; $\Omega$ denotes the Keplerian frequency of motion and $I_1$,$I_2$,$I_3$ are the principal moments of inertia. *Right* panel: phase portrait of the Poincaré map $(\varphi,I) = (\varphi(0),\dot{\varphi}(0)) \longmapsto (\varphi',I') = (\varphi(T),\dot{\varphi}(T))$, with $T = 2\pi/\Omega$. The eccentricity of the orbit is 0.02 and $\frac{3}{2}\Omega^2\frac{I_2-I_1}{I_3} = 0.075$. The phase-plane structure is qualitatively similar to the phase-plane structure of the standard map shown in Fig. 1: the phase-plane contains a mixture of invariant curves and two dimensional regions where motions seem to spread uniformly. This peculiar structure of the phase-plane qualifies this dynamical system as a non integrable system

stable and unstable manifolds of $x_*$ are curves. From the definition it is clear that the knowledge of the unstable manifold $W^u$ provides knowledge about how the orbits with initial conditions in a small neighbourhood of $x_*$ "go away" from $x_*$, while the knowledge of the stable manifold $W^s$ provides knowledge about how the orbits of the phase-space approach asymptotically $x_*$.

In the two-dimensional systems with a first integral the stable and unstable manifolds are contained in the level set of the first integral containing the fixed point. Therefore, they do not have a complicate topology. To produce a complicate topology we need different hypotheses. An hypothesis which is sufficient to prove the existence of a complicate structure for the stable and unstable manifolds due to Poincaré [18] is related to the existence of a so-called homoclinic point $x_0$, that is a point of transverse intersection of $W^s$,$W^u$. In such a case, one shows that:

- each point of the orbit of the homoclinic point $x_0$:

$$x_t = \psi^t(x_0) \quad t \in \mathbb{Z}$$

  is an homoclinic point, i.e. $W^s$ intersects $W^u$ transversely at $x_t$.
- As a consequence, the unstable manifold cuts the stable manifold transversely infinite times forming typical lobes, as it is shown in Fig. 3.

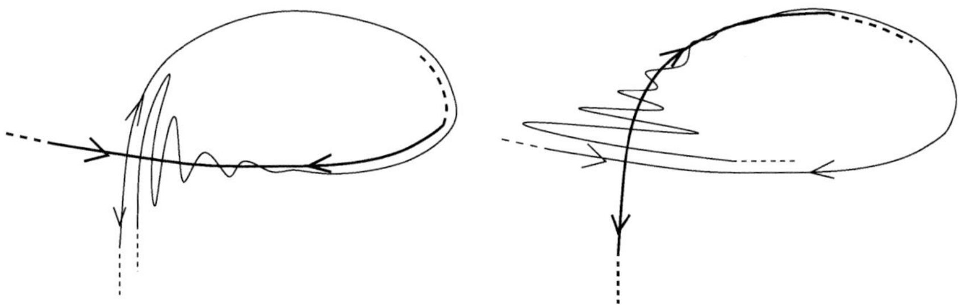

**Fig. 3** Examples of structure of stable and unstable manifolds of a saddle fixed point of a two dimensional system, with the hypothesis of existence of homoclinic points. *Left* panel: the unstable manifold of $x_*$ cuts the stable manifold (represented by the bold curve) transversely forming typical lobes. *Right* panel: the stable manifold cuts the unstable manifold (represented by the bold curve) transversely forming typical lobes

- Approaching the fixed point the base of each lobe becomes smaller and the height becomes bigger, because near the fixed point there is contraction along the stable direction and expansion along the unstable one.
- Suitably close to $x_*$, the lobes of the unstable manifold are so long that they are forced to intersect the stable manifold in points $x'$,$x''$ which are not in the orbit of $x_0$. Also the orbits of these points contain only homoclinic points (Fig. 4).

All these properties demonstrate that in the hypothesis of existence of at least an homoclinic point the structure of the stable and unstable manifolds is indeed very complicate, and is commonly called homoclinic tangle. We remark that because $W^u$ is the set of all the points whose orbit "comes from the fixed point $x_*$" (for $t \to -\infty$) and $W^s$ is the set of all the points whose orbit "goes to the fixed point $x_*$" (for $t \to +\infty$), the homoclinic points are the points whose orbit "comes from and goes to the fixed point $x_*$". Therefore, the complexity of the structure of the homoclinic tangle reflects the complexity of the dynamics related to the saddle point. A precise way of representing the complex dynamics in homoclinic tangles uses the conjugation of the map $\psi$ to special maps called "horseshoe" maps (see [20]).

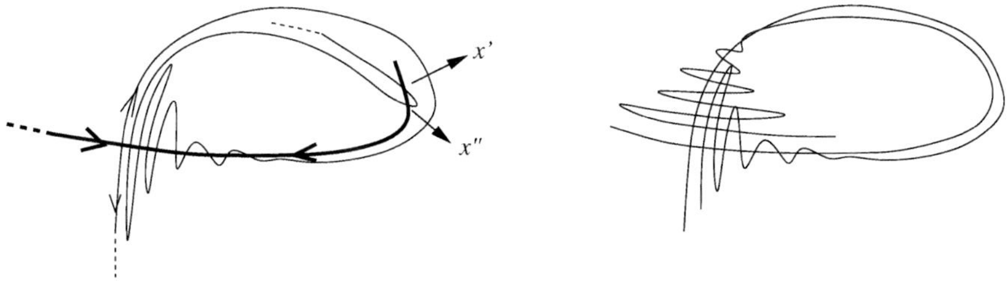


**Fig. 4** Examples of structure of stable and unstable manifolds of a saddle fixed point of a two dimensional system, with the hypothesis of existence of homoclinic points. *Left* panel: the homoclinic points $x'$ and $x''$ are not in the same homoclinic orbit. *Right* panel: homoclinic tangle near the fixed point

In general, the analytic computation of the stable and unstable manifolds, as well as the analytic determination of homoclinic points, is not straightforward. In some perturbative contexts, analytical approximations can be obtained by means of the so called Poincaré–Melnikov integrals.

Instead, the numerical localization of $W^s$, $W^u$, at least in the described case of two dimensional maps, can be obtained by numerically propagating a set of initial conditions chosen in a small neighborhood of the saddle point. In such a way, one directly constructs a neighborhood of a finite piece of the unstable manifold, while for the stable manifold one repeats the construction for the inverse map. This method gives very good results for two dimensional maps because the neighborhoods of the fixed points are two dimensional and can be propagated with reasonable CPU times. A more sophisticated method is described in [19]. In Fig. 5 we show the numerical computation of the stable and unstable manifolds of the fixed point (0,0) of the standard map: from the figures we appreciate the topological complexity of the homoclinic tangle.

## 3 From Chaos to Diffusion in two Dimensional Symplectic Maps

In the previous section we have described the structures which support chaotic motions in two-dimensional symplectic maps. However, the existence of chaotic motions does not mean that the system is characterized by macroscopic instability. For example, the region interested by chaotic motions could be localized in a small region of the phase-space. To better represent this situation, instead of the usual standard map (1), we consider a slightly different map defined by:

$$\begin{aligned} \varphi' &= \varphi + I \\ I' &= I + \varepsilon \frac{\sin\varphi'}{(\cos\varphi' + 1.1)^2}. \end{aligned} \tag{3}$$

The map (3) is more suitable than (1) to represent generic quasi-integrable maps, because the Fourier expansion of the perturbation:

$$f(\varphi') = \varepsilon \frac{\sin\varphi'}{(\cos\varphi' + 1.1)^2} \tag{4}$$

contains an infinite number of harmonics. The advantage of using maps of the form (3) has been explained in several papers [5, 7, 8, 6, 9]. We now consider the phase portraits of the map reported in Fig. 6: for $\varepsilon = 0.002$ (left panel) the phase-space contains regions characterized by chaotic motions (such as those with action around the value $I = 0.34$ and $I = 0.36$) which are disconnected by invariant curves, acting as complete barriers to the diffusion of the action variable $I$. Therefore, there do not exist orbits which connect these different chaotic regions. The invariant curves which are complete barriers to the diffusion of the action variable are KAM curves,

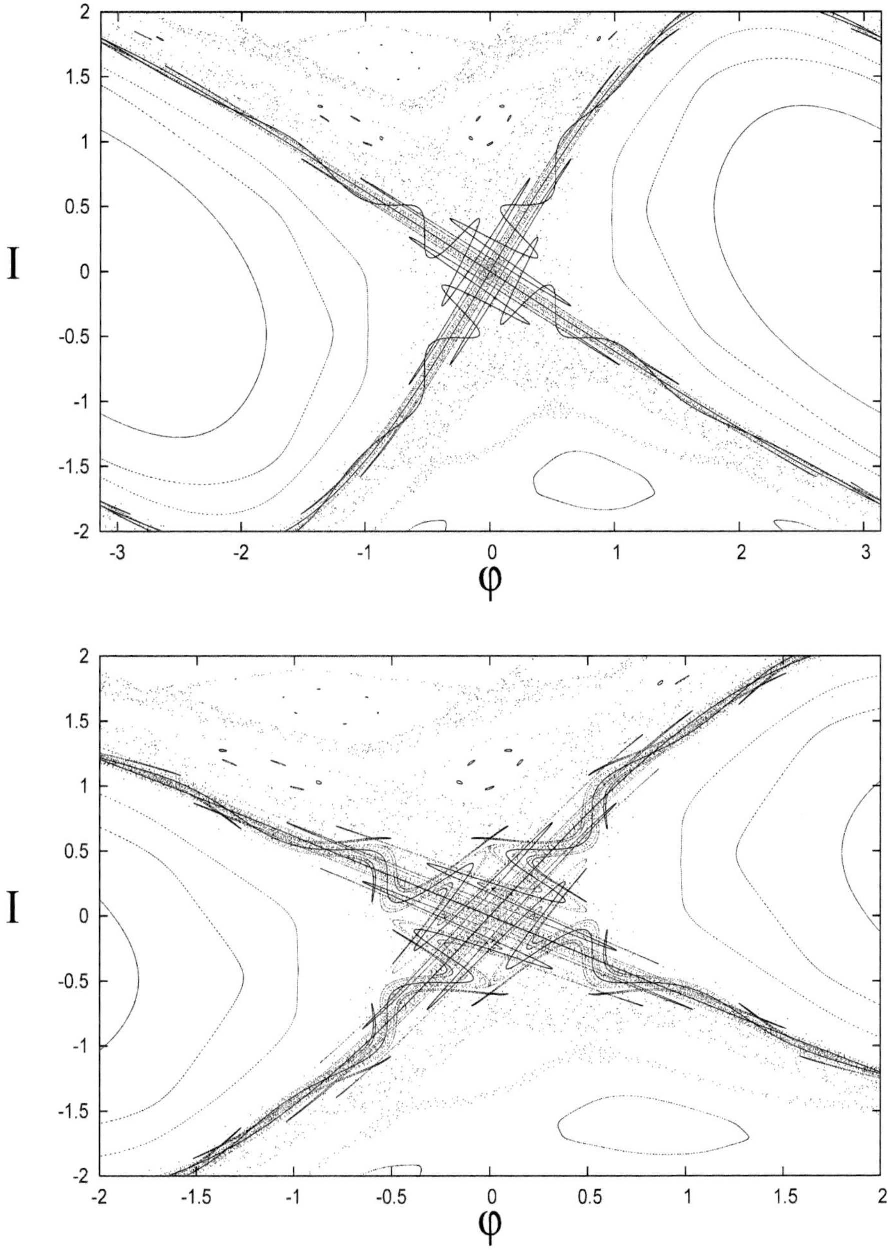


**Fig. 5** Numerical representation of finite pieces of the stable and unstable manifolds of the hyperbolic fixed point (0,0) of the standard map for $\varepsilon = 1$. On each panel the stable and unstable manifolds are plotted on the phase portrait of the map. The *top* panel represents a shorter piece of the manifold

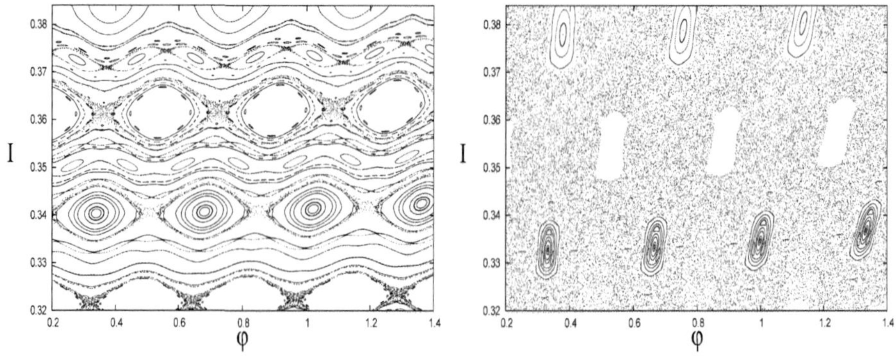


**Fig. 6** Phase portraits of the map (3) for $\varepsilon = 0.002$ (*left* panel) and $\varepsilon = 0.004$ (*right* panel). For the smallest value of $\varepsilon$ we see that there exist invariant curves which act as topological barriers to the diffusion of the action I. For the largest value of $\varepsilon$ the absence of invariant curves means that, in principle, an orbit can diffuse through all the action values

whose existence can be established by the KAM theorem (Kolmogorov [10], Arnold [2], Moser [17]). The KAM theorem proves the existence of the KAM invariant curves if the value of the perturbing parameter $\varepsilon$ is suitably small. Instead, for higher values of $\varepsilon$ the KAM curves typically are destroyed and replaced by the so called cantori, which are discontinuous Aubry–Mather invariant sets, and therefore the action variables can diffuse through their holes. In Fig. 6, right panel, we represent a the phase portrait of the map (3) for $\varepsilon = 0.004$: for such an higher value of the perturbing parameter we do not find KAM curves on the phase-portrait. The absence of the topological barriers represented by the KAM invariant curves allow the chaotic orbits to possibly diffuse through all the action values.

The possibility of an orbit of diffusing through all the action values does not mean that there exist orbits which effectively diffuse. Up to now, there does not exist a rigorous explanation of global diffusion in generic quasi-integrable systems, while there exist heuristic criteria. One of the most popular heuristic criteria for establishing the existence of global diffusion is the so called Chirikov criterion [3] of overlapping resonances. The idea behind Chirikov criterion is that global diffusion exists when the hyperbolic regions related to the hyperbolic periodic orbits of different resonances "overlap". A rigorous way of defining the overlapping of nearby resonances uses the stable and unstable manifolds of these periodic orbits. Precisely, a periodic orbit of period $k$ for a map $\psi : M \to M$ is defined by a sequence $x_j \in M, j = 0, \ldots, k-1$ such that $\psi(x_j + k) = \psi(x_j + k)$ for any $j$. In particular, any point $x_j$ is a fixed point of the map $\psi^k$. We now consider the periodic orbits of initial conditions $x_0$ such that $x_0$ is a saddle fixed point of $\psi^k$. One can therefore define stable and unstable manifolds for $x_0$ as in the case of fixed points, and look for homoclinic intersections, producing the homoclinic tangle and chaotic motions. But now one can do more and can look for the transverse intersections between the stable/unstable manifolds of the different fixed points of the map $\psi^k$ which correspond to different periodic orbits (or fixed points) of the map $\psi$.

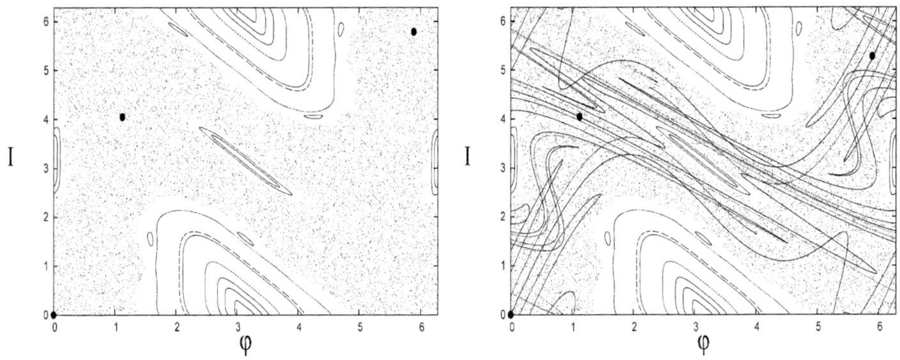


**Fig. 7** Numerical computation of heteroclinic points among the unstable manifold of a saddle fixed point and the stable manifold of a periodic orbit of period $k = 2$ of the standard map for $\varepsilon = 2$. *Left* panel: phase–portrait of $\psi^2$, where $\psi$ denotes the standard map (1). The bullet denotes the initial condition of an hyperbolic periodic orbit of period $k = 2$. *Right* panel: on the phase–portrait we report the computation of a finite piece of the unstable manifold of (0,0) and the stable manifold of the hyperbolic periodic orbit. We can appreciate the presence of many heteroclinic intersections

Let us consider the following example: let $\psi$ be the standard map (1) with $\varepsilon = 2$ and let us consider the periodic orbits of period $k = 2$. The map $\psi^2$ has (0,0) as saddle fixed point, but also a saddle fixed point $x_0 = (\varphi_0, I_0)$ which corresponds to an hyperbolic periodic orbit of the standard map of period 2 (see Fig. 7, left panel). The numerical computation of the phase portrait shows that there are not invariant KAM curves which act as topological barriers to the diffusion of the action variable in the interval $[0, I_0]$. To show that diffusion indeed occurs in this interval, we use the mechanism provided by the existence of transverse intersections among the stable manifold of one periodic orbit and the unstable manifold of the fixed point (0,0). Such transverse intersection points are called heteroclinic points, and they are used since decades to explain global diffusion in the phase-space. Establishing the existence of heteroclinic points rigorously is even more difficult than for homoclinic points. Instead, numerical methods apply without additional difficulties. In Fig. 7, right panel, we report the numerical computation of the unstable manifold of (0,0) and the stable manifold of $x_0$: the existence of transverse intersections is evident. Because of the existence of heteroclinic points one easily finds initial conditions which chaotically diffuse from a neighbourhood of the fixed points (0,0) to a neighborhood of the periodic orbit of initial condition $x_0$.

## 4 Higher Dimensional Systems: from Arnold's Model to Four Dimensional Maps

The general mechanisms which can produce drift and diffusion in the phase–space of higher dimensional systems is an interesting, and in general open, problem.

The problem becomes more difficult when the system is close to an integrable one. Interesting examples of chaos and diffusion in a three-body problem using stable/unstable manifolds is described in [16].

Many diffusive phenomena in higher dimensional systems are called Arnold diffusion, because they are more or less inspired by the pioneering example proposed by Arnold [1]. Arnold's example is defined by the Hamiltonian system:

$$H = \frac{I_1^2}{2} + \frac{I_2^2}{2} + \varepsilon \cos\varphi_1 + \varepsilon\mu(\cos\varphi_1 - 1)(\sin\varphi_2 + \sin t), \tag{5}$$

where $\varphi_1,\varphi_2 \in \mathbb{S}^1$, $I_1,I_2 \in \mathbb{R}$, and $\varepsilon,\mu$ are parameters. The Hamilton equations of (5) are:

$$\begin{aligned} \dot\varphi_1 &= I_1 \\ \dot\varphi_2 &= I_2 \\ \dot I_1 &= \varepsilon \sin\varphi_1 + \varepsilon\mu \sin\varphi_1(\sin\varphi_2 + \sin t) \\ \dot I_2 &= -\varepsilon\mu(\cos\varphi_1 - 1)\cos\varphi_2. \end{aligned} \tag{6}$$

The system depends on two small parameters $\varepsilon$ and $\mu$. For $\varepsilon = 0$ the system has only three dimensional invariant tori (considering $t$ as a periodic variable with equation $\dot t = 1$) defined by the constant value of the actions $I_1,I_2$, and the motions on these tori are quasi–periodic with three frequencies:

$$\dot\varphi_1 = I_1 \quad \dot\varphi_2 = I_2 \quad \dot t = 1. \tag{7}$$

For $\varepsilon,\mu \neq 0$ the system becomes non integrable, and understanding its dynamics is not trivial. Specifically, for suitably small $\varepsilon$ the KAM theorem (in its hysoenergetic formulation) applies to Hamiltonian (5), establishing the existence of a large volume set of invariant tori in the phase–space. However, at variance with the case of two dimensional maps, where KAM curves divided the phase–space acting as complete barriers for diffusion, in such a five dimensional system the three dimensional invariant tori do not divide the phase–space, and diffusion of orbits among the invariant tori is in principle possible.

To prove the existence of diffusion, Arnold considered the special resonance: $\dot\varphi_1 = 0$, which contains an invariant manifold $\Lambda$ defined by $I_1 = 0,\varphi_1 = 0$. The invariant manifold $\Lambda$ is fibered by the invariant two dimensional tori:

$$I_1 = 0 \quad \varphi_1 = 0 \quad \dot\varphi_2 = I_2(0) \quad \dot t = 1, \tag{8}$$

which are hyperbolic if $\mu$ is suitably small. In fact, for $\mu = 0$ the Hamiltonian of the system is:

$$H = \frac{I_1^2}{2} + \varepsilon\cos\varphi_1 + \frac{I_2^2}{2},$$

which is the Hamiltonian of a pendulum and a rotator. In this case, the stable and unstable manifolds of each invariant torus (8) are the separatrices of the pendulum.

As a consequence, for $\mu = 0$, the invariant manifold $\Lambda$ is hyperbolic. By general hyperbolic theory the manifold $\Lambda$ remains hyperbolic also for suitably small $\mu \neq 0$.

We remark that for $\mu = 0$ the action $I_2$ is a first integral of the system, while for $\mu \neq 0$ it is a first integral only for the restriction of the map to the invariant manifold $\Lambda$. Therefore, there is not diffusion of the action $I_2$ on $\Lambda$, but as soon as $\varepsilon, \mu \neq 0$ the action $I_2$ can diffuse in any small neighbourhood of $\Lambda$. To prove that this diffusion indeed exists, Arnold proved that the unstable manifolds of hyperbolic tori of $\Lambda$ intersect transversely the stable manifolds of close hyperbolic tori, thus providing the mechanism for initial conditions in the neighbourhood of $\Lambda$ to diffuse through these manifolds. This kind of diffusion is called Arnold diffusion.

Though the results on the Arnold's model have not been generalized to generic quasi–integrable systems, the ideas contained in Arnold's work have inspired in the last decades the studies of diffusion in higher dimensional systems. Below, we describe recent numerical studies of slow diffusion in four dimensional quasi–integrable systems inspired by Arnold's model of diffusion. [5, 7, 11, 8, 6, 9].

In many papers in collaboration with Froeschlé and Lega [5, 7, 8, 6, 9, 12] we considered specific quasi-integrable systems which are more generic than Arnold's model. Specifically, we considered two coupled twist maps as follows:

$$\begin{gathered} \varphi_1' = \varphi_1 + I_1, \varphi_2' = \varphi_2 + I_2 \\ I_1' = I_1 - \varepsilon \frac{\partial f}{\partial \varphi_1}(\varphi_1', \varphi_2'), I_2' = I_2 - \varepsilon \frac{\partial f}{\partial \varphi_2}(\varphi_1', \varphi_2') \end{gathered} \tag{9}$$

where $\varepsilon$ is a small parameter and the perturbation $f$ is:

$$f = \frac{1}{\cos(\varphi_1) + \cos(\varphi_2) + c}$$

with $c > 2$. This specific choice of the perturbation has been done because the Fourier expansion of $f$ contains an infinite number of harmonics, and the exponential decay of the harmonics is determined by the choice of the constant $c$.

If $\varepsilon = 0$ the map (9) is integrable: the actions of this system are constants of motion and the angles rotate at constant angular velocity. Because it is not possible to represent the orbits of (9) in the complete four dimensional phase-space, it is convenient to represent them on a two dimensional surface such as:

$$S = \{(I_1, I_2, \varphi_1, \varphi_2) \colon (\varphi_1, \varphi_2) = (0,0)\}.$$

For $\varepsilon = 0$, any orbit with initial condition $x$ on $S$ is on an invariant torus. Therefore, the orbit does not return on $S$ if the ratio of the frequencies is irrational, or it returns exactly on $S$ on the point $x$ if the ratio of the frequencies is rational. Therefore, each orbit with initial conditions on $S$ can be symbolically represented by a dot on $S$.

If $\varepsilon \neq 0$ the system is not integrable and the actions are not constants of motion, but if $\varepsilon$ is sufficiently small, the phase-space is filled by a large volume of two dimensional KAM tori. Anyone of these tori intersects transversely $S$ only on one

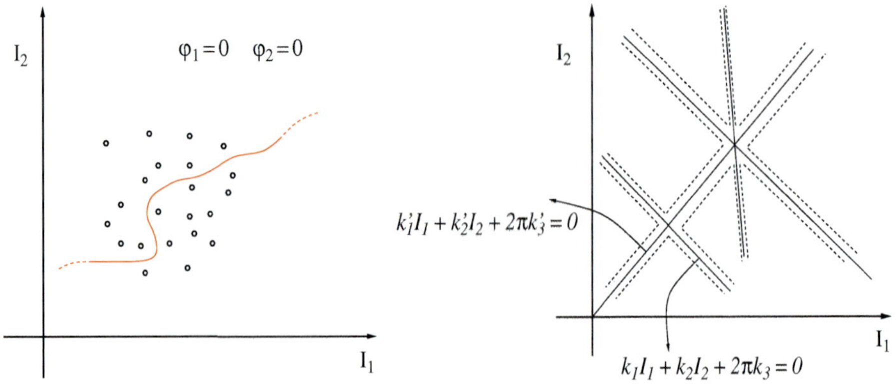


**Fig. 8** *Left* panel: the KAM tori do not trap motions in the four dimensional space. *Right* panel: the KAM tori are outside the neighborhood of the resonances defined by the Diophantine condition (12)

point (see [7]), and therefore each invariant torus is symbolically represented by a point on $S$. Therefore, the surface $S$ contains many points representing two dimensional invariant tori, which however do not trap motions in the four dimensional phase–space. There is therefore the possibility of diffusion among these invariant tori even for very small $\varepsilon \neq 0$ (see Fig. 8 for a symbolic representation of possible diffusion paths).

Diffusion, as far as we know, needs hyperbolic structures, which are related to the resonances of the system, therefore we need a method to identify the hyperbolic structures of the map. We first recall the definition of the resonances for the map (9). Any linear combination of the angles $k_1\varphi_1 + k_2\varphi_2$, with $k_1,k_2 \in \mathbb{Z}$, is resonant if there exists $k_3 \in \mathbb{Z}$ such that:

$$k_1\varphi_1' + k_2\varphi_2' = (k_1\varphi_1 + k_2\varphi_2) + (k_1I_1 + k_2I_2) = (k_1\varphi_1 + k_2\varphi_2) + 2\pi k_3 \ , \quad (10)$$

i.e. if:

$$k_1\, I_1 + k_2\, I_2 - 2\pi k_3 = 0. \quad (11)$$

From KAM theorem we know that invariant tori are located far from a suitable neighbourhood of all these resonances (see Fig. 8). In fact, a KAM torus exists near the values $(I_1,I_2)$ satisfying a non-resonance Diophantine condition of the form:

$$|k_1I_1 + k_2I_2 - 2\pi k_3| \geq \frac{\mathcal{O}(\sqrt{\varepsilon})}{|(k_1,k_2,k_3)|^{\tau}}, \quad \forall (k_1,k_2,k_3) \in \mathbb{Z}^3\backslash(0,0,0),\ \tau > 2. \quad (12)$$

The complement of the set of invariant tori, which is in the neighbourhood of the resonances, is called Arnold web, and contains the hyperbolic structures which possibly support chaotic diffusion. An efficient way of detecting numerically the Arnold web of a system is provided by the so called Fast Lyapunov Indicator, first defined in [4]. For a generic map $\psi : M \rightarrow M$ the Fast Lyapunov Indicator $FLI(x,v,T)$ is a function which depends on a point $x \in M$, on a tangent vector $v \in \mathbb{R}^n = T_xM$, and on a positive time $T$ as follows:

$$FLI(x,v,T) = \log \left\| \frac{\partial \psi^T}{\partial x}(x)v \right\|. \tag{13}$$

For a fixed vector $v$ and suitably long time $T$ the computation of the function $FLI(x,v,T)$ on the surface $S$ provides a precise detection of the Arnold web and of the hyperbolic structures of the system, as it is explained in detail in [5, 7]. Here, we report the results of the computation of the FLI for the map (9) on the surface $S$. For any point $x$ on a grid of $S$ we computed the Fast Lyapunov Indicator, and represented it with a color scale. Precisely,

- the points with the higher values of the FLI (which corresponds to white in the color scale used to represent the value of the indicator) denote motions on hyperbolic structures within the resonances of the system;
- the points with an intermediate value of the FLI (which corresponds to intermediate gray in the color scale used to represent the value of the indicator) are regular motions (including KAM tori);
- the points with lower value of the FLI (which corresponds to black or dark gray in the color scale used to represent the value of the indicator) are regular motions (including resonant tori).

Therefore, the color representation of the FLI on $S$ allows one to clearly identify the KAM tori, the resonant tori, as well as the hyperbolic structures which possibly support diffusion in the phase-space. The result of the computation is reported in Fig. 9. For $\varepsilon = 0.6$ (left panel) there is a prevalence of KAM tori in the phase-space,

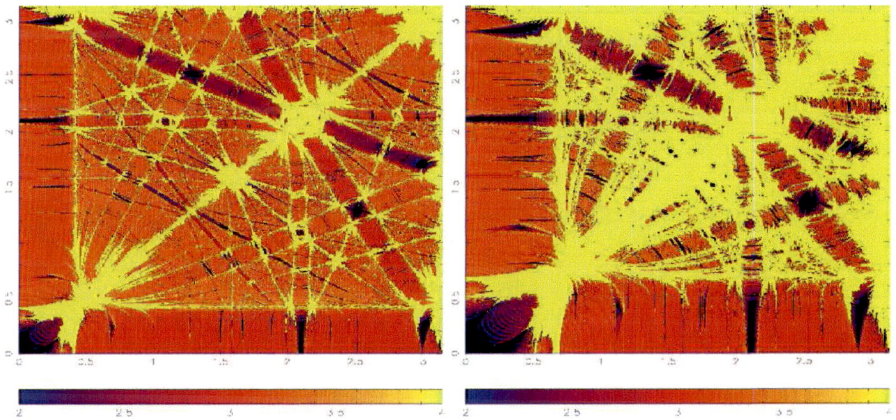

**Fig. 9** Computation of the Arnold web on the section $S$ using a color representation of the FLI. *Left* panel: $\varepsilon = 0.6$, $c = 4$. *Right* panel: $\varepsilon = 1.6$, $c = 4$.

and the hyperbolic structures are organized as a web of resonances, as predicted by the KAM theorem. Chaotic diffusion can occur only on this network of hyperbolic structures. Instead, for the higher value $\varepsilon = 1.6$ there is a prevalence of hyperbolic motions. In such a case, the hyperbolic structures are not organized in a web, we are not in a regime described by the KAM theorem, and chaotic diffusion can occur practically in any direction.

In [11, 8] we have shown how initial conditions in the hyperbolic manifolds diffuse in the Arnold web. We have chosen initial conditions in the region of the hyperbolic motions and then we computed numerically their orbits up to the very long $10^{11}$ iterations. The results are reported in Fig. 10: on the section $S$, represented

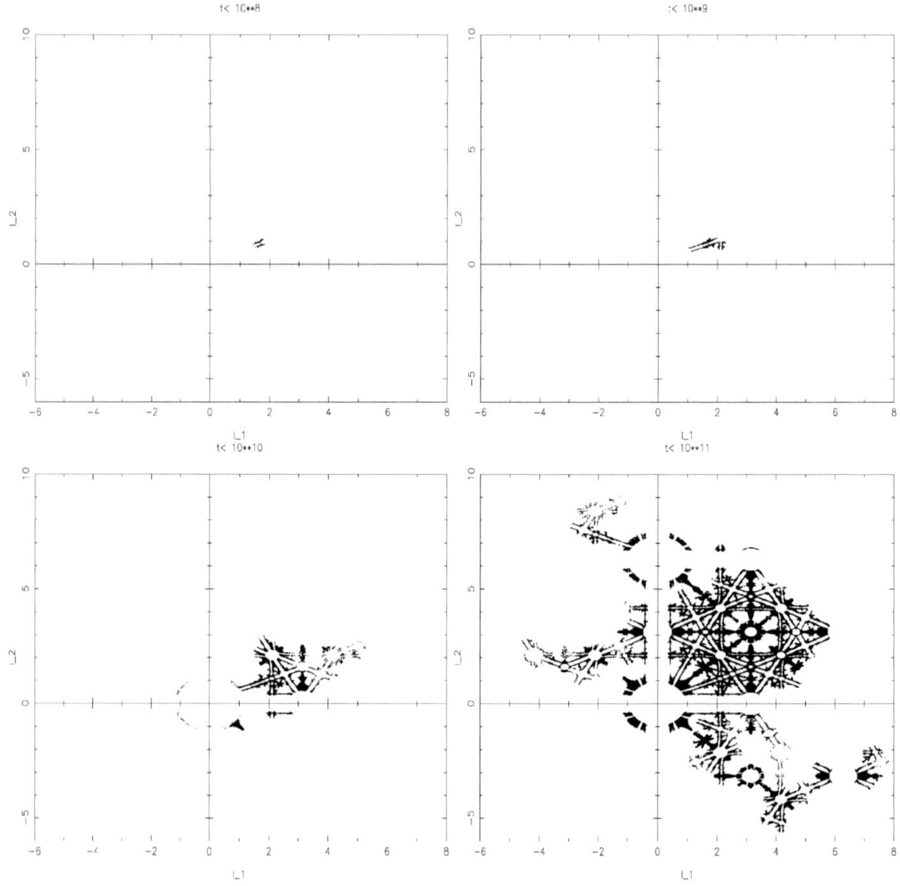


**Fig. 10** Evolution on section $S$ (*black* dots) of 20 orbits for the map (9) with hyperbolic initial conditions near $(I_1,I_2) = (1.71,0.81)$ on a time $t < 10^8$ iterations (*top left*), $t < 10^9$ iterations (*top right*), $t < 10^{10}$ iterations (*bottom left*), $t < 10^{11}$ iterations (*bottom right*) for $\varepsilon = 0.6$. The orbits fill a macroscopic region of the action plane whose structure is that of the Arnold web

by the action plane, we plotted as black dots all points of the orbits which have returned after some time near the section $S$. Because computed orbits are discrete we represented the points which enter the neighbourhood of $S$ defined by $|\varphi_1|,|\varphi_2| \leq 0.005$, (reducing the tolerance 0.005 reduces only the number of points on the section, but does not change their diffusion properties). In such a way we represent the chaotic diffusion for orbits with initial conditions in a neighborhood of $S$.It happens that the orbits fill a macroscopic region of the action plane whose structure is that of the Arnold web. The possibility of visiting all possible resonances is necessarily limited by finite computational times.

## References


1. Arnold V.I. (1964), Instability of dynamical systems with several degrees of freedom, *Sov. Math. Dokl.*, **6**, 581–585.
2. Arnold V.I. (1963a), Proof of a theorem by A.N. Kolmogorov on the invariance of quasi-periodic motions under small perturbations of the Hamiltonian. *Russ. Math. Surv.*, **18**, 9.
3. Chirikov, B.V. (1979), An universal instability of many dimensional oscillator system. *Phys. Reports*, **52**, 265.
4. Froeschlé C., Lega E. and Gonczi R. (1997), Fast Lyapunov indicators. Application to asteroidal motion. *Celest. Mech. and Dynam. Astron.*, **67**, 41–62.
5. Froeschlé C., Guzzo M. and Lega E. (2000), Graphical Evolution of the Arnold Web: From Order to Chaos, *Science*, **289**, n. 5487.
6. Froeschlé C., Guzzo M. and Lega E. (2005), Local and global diffusion along resonant lines in discrete quasi–integrable dynamical systems, *Celest. Mech. and Dynam. Astron.*, **92**, n. 1-3, 243-255.
7. Guzzo M., Lega E. and Froeschlé C. (2002), On the numerical detection of the effective stability of chaotic motions in quasi-integrable systems, *Physica D*, **163**, n. 1-2, 1-25.
8. Guzzo M., Lega E. and Froeschlé C. (2005), First Numerical Evidence of Arnold diffusion in quasi–integrable systems, *DCDS B*, **5**, n. 3.
9. Guzzo M., Lega E. and Froeschlé C. (2006), Diffusion and stability in perturbed non-convex integrable systems. *Nonlinearity*, **19**, 1049–1067.
10. Kolmogorov, A.N. (1954), On the conservation of conditionally periodic motions under small perturbation of the hamiltonian, *Dokl. Akad. Nauk. SSSR*, **98**, 524.
11. Lega E., Guzzo M. and Froeschlé C. (2003), Detection of Arnold diffusion in Hamiltonian systems, *Physica D*, **182**, 179–187.
12. Lega E., Froeschlé C. and Guzzo M. (2007), Diffusion in Hamiltonian quasi–integrable systems." In *Lecture Notes in Physics, "Topics in gravitational dynamics"*, Benest, Froeschlé, Lega eds., Springer.
13. Hirsch M.W., Pugh C.C. and Shub M. (1977), Invariant Manifolds. Lecture Notes in Mathematics, **583**. Springer-Verlag, Berlin-New York.
14. Laskar, J. (1989), A numerical experiment on the chaotic behaviour of the solar system, *Nature*, **338**, 237–238.
15. Laskar, J. (1990), The chaotic motion of the solar system - A numerical estimate of the size of the chaotic zones. Icarus 88, 266–291.
16. Llibre, J. Sim, C. (1980), Some homoclinic phenomena in the three-body problem. *J. Diff. Eq* **37**, no. 3, 444–465.
17. Moser J. (1958), On invariant curves of area-preserving maps of an annulus, *Comm. Pure Appl. Math.*, **11**, 81–114.

18. Poincaré H. (1892), Les méthodes nouvelles de la mécanique celeste, Gauthier–Villars, Paris.
19. Simo C. (1989), On the analytical and numerical approximation of invariant manifolds, in Modern Methods in Celestial Mechanics, D. Benest, Cl. Froeschlé eds, Editions Frontières, 285-329.
20. Smale S. (1967), Differentiable dynamical systems, *Bulletin of the American Mathematical Society*, **73**, 747-817.

# Regular and Chaotic Dynamics of Periodic and Quasi-Periodic Motions

**Alessandra Celletti**

**Abstract** We review some basic topics from Dynamical System theory, which are of interest in Space Manifold Dynamics. We start by recalling some notions related to equilibrium points. Floquet theorem leads to the introduction of Lyapunov exponents. Nearly–integrable systems are very common in Celestial Mechanics; their study motivated the development of perturbation theories as well as of KAM and Nekhoroshev's theorem. The Lindstedt–Poincaré technique allows to look for periodic orbits. Finally, we recall the derivation of the Lagrangian points in the circular and elliptic, planar, restricted three–body problem. Each section is *almost* self–contained and can be read independently from the others.

## 1 Introduction

This paper is intended to review basic topics of Dynamical Systems theory, which are used in Space Manifold Dynamics (hereafter, SMD). Indeed it is nowadays accepted that there is a common interplay between Dynamical Systems and Flight Dynamics: Lagrangian points, invariant manifolds, heteroclinic points now pertain to both disciplines. For example, halo orbits around one of the collinear Lagrangian points allow to place a spacecraft on the far side of the Moon, but in constant contact with the Earth. Also the so–called quasi–periodic *Lissajous orbits* are often studied in flight dynamics. Missions like ISEE-3 (International Sun/Earth Explorer 3) launched in 1978 or SOHO (Solar and Heliospheric Observatory) launched in 1995 have already benefitted from halo or Lissajous orbits. These examples corroborate the strong interaction between Orbital Dynamics and Dynamical Systems, whose mixing gives origin to SMD.

In this paper we review some topics which are essential ingredients of SMD. Each section is *almost* self–contained and can be read independently from the others. We start from the analysis of equilibrium points and their stable and unstable

A. Celletti (✉)
Dipartimento di Matematica, Università di Roma Tor Vergata, Via della Ricerca Scientifica 1, I-00133, Roma, Italy
e-mail: celletti@mat.uniroma2.it

E. Perozzi, S. Ferraz-Mello (eds.), *Space Manifold Dynamics*,
DOI 10.1007/978-1-4419-0348-8_3, 

manifolds, providing also the definition of homoclinic/heteroclinic points and that of the center manifold (Section 2). Lyapunov exponents, measuring the degree of complexity of a system, are introduced through Floquet's theorem (Section 3). Integrable and non–integrable systems (Section 4) has been often studied in connection to Celestial Mechanics, and especially with reference to the two and three–body problems. In between we find the nearly–integrable systems which motivated the development of perturbation theories (Section 5). To overcome the so–called small divisor problem appearing in perturbation theories, a breakthrough came in the middle of the twentieth century through KAM theory (Section 6), concerning the persistence of invariant tori. Arnold's diffusion and effective stability for exponential times are shortly reviewed in Section 7. Turning the attention to periodic orbits we describe the Lindstedt–Poincaré technique, which is used to compute some periodic orbits (Section 8). We conclude with an analysis of the Lagrangian points in the circular and elliptic (planar, restricted) three–body problems, and with some applications to SMD.

## 2 Around Equilibrium Points

The simplest (non trivial) example of a continuous dynamical system is provided by the so–called *harmonic oscillator*. It is described by the differential equation

$$\ddot{x} = -\omega x, \tag{1}$$

where $\omega$ is a positive real quantity. Denoting the solution of (1) at time $t$ as $(x(t),v(t))$, where $\dot{x}(t) = v(t)$ represents the velocity, one has

$$\begin{aligned} x(t) &= A \sin(\sqrt{\omega}\, t) + \alpha \\ v(t) &= A\sqrt{\omega} \cos(\sqrt{\omega}\, t + \alpha) \end{aligned} \tag{2}$$

where $A$ and $\alpha$ are real constants depending on the initial conditions. The origin is an equilibrium solution and from (2) it easily follows that

$$v(t)^2 + \omega x(t)^2 = const$$

showing that the level curves are ellipses around the origin (see Fig. 1a). The origin is called an *elliptic equilibrium point*. Let us now add a friction, so that instead of (1) we consider the equation

$$\ddot{x} = -\omega x - \gamma \dot{x},$$

where $\gamma$ is a positive constant. Then, the solution is given by

$$x(t) = A_1 e^{\lambda_1 t} + A_2 e^{\lambda_2 t},$$

where $A_1, A_2$ are real constants and $\lambda_{1,2} = -\frac{\gamma}{2} \pm \sqrt{\frac{\gamma^2}{4} - \omega}$. The origin is an attractor; according to the values of $\omega$ and $\gamma$, the solution can spiral toward the attractor (in which case it is called *spiral attractor*) or decay on the attractor without oscillations

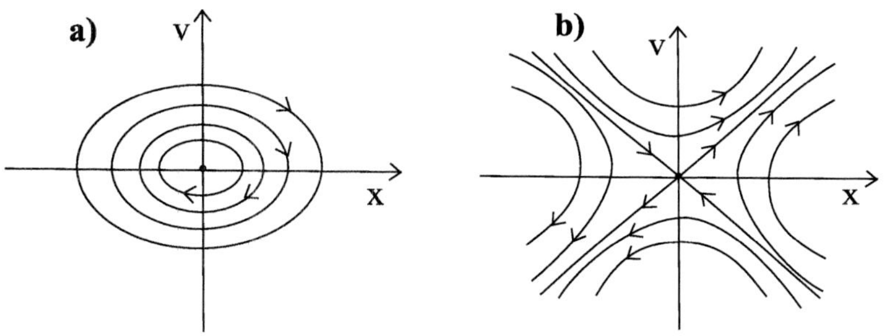


**Fig. 1** **a**) Elliptic equilibrium point. **b**) Hyperbolic equilibrium point

(*node attractor*). Since the solution converges to the equilibrium point, rather than keeping the ellipses of the conservative case, the system is said to be *structurally unstable*.

Next, we consider equation (1), but selecting the case in which $\omega$ is negative, say $\omega = -\Omega$, so that (1) can be written as

$$\ddot{x} = \Omega x.$$

The solution is given by

$$x(t) = A_1 e^{\sqrt{\Omega}t} + A_2 e^{-\sqrt{\Omega}t}$$
$$v(t) = A_1 \sqrt{\Omega} e^{\sqrt{\Omega}t} - A_2 \sqrt{\Omega} e^{-\sqrt{\Omega}t},$$

where $A_1$, $A_2$ are real constants. The origin is an *hyperbolic equilibrium* point and the level curves are hyperbolae (see Fig. 1b). The asymptotes are the stable and unstable curves. In particular, the points of the phase space which converge to the hyperbolic point form the *stable manifold*, while those getting far from the equilibrium position make the *unstable manifold*. The intersections between the stable and unstable manifolds of an equilibrium position are called *homoclinic points*; if the stable and unstable manifolds pertain to different equilibrium positions, their intersections are called *heteroclinic points* (see [15] for applications to the three–body problem). The model with friction keeps the same hyperbolic features, thus showing an overall structural stability.

It is also useful to introduce the *center manifolds*, which allow to analyze the behavior of an equilibrium position using another dynamical system with a reduced dimension. The key result is provided by the so–called *Center Manifold Theorem*, which can be stated as follows. Assume that the equations of motion are given by

$$\dot{\underline{x}} = \underline{f}(\underline{x}) \qquad \underline{x} \in \mathbf{R}^n$$

and assume that $\underline{f}(\underline{0}) = \underline{0}$.

**Center Manifold Theorem**. Let $J$ be the Jacobian of $\underline{f}$; decompose the spectrum of $J$ into three parts: $\sigma_s$ is composed by the eigenvalues with negative real part, $\sigma_c$ by the eigenvalues with zero real part, $\sigma_u$ by the eigenvalues with positive real part. Let $E^s$, $E^c$, $E^u$ be the eigenspaces associated to, respectively, $\sigma^s, \sigma^c, \sigma^u$. Then, there exist stable, unstable and center manifolds tangent, respectively, to $E^s$, $E^u$, $E^c$ at the origin. The center manifold is invariant and every solution in the proximity of the origin approaches the center manifold.

To be concrete, let us consider the dynamical system described by the differential equations

$$\dot{\underline{x}}_1 = A_1\underline{x}_1 + \underline{f}_1(\underline{x}_1,\underline{x}_2), \quad \underline{x}_1 \in \mathbf{R}^{m_1}$$
$$\dot{\underline{x}}_2 = A_2\underline{x}_2 + \underline{f}_2(\underline{x}_1,\underline{x}_2), \quad \underline{x}_2 \in \mathbf{R}^{m_2},$$

where $A_1$ is an $m_1 \times m_1$ matrix and $A_2$ is an $m_2 \times m_2$ matrix, $\underline{f}_1, \underline{f}_2$ are regular vector functions. Assume that all eigenvalues of $A_1$ have zero real part, while all eigenvalues of $A_2$ have negative real part. Moreover, assume that $\underline{f}_1$ and $\underline{f}_2$, as well as their Jacobians, are zero at the origin. The stable manifold at the origin is $\underline{x}_1 = \underline{0}$, while $\underline{x}_2 = \underline{0}$ is the center manifold. In a small neighborhood of the origin, the invariant manifold described by the equation $\underline{x}_2 = \underline{h}(\underline{x}_1)$ for some vector function $\underline{h}$ is a center manifold, if $\underline{h}$ and its Jacobian are zero at the origin. The equation on the center manifold is

$$\dot{\underline{y}} = A_1\underline{y} + \underline{f}_1(\underline{y},\underline{h}(\underline{y}))\,, \qquad \underline{y} \in \mathbf{R}^{m_1}$$

which determines the dynamics in a neighborhood of the origin.

## 3 From Floquet to Lyapunov

A tool for studying the dynamical character of a trajectory is provided by *Floquet theory*, which allows to introduce the *Lyapunov exponents*, yielding an indication of the regular or chaotic behavior of the system.

Let us consider a dynamical system described by the differential equations

$$\dot{\underline{x}} = A(t)\underline{x}\,, \qquad \underline{x} \in \mathbf{R}^n\,, \tag{3}$$

where $A = A(t)$ is an $n \times n$ periodic matrix with period $T$. Floquet theory consists in constructing a coordinate change, so that the study of (3) is reduced to the analysis of a system with constant real coefficients. To this end, we introduce the *principal monodromy* or *fundamental* matrix $\Phi(t)$, whose columns are linear independent solutions of (3) and such that $\Phi(0)$ is the identity matrix. It follows that after a period $T$:

$$\Phi(t+T) = \Phi(t)\Phi(T).$$

By Floquet theorem, there exists a matrix $B$ and a periodic symplectic matrix $C(t)$ such that for any time

$$\Phi(t) = e^{Bt}C(t).$$

Performing the change of variables $\underline{y} = C^{-1}(t)\underline{x}$, one is reduced to study the linear system

$$\dot{\underline{y}} = B\underline{y}$$

which reduces to the study of a differential equation with real constant coefficients. The eigenvalues of $\Phi(T)$ are called the *characteristic multipliers*, which measure the rate of expansion or contraction of a solution. A *characteristic exponent* is a quantity $\ell$ such that $e^{\ell T}$ is a characteristic multiplier. The real parts of the characteristic exponents are the Lyapunov exponents; if all Lyapunov exponents are negative, then the zero solution is asymptotically stable; if they are positive, the solution is unstable.

# 4 Integrable versus Non–Integrable

Within real physical models it is rather uncommon to find closed mathematical formulae which allow to describe the solution of the equations of motion. The models that admit an analytical solution are typically very simplified versions of the physical problem. For example, the mathematical description of a pendulum, which is an elementary example described in many textbooks of classical mechanics, is based on the idealization of a mass attached to a perfectly rigid rod in the absence of any friction and subject only to a gravitational force. However, the grim reality is that of a mass attached to a rod which is typically extensible and not exactly rigid, that mass and rod move in a medium (like the air) which exerts a small friction, and that many different forces can act on the system (for example, also the gravitational attraction of the Moon beside that of the Earth). In the realistic case it is difficult to provide an exact mathematical solution for the dynamics of the problem. In the language of Dynamical Systems the ideal model is called *integrable*, whenever it is possible to provide mathematical formulae which provide the description of the motion; the realistic problem is usually *non–integrable*, since in general it is not possible to provide a mathematical description of the dynamics. Paradigmatic examples of integrable and non–integrable systems come from the study of the celestial motions, in particular from the two and three body problems. Let us start by considering the motion of an asteroid moving around the Sun and assume to neglect all other forces, including the gravitational influence exerted by the other planets. Kepler's laws provide the solution of the two–body problem, according to which the asteroid moves around the Sun on an elliptic orbit which can be described by elementary formulae. The two–body problem is the archetype of an integrable system, whose solution can be explicitly found. Next we pass to consider a more realistic problem by adding the attraction of a third body, for example Jupiter. Since the eighteenth century many efforts have been devoted to the study of the three–body problem, though an explicit solution was never found. Indeed, H. Poincaré [26] proved that the three–body problem is non–integrable and it is not possible to provide explicit mathematical formulae which describe the solution. However, in the example considered here, namely the Sun-asteroid-Jupiter problem, it is important to stress that the gravitational force exerted by Jupiter on the asteroid is much smaller than that due to the Sun. In fact, the mass of Jupiter amounts to $10^{-3}$ times the mass of the Sun. For this reason we classify this problem as *nearly–integrable*: the two–body Keplerian solution describing the Sun–asteroid motion is only slightly perturbed by the presence of Jupiter. Though a complete mathematical solution cannot be achieved, one can implement analytical techniques which allow to find an *approximate* solution of the equations of motion. This approach, known as *perturbation theory*, will be described in Section 5.

Rigorously speaking, let us consider a mechanical system with $n$ degrees of freedom, described by the Hamiltonian function

$$\tilde{H} = \tilde{H}(\underline{p},\underline{q}) , \qquad \underline{p} \in \mathbf{R}^n , \quad \underline{q} \in \mathbf{R}^n .$$

According to the Liouville–Arnold theorem [4], integrable systems admit a canonical change of variables, say $\mathcal{C}:(\underline{A},\underline{\varphi}) \in \mathbf{R}^n \times \mathbf{T}^n \to (\underline{p},\underline{q}) \in \mathbf{R}^{2n}$ (with $\mathbf{T} \equiv \mathbf{R}/2\pi\mathbf{Z}$), such that the new Hamiltonian depends just on the variables $\underline{A}$:

$$H \circ \mathcal{C}(\underline{A},\underline{\varphi}) = h(\underline{A}) \tag{4}$$

The set of coordinates $(\underline{A},\underline{\varphi})$ are known as *action–angle* variables [4]. The fact that the Hamiltonian system (4) is integrable can be immediately recognized by writing Hamilton's equations. In fact, denoting by

$$\underline{\omega} = \underline{\omega}(\underline{A}) = \frac{\partial h(\underline{A})}{\partial \underline{A}} \tag{5}$$

the *frequency* or *rotation number* of the system, one has

$$\begin{aligned} \dot{\underline{A}} &= \frac{\partial h(\underline{A})}{\partial \underline{\varphi}} = \underline{0} \\ \dot{\underline{\varphi}} &= \frac{\partial h(\underline{A})}{\partial \underline{A}} = \underline{\omega}(\underline{A}) . \end{aligned}$$

Therefore, $\underline{A}(t) = \underline{A}_0$ is constant along the motion, while from the second equation we obtain $\underline{\varphi}(t) = \underline{\omega}(\underline{A}_0)t + \underline{\varphi}_0$, where $(\underline{A}_0,\underline{\varphi}_0)$ denote the initial conditions. Introducing a small parameter, say $\varepsilon > 0$, a nearly–integrable system can be described by a Hamiltonian of the form

$$H(\underline{A},\underline{\varphi}) = h(\underline{A}) + \varepsilon f(\underline{A},\underline{\varphi}) , \tag{6}$$

where $h(\underline{A})$ is the integrable part, while $\varepsilon f(\underline{A},\underline{\varphi})$ is the perturbing function. Going back to the three–body example, we can identify the integrable part, described by $h(\underline{A})$, with the two–body problem (the asteroid–Sun interaction), the parameter $\varepsilon$ represents the Jupiter–Sun mass ratio and the term $\varepsilon f(\underline{A},\underline{\varphi})$ corresponds to the asteroid–Jupiter gravitational interaction. Many other problems in Celestial Mechanics are described in terms of nearly–integrable systems. For example, the motion of a rigid satellite rotating about an internal spin–axis and revolving around a central planet. The system is integrable whenever the satellite is assumed to be spherical; the non–integrability comes from the oblateness of the satellite which can be identified with the perturbing parameter.

## 5 Getting Started with Perturbation Theory

Perturbation theories greatly influenced the advances in Celestial Mechanics. The most famous example is provided by the computations made by J.C. Adams and U.J.J. Leverrier, which led to the discovery of Neptune. The prediction of the position of Neptune was computed with astonishing accuracy to recover unexplained perturbations on the motion of Uranus.

The fundamentals of perturbation theory [5, 6, 11, 27] can be summarized as follows. Consider a nearly–integrable Hamiltonian function of the form (6). Neglecting the perturbation, the equations of motion are

$$\begin{aligned} \dot{\underline{A}} &= \underline{0} \\ \dot{\underline{\varphi}} &= \underline{\omega}(\underline{A}) \,, \end{aligned}$$

where $\underline{\omega}$ has been defined in (5). The corresponding solution shows that the actions remain constant, say $\underline{A} = \underline{A}_0$, while the angles rotate with frequency $\underline{\omega}_0 \equiv \underline{\omega}(\underline{A}_0)$, i.e. $\underline{\varphi}(t) = \underline{\omega}(\underline{A}_0) + \underline{\varphi}(0)$. This solution represents an approximation of the true solution, valid up to a time of the order of $1/\varepsilon$. Through suitable canonical changes of variables, perturbation theory allows to get better approximations of the equations of motion by removing the perturbation to higher order in the perturbing parameter. In particular, let us look for a change of variables $\mathcal{C}:(\underline{A}',\underline{\varphi}') \to (\underline{A},\underline{\varphi})$ such that the transformed Hamiltonian $H'(\underline{A}',\underline{\varphi}')$ takes the form

$$H'(\underline{A}',\underline{\varphi}') = H \circ \mathcal{C}(\underline{A}',\underline{\varphi}') = h'(\underline{A}') + \varepsilon^2 f'(\underline{A}',\underline{\varphi}') \,, \tag{7}$$

for suitable regular functions $h'$ and $f'$. The perturbing function has now been removed to orders $\varepsilon^2$; Hamilton's equations associated to (7) can be integrated up to times of order $1/\varepsilon^2$. Perturbation theories have been developed since the eighteenth century with the aim to compute accurate solutions for the motion of the celestial bodies. Classical perturbation theory can be implemented under very general assumptions, namely the non–degeneracy of the unperturbed Hamiltonian - the determinant of the Hessian matrix associated to $h$ is different from zero for any $\underline{A} \in \mathbf{R}^n$ – and a non–resonance condition on the frequency, namely $|\underline{\omega}(\underline{A}) \cdot \underline{k}| > 0$ for any $\underline{k} \in \mathbf{Z}^n$ (or, at least, for any $\underline{k} \in \mathbf{Z}^n$ with $|\underline{k}| \le N$ for a suitable positive integer $N$). The recipe is based on a coordinate change of variables, a Taylor series expansion in terms of the perturbing parameter $\varepsilon$ and a Fourier series to determine explicitely the canonical transformation. We report the algorithm which allows to compute a first order perturbation theory. We emphasize that such algorithm is completely constructive, in the sense that the change of variables and the new Hamiltonian function can be explicitly computed. Let $(\underline{A},\underline{\varphi}) \rightarrow (\underline{A}',\underline{\varphi}')$ be a canonical change of variables such that

$$\underline{A} = \underline{A}' + \varepsilon \frac{\partial \Phi(\underline{A}',\underline{\varphi})}{\partial \underline{\varphi}}$$

$$\underline{\varphi}' = \underline{\varphi} + \varepsilon \frac{\partial \Phi(\underline{A}',\underline{\varphi})}{\partial \underline{A}'} ,$$

where the *unknown* function $\Phi(\underline{A}',\underline{\varphi})$ is referred to as the *generating function*. Notice that $\underline{\varphi}$ depends on the old angles $\underline{\varphi}$ and on the new actions $\underline{A}'$. Let us start by splitting the perturbing function $f(\underline{A},\underline{\varphi})$ into its average $f_0(\underline{A}) = \frac{1}{(2\pi)^n} \int_{\mathbf{T}^n} f(\underline{A},\underline{\varphi}) d\underline{\varphi}$ and a remainder function $\tilde{f}(\underline{A},\underline{\varphi})$ defined as $\tilde{f}(\underline{A},\underline{\varphi}) \equiv f(\underline{A},\underline{\varphi}) - f_0(\underline{A})$. Inserting the transformation in (6) and expanding in Taylor series up to the second order around $\varepsilon = 0$, one obtains

$$h(\underline{A}' + \varepsilon \frac{\partial \Phi(\underline{A}',\underline{\varphi})}{\partial \underline{\varphi}}) + \varepsilon f(\underline{A}' + \varepsilon \frac{\partial \Phi(\underline{A}',\underline{\varphi})}{\partial \underline{\varphi}},\underline{\varphi})$$

$$= h(\underline{A}') + \underline{\omega}(\underline{A}') \cdot \varepsilon \frac{\partial \Phi(\underline{A}',\underline{\varphi})}{\partial \underline{\varphi}} + \varepsilon f_0(\underline{A}') + \varepsilon \tilde{f}(\underline{A}',\underline{\varphi}) + O(\varepsilon^2) .$$

Imposing that the new Hamiltonian does not depend on the angles up to $O(\varepsilon^2)$, we need to require that the function $\Phi$ satisfies the equation

$$\underline{\omega}(\underline{A}') \cdot \frac{\partial \Phi(\underline{A}',\underline{\varphi})}{\partial \underline{\varphi}} + \tilde{f}(\underline{A}',\underline{\varphi}) = \underline{0} . \tag{8}$$

Notice that equation (8) is well defined, since its average over the angles is zero. Once a solution of (8) is found, we conclude that the new integrable Hamiltonian is given by

$$h'(\underline{A}') = h(\underline{A}') + \varepsilon f_0(\underline{A}'),$$

whose associated Hamilton's equations provide the solution of the motion up to $O(\varepsilon^2)$. In order to derive an explicit expression for the generating function, let us expand $\Phi$ and $\tilde{f}$ in Fourier series as

$$\Phi(\underline{A}',\underline{\varphi}) = \sum_{\underline{k} \in \mathbf{Z}^n \setminus \{\underline{0}\}} \hat{\Phi}_{\underline{k}}(\underline{A}') e^{i\underline{k}\cdot\underline{\varphi}} ,$$

$$\tilde{f}(\underline{A}',\underline{\varphi}) = \sum_{\underline{k} \in \mathbf{Z}^n \setminus \{\underline{0}\}, |\underline{k}| \leq N} \hat{f}_{\underline{k}}(\underline{A}') \, e^{i\underline{k}\cdot\underline{\varphi}} ,$$

where we assumed that $\tilde{f}$ is composed by a finite number of terms, precisely those with $|\underline{k}| \leq N$ for a suitable integer $N$. Inserting the Fourier expansion in (8) we get

$$i \sum_{\underline{k} \in \mathbf{Z}^n \setminus \{\underline{0}\}} \underline{\omega}(\underline{A}') \cdot \underline{k} \, \hat{\Phi}_{\underline{k}}(\underline{A}') \, e^{i\underline{k}\cdot\underline{\varphi}} = - \sum_{\underline{k} \in \mathbf{Z}^n \setminus \{\underline{0}\}, |\underline{k}| \leq N} \hat{f}_{\underline{k}}(\underline{A}') \, e^{i\underline{k}\cdot\underline{\varphi}} ,$$

which implies

$$\hat{\Phi}_{\underline{k}}(\underline{A}') = -\frac{\hat{f}_{\underline{k}}(\underline{A}')}{i\,\underline{\omega}(\underline{A}')\cdot\underline{k}}\ .$$

Casting together the above formulae, we obtain that the generating function takes the form

$$\Phi(\underline{A}',\underline{\varphi}) = i \sum_{\underline{k}\in\mathbf{Z}^n\setminus\{\underline{0}\},|\underline{k}|\le N} \frac{\hat{f}_{\underline{k}}(\underline{A}')}{\underline{\omega}(\underline{A}')\cdot\underline{k}}\, e^{i\underline{k}\cdot\underline{\varphi}}\ . \qquad (9)$$

Due to the divisor appearing in (9), it is necessary to impose the non–resonance condition on the frequency $\underline{\omega}$ at least for some Fourier indexes, say $\underline{k} \in \mathbf{Z}^n$ with $|\underline{k}| \le N$. When a *resonance* is met, namely $|\underline{\omega}(\underline{A})\cdot\underline{k}| = 0$ for some $\underline{k} \in \mathbf{Z}^n$, then the algorithm fails and classical perturbation theory cannot be implemented. Nevertheless, even if the divisors are not zero thanks to the fact that the frequency vector $\underline{\omega}$ is rationally independent, the terms $\underline{\omega}(\underline{A})\cdot\underline{k}$ can become arbitrarily small, thus leading to the divergence of the series defining the generating function. This obstacle is called the *small divisor* problem and it prevents the iteration of the above procedure to higher orders in $\varepsilon$. To overcome this problem a breakthrough came in the middle of the twentieth century through Kolmogorov's theorem, which allowed to establish the persistence of invariant tori under very mild assumptions.

## 6 Birth and Death of Invariant Tori

Let us consider a nearly–integrable system with $n$ degrees of freedom. Assume that the frequency vector $\underline{\omega} = (\omega_1,\dots,\omega_n)$ is rationally independent; a *quasi–periodic* motion is a solution that can be expressed (as time $t$ varies) as a function of the form $F = F(\omega_1 t,\dots,\omega_n t)$, where $F(\varphi_1,\dots,\varphi_n)$ is a multiperiodic function, $2\pi$–periodic in each component $\varphi_i$. We have seen that classical perturbation theory allows to approximate the solutions of nearly–integrable systems; once the approximation is constructed, one can fix the initial conditions and proceed to investigate the subsequent evolution. The point of view of Kolmogorov's theorem is quite different, since instead of investigating the motion with preassigned initial conditions, it explores the dynamics on which a quasi–periodic motion with fixed frequency $\underline{\omega}$ takes place [21]. Starting with a nearly–integrable Hamiltonian function of the form (6) and having fixed a rationally independent frequency vector $\underline{\omega}_0$, under suitable assumptions on the unperturbed Hamiltonian $h$ and on the frequency $\underline{\omega}_0$, Kolmogorov's theorem states that if the perturbing parameter $\varepsilon$ is sufficiently small, there exists an invariant torus on which a quasi–periodic motion with frequency $\underline{\omega}_0$ takes place. The set of such invariant tori has positive measure in the phase space. Short afterwards, V.I. Arnold [1, 2] and J. Moser [24] provided alternative proofs in different contexts. Since then, the overall theory is known as *KAM theory*.

Starting from (6), we know that in the integrable case $\varepsilon = 0$ an invariant torus with frequency $\underline{\omega}_0$ is the set $\mathcal{T}_{\underline{\omega}_0} \equiv \{\underline{A}_0\} \times \mathbf{T}^n$ where $\underline{\omega}_0 = \underline{\omega}(\underline{A}_0)$. The hypotheses under which KAM theory can be applied are the following:

i) the non–degeneracy of the unperturbed Hamiltonian, i.e.

$$\det \frac{\partial^2 h(\underline{A})}{\partial \underline{A}^2} \neq 0\,, \qquad \forall\, \underline{A} \in \mathbf{R}^n\,;$$

ii) the frequency vector $\underline{\omega}_0$ satisfies the *diophantine* inequality

$$|\underline{\omega}_0 \cdot \underline{k}|^{-1} \le C|\underline{k}|^{\tau}\,, \qquad \forall\, \underline{k} \in \mathbf{Z}^n \backslash \{\underline{0}\}\,, \tag{10}$$

for some positive constants $C$ and $\tau$. Notice that this assumption is stronger than the non–resonance condition.

Looking at the Hamiltonian system described by (6), KAM theory concerns the persistence of an invariant torus on which a quasi–periodic motion with frequency $\underline{\omega}_0$ takes place (see Appendix A for a link between periodic and quasi-periodic motions). The answer is positive provided the size of the perturbing parameter $\varepsilon$ is sufficiently small, say $\varepsilon \le \varepsilon_{KAM}(\underline{\omega}_0)$. This is necessary to infer the convergence of the series expansions involved in the proof. The qualitative behaviour of a KAM torus as the perturbing parameter is varied is the following. If $\varepsilon$ is very small, the invariant torus is located close to the unperturbed torus; as the perturbing parameter increases, the torus becomes more and more displaced and deformed until a critical value of $\varepsilon$ is reached, say $\varepsilon = \varepsilon_c(\underline{\omega}_0)$, at which the invariant torus ceases to be regular and breaks down. For applications of KAM theory to Celestial Mechanics we refer the reader to [7] and references therein.

The idea at the basis of Kolmogorov's theorem is to clear the hurdle due to small divisors by performing a *superconvergent* sequence of canonical transformations. Classical perturbation theory allows to transform the initial Hamiltonian, say $H_1 = h_1 + \varepsilon f_1$, to one of the form $H_2 = h_2 + \varepsilon^2 f_2$, where the perturbation is, roughly speaking, of order $\varepsilon^2$. The $j$–th step of this iteration produces the Hamiltonian $H_j = h_j + \varepsilon^j f_j$. The main problem is then to show the convergence of the sequence of canonical transformations. Kolmogorov's theorem is based on the implementation of a *superconvergent* or *quadratic* method such that the initial Hamiltonian is transformed to $H_2 = h_2 + \varepsilon^2 f_2$ and then to $H_3 = h_3 + \varepsilon^4 f_3$, while the $j$–th step provides $H_j = h_j + \epsilon^{2^{j-1}} f_j$. The fact that the perturbation decreases with $\varepsilon$ faster than linearly allows to control the small divisors appearing in the sequence of canonical transformations. An instructive way to understand the difference between classical perturbation theory and KAM theory is provided by Newton's method for finding the real root of an equation $f(\xi) = 0$ (see Fig. 2). Let $\xi_0$ be the initial approximation and let $e_0 \equiv |\xi - \xi_0|$ be an estimate on the initial error. Implement a linear method to find the next approximation $\xi_1$, which is determined as the intersection of the tangent to the curve at the point $(\xi_0, f(\xi_0))$ with the $\xi$–axis. Let $\eta_0 \equiv f'(\xi_0)$;

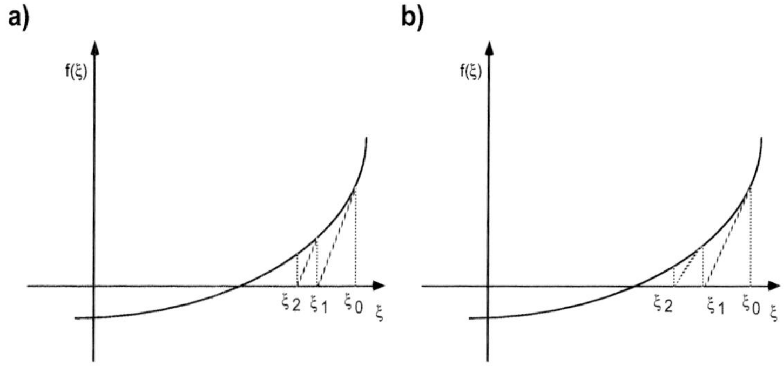


**Fig. 2** Newton's method for finding the root of the equation $f(\xi) = 0$. **a**) Linear convergence; **b**) quadratic convergence

next approximation $\xi_2$ is computed as the abscissa of the line through $(\xi_1, f(\xi_1))$ with slope $\eta_0$. Iterating this procedure, one finds that the error at the $j$–th step is $e_j = |\xi - \xi_j| = O(e_0^{j+1})$. This is the kind of results usually found in perturbation theory. Using a quadratic procedure, the successive approximations are given by the intersection between the $\xi$–axis and the tangent to the function computed at the previous step, i.e.

$$\xi_{j+1} = \xi_j - \frac{f(\xi_j)}{f'(\xi_j)} \qquad j = 0,1,2\dots$$

Let us expand $f(\xi)$ around $\xi_j$; the second order expansion is equal to

$$0 = f(\xi) = f(\xi_j) + f'(\xi_j)(\xi - \xi_j) + \frac{f''(\xi_j)}{2!}(\xi - \xi_j)^2 + O(|\xi - \xi_j|^3)\,,$$

from which we obtain that

$$\xi_{j+1} - \xi = \frac{1}{2!}\frac{f''(\xi_j)}{f'(\xi_j)}(\xi - \xi_j)^2\,.$$

The error at the $j$–th step goes quadratically as

$$e_{j+1} = O(e_j^2) = O(e_0^{2^j})\,.$$

This procedure is adopted for proving KAM theorem. For complete details on the KAM proof we refer the reader to [21, 1, 24, 5, 22, 27].

# 7 Diffusion and Exponential Stability

The existence of invariant tori is particularly relevant in low–dimensional Hamiltonian systems. In fact, if $n = 2$ the phase space has dimension 4, the constant energy surfaces have dimension 3 and the KAM tori have dimension 2, thus providing a separation of the constant energy phase space into invariant regions. This stability result is no more valid for higher dimensions. For example, if $n = 3$ the phase space has dimension 6, the constant energy surfaces have dimension 5 and the 3–dimensional invariant tori do not separate anymore the constant energy phase space. In this case invariant tori form the majority of the solutions; however, the resonances generate gaps between the invariant tori, so that the trajectories can leak out to arbitrarily far regions of the phase space. Arnold provided an example of such phenomenon [3], which is known as *Arnold's diffusion*. In particular, he considered the Hamiltonian function

$$H(A_1,A_2,\varphi_1,\varphi_2,t) = \frac{1}{2}(A_1^2 + A_2^2) + \varepsilon[(\cos\varphi_1 - 1)(1 + \mu\sin\varphi_2 + \mu\cos t)] ,$$

where $A_1, A_2 \in \mathbf{R}$, $\varphi_1, \varphi_2 \in \mathbf{T}$ and $\varepsilon$, $\mu$ are positive parameters. For suitable values of $\varepsilon$, $\mu$, and for given values of the actions such that $0 < A_2^{(a)} < A_2^{(b)}$, it is shown that there exists an orbit connecting regions where the values of the action $A_2$ are far from each other, say $A_2 < A_2^{(a)}$ and $A_2 > A_2^{(b)}$. As shown by Nekhoroshev [25] the diffusion time can be exponentially large. In order to get diffusion it is necessary to build up unstable tori, called *whiskered* tori; a chain of heteroclinic intersections provides the transport of the trajectories from one neighborhood of a torus to the neighborhood of another torus. On the other hand, Nekhoroshev's theorem [25] shows that, under quite general assumptions on the Hamiltonian, the action variables remain confined for an exponentially long time, i.e. if $\varepsilon$ is sufficiently small, say $\varepsilon < \varepsilon_0$, then

$$\|\underline{A}(t) - \underline{A}_0\| \le C_1(\frac{\varepsilon}{\varepsilon_0})^\alpha \qquad \text{for } |t| \le C_2\, e^{C_3(\frac{\varepsilon}{\varepsilon_0})^\beta} ,$$

for suitable positive real constants $\alpha$, $\beta$, $C_1$, $C_2$, $C_3$.

## 8 Hunting for Periodic Orbits

Consider a dynamical system described by the second–order differential equation

$$\ddot{x} + \omega_0^2 x = \varepsilon f(x,\dot{x}) , \qquad x \in \mathbf{R} , \tag{11}$$

where $\varepsilon \ge 0$ is a small real parameter and $f: \mathbf{R}^2 \to \mathbf{R}$ is a regular function. For $\varepsilon = 0$ the system reduces to a harmonic oscillator, whose solution is periodic with period $T_0 = \frac{2\pi}{\omega_0}$. The *Lindstedt–Poincaré technique* allows to find periodic solutions for $\varepsilon \neq 0$ by taking into account that the frequency of the motion can change due to the nonlinear terms. Indeed, for $\varepsilon \neq 0$ the period $T$ coincides with $T_0$ only up to terms of order $\varepsilon$. The basic idea consists in expanding the solution $x(t)$ and the (unknown) frequency $\omega$ as a function of $\varepsilon$:

$$\begin{aligned} x(t) &= x_0(t) + \varepsilon x_1(t) + \varepsilon^2 x_2(t) + \dots \\ \omega &= \omega_0 + \varepsilon \omega_1 + \varepsilon^2 \omega_2 + \dots , \end{aligned} \tag{12}$$

where it is assumed that $x_j(T) = x_j(0)$, being the quantities $x_j(t)$, $\omega_j$, $j \geq 0$, unknown. Let us perform the change of variables $s = \omega t$, so that (11) becomes

$$\omega^2 x'' + \omega_0^2 x = \varepsilon f(x,\omega x') , \tag{13}$$

where $x'$ and $x''$ denote the first and second derivatives with respect to $s$. Let us expand the perturbation as powers of $\varepsilon$:

$$f(x,\omega x') = f(x_0,\omega_0 x_0') + \varepsilon \left[ x_1 \frac{\partial f(x_0,\omega_0 x_0')}{\partial x} + x_1' \frac{\partial f(x_0,\omega_0 x_0')}{\partial x'} + \omega_1 \frac{\partial f(x_0,\omega_0 x_0')}{\partial \omega} \right] + O(\varepsilon^2) .$$

Inserting the series expansion (12) in (13) and equating terms of the same order of $\varepsilon$, one obtains the equations

$$\begin{aligned} \omega_0^2 x_0'' + \omega_0^2 x_0 &= 0 \\ \omega_0^2 x_1'' + \omega_0^2 x_1 &= f(x_0,\omega_0 x_0') - 2\omega_0 \omega_1 x_0'' \\ &\dots \end{aligned}$$

These equations can be solved in sequence and the quantities $\omega_j$ can be found by requiring the periodicity condition $x_j(s + 2\pi) = x_j(s)$, $j = 0,1,2,\dots$

As an example, we study the Duffing equation

$$\ddot{x} + x = -\varepsilon x^3 ,$$

where $x(0) = 1$ and $\dot{x}(0) = 0$. Let us expand the solution $x(t)$ and the unknown frequency $\omega$ as in (12); transforming time as $s = \omega t$ one gets the equation

$$\omega^2 x''(s) + x(s) = -\varepsilon x(s)^3 .$$

To the zeroth order in $\varepsilon$ one gets the equation

$$\omega_0^2 x_0'' + x_0 = 0$$

and, taking into account the initial conditions, one obtains $\omega_0 = 1$ and $x_0(s) = A\cos s$ for some real constant $A$. To the first order in $\varepsilon$ one obtains the equation

$$x_1'' + x_1 = A\left(2\omega_1 - \frac{3}{4}A^2\right)\cos s - \frac{1}{4}A^3 \cos 3s ;$$

secular terms are avoided provided $\omega_1 = \frac{3}{8}A^2$, thus yielding the first order solution $x_1(s) = \frac{1}{32}A^3 \cos 3s$. The solution at the successive orders is obtained in a similar way.

# 9 The Lagrangian Solutions

## 9.1 The Restricted, Planar, Circular Lagrangian Solutions

Let us consider the restricted, circular three–body problem; in the rotating reference frame, there exist equilibrium solutions, known as the *collinear* and *triangular* equilibrium points. A number of asteroids, Trojan and Greek asteroids, is observed to form an equilateral triangle with Jupiter and the Sun. We present here the mathematical derivation of the equilibrium positions and a discussion of their stability.

Let the primary bodies have masses $m_1$ and $m_2$ with $m_1 > m_2$ and let

$$\overline{\mu} \equiv \frac{m_2}{m_1 + m_2} \ ,$$

so that $\mu_1 \equiv Gm_1 = 1-\overline{\mu}$, $\mu_2 \equiv Gm_2 = \overline{\mu}$ (notice that $\mu_1+\mu_2 = 1$). We study the motion of a body with mass $m_3$, which is assumed to be much smaller than $m_1$, $m_2$. Let $O$ be the barycenter of the primaries and normalize to unity the angular velocity of the primaries. We introduce a synodic reference frame $(O,x,y,z)$, rotating with the angular velocity of the primaries; fix the axes so that the coordinates of the primaries are $(x_1,y_1,z_1) = (-\mu_2,0,0)$, $(x_2,y_2,z_2) = (\mu_1,0,0)$. The equations of motion of the body with mass $m_3$ are given by

$$\begin{aligned} \ddot{x} - 2\dot{y} &= \frac{\partial U}{\partial x} \\ \ddot{y} + 2\dot{x} &= \frac{\partial U}{\partial y} \\ \ddot{z} &= \frac{\partial U}{\partial z} \ , \end{aligned} \tag{14}$$

where $U$ is defined as

$$U = U(x,y,z) \equiv \frac{1}{2}(x^2 + y^2) + \frac{\mu_1}{r_1} + \frac{\mu_2}{r_2} \ , \tag{15}$$

being

$$r_1 = \sqrt{(x + \mu_2)^2 + y^2 + z^2} \ , \qquad r_2 = \sqrt{(x - \mu_1)^2 + y^2 + z^2} \ .$$

Multiplying (14) by $\dot{x}$, $\dot{y}$, $\dot{z}$, adding the results and integrating with respect to time, one obtains

$$\dot{x}^2 + \dot{y}^2 + \dot{z}^2 = 2U - C_J \ , \tag{16}$$

where $C_J$ is the *Jacobi constant*. From (15) one obtains

$$C_J = x^2 + y^2 + 2\frac{\mu_1}{r_1} + 2\frac{\mu_2}{r_2} - (\dot{x}^2 + \dot{y}^2 + \dot{z}^2) \ .$$

From (16) it follows that $2U - C_J \geq 0$; the zero velocity curve $C_J = 2U$ provides a boundary called *Hill's surface*, which separates regions where the motion is allowed or forbidden.

The location of the equilibrium points can be found as follows. Assuming a *planar* motion and using that $\mu_1 + \mu_2 = 1$, one finds

$$U = \mu_1 \left( \frac{1}{r_1} + \frac{r_1^2}{2} \right) + \mu_2 \left( \frac{1}{r_2} + \frac{r_2^2}{2} \right) - \frac{1}{2} \mu_1 \mu_2 \, .$$

The equilibrium points are the solutions of the system formed by the derivatives of $U$ with respect to $x$ and $y$, i.e.

$$\begin{aligned} \frac{\partial U}{\partial x} &= \mu_1 \left( -\frac{1}{r_1^2} + r_1 \right) \frac{x + \mu_2}{r_1} + \mu_2 \left( -\frac{1}{r_2^2} + r_2 \right) \frac{x - \mu_1}{r_2} = 0 \\ \frac{\partial U}{\partial y} &= \mu_1 \left( -\frac{1}{r_1^2} + r_1 \right) \frac{y}{r_1} + \mu_2 \left( -\frac{1}{r_2^2} + r_2 \right) \frac{y}{r_2} = 0 \, . \end{aligned} \tag{17}$$

We immediately remark that $r_1 = r_2 = 1$ is a solution of (17), which is equivalent to

$$(x + \mu_2)^2 + y^2 = 1 \, , \qquad (x - \mu_1)^2 + y^2 = 1 \, ;$$

these equations yield the *triangular* Lagrangian solutions $L_4$ and $L_5$ whose coordinates are

$$(x,y) = \left( \frac{1}{2} - \mu_2, \frac{\sqrt{3}}{2} \right) \, , \qquad (x,y) = \left( \frac{1}{2} - \mu_2, -\frac{\sqrt{3}}{2} \right) \, .$$

Since $y = 0$ is a solution of the second of (17), it can be proved that there exist three *collinear* equilibrium positions, denoted as $L_1$, $L_2$, $L_3$, with $L_1$ located between the primaries, while $L_2$ and $L_3$ are on opposite sides with respect to the primaries (see Fig. 3).

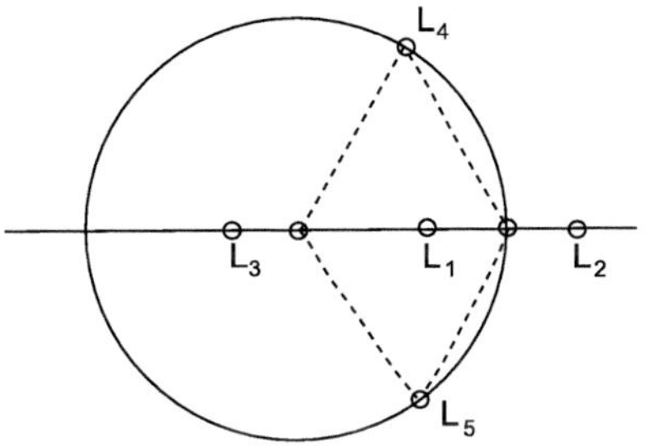


**Fig. 3** The triangular and collinear equilibrium points

Let us discuss the linear stability of one of the above (collinear or triangular) equilibrium positions, that we generically indicate as $(x_0,y_0)$. Let $(\xi,\eta)$ be a small displacement from the equilibrium, i.e. $(x,y) = (x_0+\xi, y_0+\eta)$. In a neighborhood of the equilibrium, using the first two equations in (14) and setting

$$U_{xx} = \frac{\partial^2 U(x_0,y_0)}{\partial x^2}\,, \qquad U_{xy} = \frac{\partial^2 U(x_0,y_0)}{\partial x \partial y}\,, \qquad U_{yy} = \frac{\partial^2 U(x_0,y_0)}{\partial y^2}\,,$$

the variational equations are given by

$$\begin{pmatrix} \dot\xi \\ \dot\eta \\ \ddot\xi \\ \ddot\eta \end{pmatrix} = A \begin{pmatrix} \xi \\ \eta \\ \dot\xi \\ \dot\eta \end{pmatrix}\,,$$

where

$$A \equiv \begin{pmatrix} 0 & 0 & 1 & 0 \\ 0 & 0 & 0 & 1 \\ U_{xx} & U_{xy} & 0 & 2 \\ U_{xy} & U_{yy} & -2 & 0 \end{pmatrix}.$$

The eigenvalues of the matrix $A$ are the solutions of the secular equation

$$\lambda^4 + (4 - U_{xx} - U_{yy})\lambda^2 + (U_{xx}U_{yy} - U_{xy}^2) = 0\,.$$

This equation admits four roots:

$$\lambda_{1,2} = \pm\left[\frac{1}{2}\left(U_{xx}+U_{yy}-4\right) - \frac{1}{2}\left[\left(4-U_{xx}-U_{yy}\right)^2 - 4\left(U_{xx}U_{yy}-U_{xy}^2\right)\right]^{\frac{1}{2}}\right]^{\frac{1}{2}}$$

$$\lambda_{3,4} = \pm\left[\frac{1}{2}\left(U_{xx}+U_{yy}-4\right) + \frac{1}{2}\left[\left(4-U_{xx}-U_{yy}\right)^2 - 4\left(U_{xx}U_{yy}-U_{xy}^2\right)\right]^{\frac{1}{2}}\right]^{\frac{1}{2}}.$$

The equilibrium solution is linearly stable[1], if the eigenvalues are purely imaginary. One can show that the collinear points are unstable for any value of the masses, while triangular equilibrium positions are stable provided

$$1 - 27(1-\mu_2)\mu_2 \geq 0\,.$$

[1] We remark that effective estimates based on Nekhoroshev's theorem has been largely developed to analyze the behavior of the Lagrangian points (see, e.g., [8, 12, 13, 16, 17, 23, 28] and references therein).

## 9.2 The Restricted, Planar, Elliptic Lagrangian Solutions

We remove the assumption that the primaries orbit on a circular trajectory and we let them move on an elliptic trajectory around their common center of mass $O$. Nevertheless we keep the hypothesis that the motion of the three bodies takes place on the same plane. Let $v$ denote the true anomaly of the common ellipse and let $r = \frac{1-e^2}{1+e\cos v}$ be the distance between $P_1$ and $P_2$, where $e$ denotes the eccentricity of the ellipse whose semimajor axis has been normalized to one. We want to describe the equations of motion in a *rotating–pulsating* reference frame. To this end, we first we introduce an inertial barycentric reference frame $(O,x,y)$. Then we define a barycentric reference frame $(O,\xi,\eta)$ rotating with variable angular velocity, such that the rotation angle coincides with the true anomaly $v$:

$$
\begin{aligned}
x &= \xi \cos v - \eta \sin v \\
y &= \xi \sin v + \eta \cos v \; .
\end{aligned}
$$

We introduce rotating–pulsating coordinates $(X,Y)$ defined by

$$
\begin{aligned}
\xi &= rX \\
\eta &= rY.
\end{aligned}
$$

Finally, we adopt the true anomaly as independent variable, transforming the time as

$$
dt = \frac{1}{\sqrt{p}} r^2 \, dv \; ,
$$

where $p$ is the parameter of the ellipse. Denoting by a prime the derivative with respect to the true anomaly, the equations of motion are given by

$$
\begin{aligned}
X'' - 2Y' &= \Omega_X \\
Y'' + 2X' &= \Omega_Y \; ,
\end{aligned}
$$

where

$$
\Omega = \frac{1}{1+e\cos v} \left[ \frac{1}{2}(X^2+Y^2) + \frac{1-\mu}{r_1} + \frac{\mu}{r_2} + \frac{1}{2}\mu(1-\mu) \right] .
$$

The equilibrium positions are the solutions of $\Omega_X = \Omega_Y = 0$ and it is readily seen that they coincide with those of the circular model. In particular, the triangular solutions located at $(\mu - \frac{1}{2}, \pm \frac{\sqrt{3}}{2})$ pulsate as the coordinates. A discussion of the linear stability in the parameter plane $(\mu,e)$ is provided in [9].

## 9.3 The Lagrangian Points in Flight Dynamics

A careful exploration of the dynamical features of the Lagrangian points has been performed in the framework of flight dynamics. In 1968 R. Farquhar [10] christened *halo orbits*, some three–dimensional periodic trajectories close to the collinear Lagrangian points (see, e.g., [14]). The stable manifold of a halo orbit can be run to transfer a spacecraft from the Earth to the proximity of the Lagrangian point. Halo orbits located around the $L_2$ points on the far side of the Moon allow a continuous communication with the Earth.

Since the triangular Lagrangian points of the Earth–Moon three–body problem are stable, there exists a neighborhood such that any initial condition within it remains close to the equilibrium position for a long time. By a theorem due to Lyapunov, for any value of the mass ratio there exists a vertical family of periodic orbits emanating from the triangular equilibrium position. Interesting studies about equilateral libration points were performed in [18, 20]; a spacecraft in the Earth–Moon system is studied using a restricted three–body problem subject to a perturbation modeled by quasi–periodic functions to describe the real motion of Earth and Moon under the gravitational attraction of the Sun. The preservation of maximal dimension tori has been exploited in [19]. The persistence of lower dimensional tori has been studied in [20]: some of these tori are not destroyed, but just deformed by the perturbation. In the context of space flight dynamics, lower dimensional tori related to the $L_2$ Lagrangian point of the Earth–Sun system have been studied for a mission named *Terrestrial Planet Finder*, whose aim is the investigation of exoplanetary systems.

## Appendix A. Periodic and Quasi–Periodic Motions

The difference between periodic and quasi–periodic motions relies on the choice of the rotation number. To make a concrete example, we consider a Hamiltonian system with 2 degrees of freedom. A quasi–periodic motion takes place on a two–dimensional torus, whose points are determined by assigning the longitude and the latitude, say $\varphi_1$, $\varphi_2$ with $0 \le \varphi_1 \le 2\pi$, $0 \le \varphi_2 \le 2\pi$. Let $\underline{\omega} \equiv (\omega_1,\omega_2)$ be the frequency of the motion; in the quasi–periodic case it is assumed that $\frac{\omega_1}{\omega_2}$ is irrational, so that the evolution never retraces the initial position and the trajectory is everywhere dense on the torus. On the contrary, if $\frac{\omega_1}{\omega_2}$ is rational, say $\frac{\omega_1}{\omega_2} = \frac{p}{q}$ with $p,q \in \mathbf{Z}_+$, then the motion is periodic and it retraces the same steps. Let us assume that the frequency vector is of the form $\underline{\omega_0} \equiv (\gamma,1)$ with $\gamma$ being a real number. We define the *continued fraction* representation of $\gamma$ as the sequence of positive integers $[a_0;a_1,a_2,a_3\dots]$, such that

$$\gamma = a_0 + \cfrac{1}{a_1 + \cfrac{1}{a_2 + \frac{1}{a_3+\dots}}} .$$

Using a standard notation, we can write $\gamma = [a_0;a_1,a_2,\dots]$. If $\gamma$ is a rational number, its continued fraction representation is composed by a finite number of terms, i.e. there exists a positive integer $N$ such that $\gamma = [a_0;a_1,a_2,\dots,a_N]$; if $\gamma$ is irrational, the continued fraction expansion is composed by an infinite number of terms. *Noble* numbers are irrational numbers whose continued fraction is definitely equal to one. Number theory guarantees that noble numbers satisfy the diophantine condition (10) with $\tau = 1$. The *golden ratio* $\omega_g \equiv \frac{\sqrt{5}+1}{2}$ has the property that its continued fraction representation is composed by all ones: $\omega_g = [1;1,1,1\dots]$; moreover it satisfies condition (10) with the smallest diophantine constant $C$, being $C = \frac{3+\sqrt{5}}{2}$. Successive truncations of the continued fraction expansion of an irrational number provide the sequence of rational approximants $\{\frac{p_k}{q_k}\}_{k\in\mathbf{Z}}$ given by

$$\frac{p_1}{q_1} = a_0 + \frac{1}{a_1}, \qquad \frac{p_2}{q_2} = a_0 + \frac{1}{a_1 + \frac{1}{a_2}}, \qquad \frac{p_3}{q_3} = a_0 + \frac{1}{a_1 + \frac{1}{a_2 + \frac{1}{a_3}}} \quad \ldots$$

In the case of the golden ratio the rational approximants coincide with the ratio of the so–called Fibonacci's numbers; indeed, the sequence $\left\{1, \frac{2}{1}, \frac{3}{2}, \frac{5}{3}, \frac{8}{5}, \frac{13}{8}, \frac{21}{13}\right\}, \ldots$ converges to $\frac{\sqrt{5}+1}{2}$.

## References


1. V.I. Arnold, *Proof of a Theorem by A.N. Kolmogorov on the invariance of quasi-periodic motions under small perturbations of the Hamiltonian*, Russ. Math. Surv. **18**, 13–40 (1963)
2. V.I. Arnold, *Small divisor problems in classical and Celestial Mechanics*, Russian Math. Surv. **18**, 85–191 (1963)
3. V.I. Arnold, *Instability of dynamical systems with several degrees of freedom*, Sov. Math. Dokl. **5**, 581–585 (1964)
4. V.I. Arnold, *Mathematical Methods of Classical Mechanics*, Springer–Verlag, New York (1978) (Russian original, Moscow, 1974)
5. V.I. Arnold (editor), *Encyclopaedia of Mathematical Sciences*, Dynamical Systems III, Springer–Verlag **3** (1988)
6. A. Celletti, *Perturbation Theory in Celestial Mechanics*, Encyclopedia of Complexity and System Science, R.A. Meyers ed., Springer–Verlag (2009)
7. A. Celletti, L. Chierchia, *KAM stability and celestial mechanics*, Memoir. Am. Math. Soc. **187**, n. 878 (2007)
8. A. Celletti, A. Giorgilli, *On the stability of the Lagrangian points in the spatial restricted problem of three bodies*, Cel. Mech. Dyn. Astr. **50**, 31–58 (1991)
9. J.M.A. Danby, *Stability of the triangular points in the elliptic restricted problem of three bodies*, Astron. J. **69**, 165–172 (1974)
10. R.W. Farquhar, *The Control and Use of Libration–Point Satellites*, Ph.D. Dissertation, Dept. of Aeronautics and Astronautics, Stanford University, Stanford, CA (1968)
11. S. Ferraz–Mello, *Canonical Perturbation Theories*, Springer–Verlag, Berlin, Heidelberg, New York (2007)
12. A. Giorgilli, C. Skokos, *On the stability of the trojan asteroids*, Astron. Astroph. **317**, 254–261 (1997)
13. A. Giorgilli, A. Delshams, E. Fontich, L. Galgani, C. Simó, *Effective stability for a Hamiltonian system near an elliptic equilibrium point, with an application to the restricted three–body problem*, J. Diff. Eq. **77**, 167–198 (1989)
14. G. Gomez, A. Jorba, J. Masdemont, C. Simó, *Study of the transfer from the Earth to a halo orbit around the equilibrium point* $L_1$, Cel. Mech. Dyn. Astr. **56**, 541–562 (1993)
15. G. Gomez, J. Llibre, J. Masdemont, *Homoclinic and heteroclinic solutions in the restricted three–body problem*, Cel. Mech. **44**, 239–259 (1988)
16. M. Guzzo, Z. Knezevic, A. Milani, *Probing the Nekhoroshev stability of asteroids*, Cel. Mech. Dyn. Astr. **83**, 121–140 (2002)
17. M. Guzzo, A. Morbidelli, *Construction of a Nekhoroshev like result for the asteroid belt dynamical system*, Cel. Mech. Dyn. Astr. **66**, 255–292 (1997)

18. À. Jorba, *On Practical Stability Regions for the Motion of a Small Particle Close to the Equilateral Points of the Real Earth–Moon System*, Proc. Hamiltonian Systems and Celestial Mechanics, J. Delgado, E.A. Lacomba, E. Perez–Chavela, J. Llibre eds., World Scientific Publishing, Singapore (1998)
19. À. Jorba, C. Simó, *On quasi–periodic perturbations of elliptic equilibrium points*, SIAM J. Math. Anal. **27**, 1704–1737 (1996)
20. À. Jorba, J. Villanueva, *On the persistence of lower dimensional invariant tori under quasi–periodic perturbations*, J. Nonlinear Sci. **7**, 427–473 (1997)
21. A.N. Kolmogorov, *On the conservation of conditionally periodic motions under small perturbation of the Hamiltonian*, Dokl. Akad. Nauk SSR **98**, 527–530 (1954)
22. R. de la Llave, A. Gonzàlez, À. Jorba, J. Villanueva, *KAM theory without action-angle variables*, Nonlinearity **18**, n. 2, 855–895 (2005)
23. Ch. Lhotka, C. Efthymiopoulos, R. Dvorak, *Nekhoroshev stability at $L_4$ or $L_5$ in the elliptic restricted three body problem - application to Trojan asteroids*, MNRAS **384**, 1165–1177 (2008)
24. J. Moser, *On invariant curves of area-preserving mappings of an annulus*, Nach. Akad. Wiss. Göttingen, Math. Phys. Kl. II, 1–20 (1962)
25. N.N. Nekhoroshev, *An exponential estimate of the time of stability of nearly–integrable Hamiltonian systems*, Russ. Math. Surv. **32**, 1–65 (1977)
26. H. Poincaré, *Les méthodes nouvelles de la méchanique céleste*, Gauthier Villars, Paris (1899)
27. C.L. Siegel, J.K. Moser, *Lectures on Celestial Mechanics*, Springer–Verlag, Berlin, Heidelberg, New York (1971)
28. K. Tsiganis, H. Varvoglis, R. Dvorak, *Chaotic diffusion and effective stability of Jupiter Trojans*, Cel. Mech. Dyn. Astr. **92**, 71–87 (2005)

# Survey of Recent Results on Weak Stability Boundaries and Applications

**Edward Belbruno**

**Abstract** A region of transitional stability in the three-body problem has proved to have interesting mathematical properties and also important applications to several fields. This region, called the weak stability boundary, was first discovered for its use in providing a methodology for computing new types of low energy transfers. Up to recently, understanding it's underlying structure has been elusive. Recent results on understanding the mathematical structure of this region are presented as well as associated dynamics. This includes both numerical and theoretical results indicating the underlying complicated structure of invariant hyperbolic manifolds. Associated resonance dynamics are described. Recent applications are described within the field of astronomy on a theory for the origin of the Moon and also on the minimal energy transfer of solid material between planetary systems.

## 1 Introduction

The search for a new type of transfer from the Earth to the Moon for spacecraft in 1986 led to the discovery of an interesting region of unstable motion about the Moon [3]. The motivation was to find a way for a spacecraft to arrive near the Moon with a substantially reduced relative velocity as compared to the standard Hohmann transfer. A Hohmann transfer approaches the Moon with a relative velocity of approximately 1 km/s, resulting in a significant amount of fuel being required to slow down and go into lunar orbit. It was desired to reduce the approach velocity to 0 km/s. This is called ballistic capture. Although such a capture was conjectured in the 1960s by C. Conley [12, 3] in the three-body problem between the Earth-Moon-spacecraft for transfers starting from arbitrarily near to the Earth to near the Moon, it had never been demonstrated. It was suggested from Conley's work that the invariant manifold structure associated to the unstable collinear Lagrange points $L_1$,$L_2$ near the Moon would have to somehow play a role, but this was not understood.

E. Belbruno (✉)
Department of Astrophysical Sciences, Princeton University, Princeton, NJ, 08544, USA
e-mail: belbruno@princeton.edu

E. Perozzi, S. Ferraz-Mello (eds.), *Space Manifold Dynamics*,
DOI 10.1007/978-1-4419-0348-8_4, 

A way to achieve ballistic capture was numerically demonstrated in 1986 for transfers starting sufficiently far from the Earth [5, 3]. The solution to this problem, yielding a transfer to the Moon for a spacecraft using low thrust, with a flight time of 2 years, utilized a region about the Moon where the stability of motion is in transition. As we will describe below, this is where a particle, say a spacecraft, is in between capture and escape with respect to the Moon. The capture, defined by a negative Kepler energy, is temporary, termed weak capture. The region about the Moon where weak capture occurs is defined by the weak stability boundary (WSB). As we will describe, it can be estimated by a numerical algorithm which determines the transition between "stable" and "unstable" motion about the Moon. Stable and unstable motion in this case are associated to whether or not a spacecraft can, or cannot, respectively, perform a complete cycle about the Moon.

A transfer to the Moon arriving in ballistic capture can be achieved by arriving in weak capture, or equivalently at the WSB. It turned out that the solution obtained in 1986 did utilize the dynamics near the invariant manifolds associated with the Lyapunov orbits near the collinear Lagrange points. The full solution to Conley's conjecture was to find a transfer arriving at the lunar WSB starting from an *arbitrary* distance from the Earth instead of sufficiently far away. This was accomplished in 1991 with the rescue of a Japanese lunar mission and getting it's spacecraft, *Hiten*, to the Moon with very little fuel on a new type of transfer. This solution utilized a four-body problem between the Earth-Moon-Sun-spacecraft [7, 5, 3]. The manifold structure associated to the dynamics of that transfer was partially uncovered in 1994 [6]. This was further explored in 2000 by Marsden et al. [16]. The transfer that *Hiten* used promises to play an important role in future lunar missions. Also, see the work by Circi and Teofillato [11].

Up to recently, the nature of the weak stability boundary, and associated dynamics, has not been well understood. One of the main goals of this paper is to describe recent work which sheds light on this problem. We will consider the restricted three-body problem for the motion of a particle $P_3$ of zero mass in a gravitational field generated by two primary particles $P_1$,$P_2$ in mutual circular motion, where the mass of $P_1$ is much larger than the mass of $P_2$. The motion of $P_3$ is studied relative to $P_2$, and the weak stability boundary exists about $P_2$.

The first results we describe give a much better understanding of the weak stability boundary by visualizing this region obtained from it's algorithmic definition. This was described in an interesting paper by Garcia and Gomez [13] in 2007. Prior to this result, the region that the algorithmic definition gave rise to was only roughly determined. The work in [13] precisely determines this set and goes considerably beyond it by generalizing the definition of the WSB. Work by Topputo and Belbruno [19] provides some refinements and further insights in their results. We will also describe how this work suggests a connection between the weak stability boundary and the limit set obtained from the invariant manifolds associated to the Lyapunov orbits about $L_1$,$L_2$ near $P_2$. This is an interesting new result.

An earlier result that shed some light on the structure of this region, but using different methods, utilized a rough approximation for the location of this boundary. Using this, it was proven that ballistic capture is an unstable and chaotic process.

This was done when the motion of $P_3$ is near parabolic with respect to $P_1$ [5]. This result will be briefly described. The approximation for the location of the boundary is used in [8] to reveal a rich resonance structure.

The WSB region is not an invariant set, nor does it lie on individual energy levels which has made its study difficult. However, these results promise to help better understand this region.

The last topic briefly discussed in this paper is on application of the WSB region to two problems in astrophysics. The first application is on a theory on the origin of the Moon by Belbruno and Gott [4]. This theory uses the WSB region to construct collision trajectories with the Earth that are used to explain how the Earth may have been impacted by a large object to form the Moon. A more recent work utilizes the WSB regions in two different three-body problems to provide a mechanism for weak capture of solid material from one planetary system into another viewed as a five-body problem. This result has implications on the question on the transfer of life bearing material into our own solar system in its early stages of development [9].

The general goal of this paper is to give a sense of selected recent work, since 2004, on the nature of the weak stability boundary and some applications. The results we have chosen to describe are not meant to be exhaustive.

## 2 Restricted Three-Body Problem Model

The model we will use until further notice is the restricted three-body problem mentioned in the Introduction between particles $P_1$,$P_2$,$P_3$. This is a model that is well known and provides a way to study dynamics in the three-body problem where the problem has been simplified as much as possible and still preserve the three-body interaction. Although this more simplified version of the three-body problem is being used, it is found that the results obtained are close to a more realistic three-body modeling.

It is assumed: 1. $P_1$,$P_2$ move in mutual Keplerian circular orbits about their common center of mass which is placed at the origin of an inertial coordinate system $X,Y$. 2. The mass of $P_3$ is zero. Letting $m_k$ represent the masses of $P_k$, $k = 1,2,3$, then $m_3 = 0$. We also assume $m_2 << m_1$. Let $\omega$ be the constant frequency of circular motion of $m_1$ and $m_2$, $\omega = 2\pi/P$, where $P$ is the period of the motion. We consider a rotating coordinate system $(x,y)$ which rotates with the constant frequency $\omega$ as $P_1$ and $P_2$. In the $x$–$y$ coordinate system the positions of $P_1$ and $P_2$ are fixed. Without loss of generality, we can set $\omega = 1$ and place $P_1$ at $(\mu,0)$ and $P_2$ at $(-1+\mu,0)$. Here we normalize the mass of $m_1$ to $1-\mu$ and $m_2$ to $\mu$, $\mu = m_2/(m_1+m_2)$. The equations of motion for $P_3$ are

$$\begin{aligned} \ddot{x} - 2\dot{y} &= x + \Omega_x \\ \ddot{y} + 2\dot{x} &= y + \Omega_y, \end{aligned} \tag{1}$$

where $\dot{} \equiv \frac{d}{dt}, \Omega_x \equiv \frac{\partial\Omega}{\partial x}$,

$$\Omega = \frac{1-\mu}{r_1} + \frac{\mu}{r_2},$$

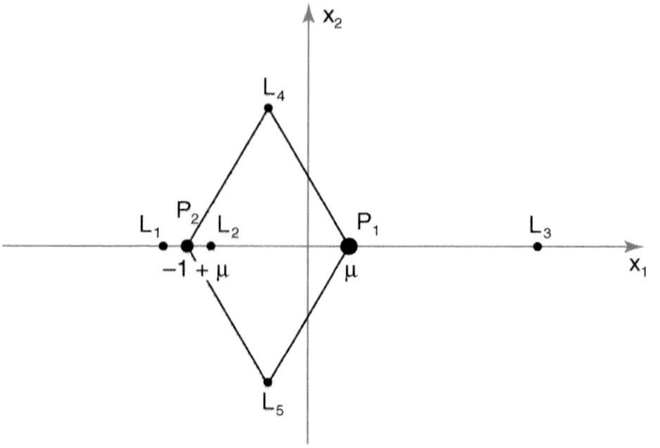


**Fig. 1** Rotating coordinate system and locations of the Lagrange points

$r_1$ = distance of $P_3$ to $P_1$ $= [(x-\mu)^2 + y^2]^{\frac{1}{2}}$, and $r_2$ = distance of $P_3$ to $P_2 = [(x+1-\mu)^2 + y^2]^{\frac{1}{2}}$, see Fig. 1. We note that the units of position, velocity and time are dimensionless. To obtain position in kilometers, and the velocity in kilometers per second, the dimensionless position $(x,y)$ is multiplied by the distance, $d$, in kilometers between $P_1 - P_2$ and the velocity $\dot{x},\dot{y}$ is multiplied by the circular velocity, $vc$, in kilometers per second of $P_2$ about $P_1$. For example in the case of $P_1$, $P_2$ = Sun, Earth, respectively, then $d$ = 149,600,000 km, $vc$ = 29.78 km/s, and $\mu = 0.000003$. Also, $t = 2\pi$ corresponds to 1 year. Analogously, if $P_1$,$P_2$ = Earth, Moon, respectively, then $d$ = 386,000 km, $vc$ = 1 km/s, $\mu = 0.012$.

System (1) of differential equations has five equilibrium points at the well known Euler-Lagrange points $L_k$,$k$ = 1,2,3,4,5, where $\ddot{x} = \ddot{y} = 0$ and $\dot{x} = \dot{y} = 0$. Placing $P_3$ at any of these locations implies it will remain fixed at these positions for all time. The relative positions of $L_k$ are shown in Fig. 1. Three of these points are collinear and lie on the $x$-axis, and the two that lie off of the $x$-axis are called equilateral points.

The three collinear Lagrange points $L_k$,$k$ = 1,2,3 lying on the $x$-axis are unstable. They are saddle-center points [5]. This implies that a gravitational perturbation of $P_3$ at any of the collinear Lagrange points will cause $P_3$ to move away from these points as time progresses. The two equilateral points are stable, so that if $P_3$ were place at these points and gravitationally perturbed a sufficiently small amount, it will remain in motion near these points for all time.

This stability result for $L_4$,$L_5$ is subtle and was a motivation for the development of the so called Kolmogorov-Arnold-Moser(KAM) theorem on the stability of motion of quasi-periodic motion in general Hamiltonian systems of differential equations [1]. A variation of this theorem was applied to the stability problem of $L_4$,$L_5$ by Deprit & Deprit-Bartolomé in 1967 [18]. Since we will refer to this result later in the paper, we record it here for reference,

**Theorem 1** *$L_4$,$L_5$ are locally stable if* $0 < \mu < \mu_1$, $\mu_1 = \frac{1}{2}(1 - \frac{1}{9}\sqrt{69}) \approx 0.0385$, *and* $\mu \neq \mu_k$, $k = 2,3,4$ $\mu_2 = \frac{1}{2}(1 - \frac{1}{45}\sqrt{1833}) \approx 0.0243$, $\mu_3 = \frac{1}{2}(1 - \frac{1}{15}\sqrt{213}) \approx 0.0135$, $\mu_4 \approx 0.0109$.

As an example, in the case of the Earth-Sun system, $\mu = 0.000003$ which is substantially less than $\mu_1$ and the exceptional values $\mu_k, k = 2,3,4$ so that $L_4$ is clearly stable in this case.

An integral of motion for (1) is the Jacobi energy given by

$$J = -(\dot{x}^2 + \dot{y}^2) + (x^2 + y^2) + \mu(1 - \mu) + 2\Omega. \tag{2}$$

Thus $\Sigma(C) = \{(x,y,\dot{x},\dot{y}) \in \mathbb{R}^4 \mid J = C,\ C \in \mathbb{R}\}$ is a three-dimensional surface in the four-dimensional phase space $(x,y,\dot{x},\dot{y})$, such that the solutions of (1) which start on $\Sigma(C)$ remain on it for all time. $C$ is called the Jacobi constant. The manifold $\Sigma(C)$ exists in the four-dimensional phase space. It's topology changes as a function of the energy value $C$. This can be seen if we project $\Sigma$ into the two-dimensional position space $(x,y)$. This yields the Hill's regions $\mathcal{H}(C)$ where $P_3$ is constrained to move. The qualitative appearance of the Hill regions $\mathcal{H}(C)$ for different values of $C$ are described in [5]. As $C$ decreases in value, $P_3$ has a higher velocity magnitude at a given point in the $(x,y)$-plane. If we let $C_k = J(L_k)$, $k = 1,2,3,4,5$, then $3 = C_4 = C_5 < C_3 < C_1 < C_2$. For $C < 3$, the Hill's region becomes the entire $(x,y)$-plane.

## 3 Determination of the Weak Stability Boundary

The weak stability boundary region was first estimated in 1986 as a way for spacecraft to be ballistically captured into orbit about the Moon, defined where the Kepler energy, $H_2$, with respect to the Moon, $P_2$, is negative,

$$H_2 = \frac{1}{2}v_2^2 - \frac{\mu}{r_2} < 0, \tag{3}$$

$v_2$ is the magnitude of the velocity of $P_3$ with respect to $P_2$. This gives rise to weak capture where $P_3$ will remain captured about $P_2$ for a finite time, $t$, $t_1 \leq t \leq t_2$, and where $H_2 > 0$ for $t < t_1, t > t_2$, i.e. where it escapes $P_2$. Until further notice, $P_2$ need not be the Moon, but any primary of mass $\mu$, and $P_1$ any primary of mass $1 - \mu$.

The estimation of this region was originally accomplished by a numerical algorithm which measured when $P_3$ was able to perform a complete cycle about $P_2$ with initial elliptic conditions on a radial line $l$ centered at $P_2$ and returning to $l$. This was first done in 1986 [3] then more rigorously in [5]. More precisely, the initial conditions on $l$ assume, therefore, that $H_2 < 0$, or equivalently, where the initial eccentricity $e_2$ of $P_3$ with respect to $P_2$ satisfies $e_2 < 1$ at the initial time $t = 0$. A value of $e_2 \in [0,1)$ is fixed. It is assumed that the initial velocity vector on $l$ is normal to the line, in the posigrade direction, and that the initial state is at the periapsis of an osculating ellipse. Thus,

$$v_2 = \sqrt{\frac{\mu(1 + e_2)}{r_2}}. \tag{4}$$

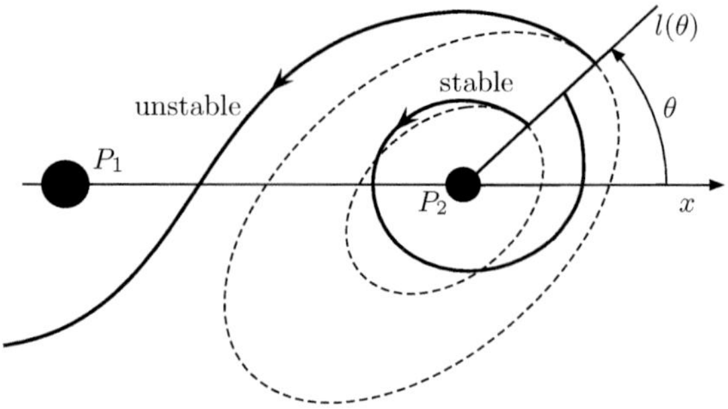


**Fig. 2** Stable and unstable motion after one cycle about $P_2$

We assume $l$ makes an angle $\theta_2 \in [0,2\pi]$ with the $x-$axis, indicated by $l(\theta_2)$, which is fixed. With a given initial state for $P_3$ at $t = 0$, the differential equations given by (1) are numerically integrated for $t > 0$. If the trajectory for $P_3$, $\psi(t) = (x(t),y(t))$, performs a full cycle about $P_2$ and returns to $l$ with $H_2 < 0$, then the motion is called *stable*. If, on the other hand, $P_3$ returns to $l$ with $H_2 \geq 0$, or if $\psi(t)$ moves away from $P_2$ and makes a full cycle about $P_1$, then the motion is called *unstable*. (See Fig. 2.)

By iterating between stable and unstable motion, one finds a critical distance $r^*$ on $l$ with the property that for $r_2 < r^*$ the motion is stable and for $r_2 > r^*$ the motion is unstable. Since $r^*$ depends on $\theta_2$ and $e_2$, that are held fixed during the iteration process, we obtain a functional relationship

$$r^* = f(e_2,\theta_2) \tag{5}$$

which defines the weak stability boundary about $P_2$, we label $\mathcal{W}$.

The definition of the weak stability boundary given by $\mathcal{W}$ can be generalized to find a more accurate definition of this transition region. This is because the transition distance given by (5) is not unique. This generalization was given by Garcia and Gomez [13] and shows that the weak stability boundary is much more complicated than originally thought. Their analysis was studied by Topputo and Belbruno and additional refinements were obtained [19]. Some of the results of these studies are summarized here and the reader can find many more details in these papers.

It was found that for a given value of $\theta_2$ and $e_2$, there are a countable number of open intervals, $I_k = (r^*_{2k-1}, r^*_{2k})$, $k = 1,2,3,\ldots$, $r^*_1 = 0$, containing stable points along $l$ and that the points defining the transition between stable and unstable motion, which define the weak stability boundary, lie at the boundaries of these open intervals. The stable set of points, $U_1(e_2,\theta_2)$, is therefore given by,

$$U_1(e_2,\theta_2) = \bigcup_{k\geq 1} I_k. \tag{6}$$

The more general definition of the weak stability boundary as $e_2,\theta_2$ vary is given by the set of boundary points of this set, except $r^*_1$. We label this

$$\mathcal{W}_1 = \partial \bigcup_{e_2\in[0,1),\theta_2\in[0,2\pi]} U_1(e_2,\theta_2), \tag{7}$$

where, for a set $A$, $\partial A$ represents the boundary of $A$.

Equation 7 yields a definition of the weak stability boundary when studying trajectories that make one cycle about $P_2$. We refer to this as the weak stability boundary of order one. Analogously, $U_1$ is referred to as a set where the points are 1-stable. More generally, a similar definition can be made after analyzing $n$ cycles about $P_2$ before returning to $l$, $n = 1,2,3,\ldots$. Thus, the weak stability boundary of order $n = 1,2,3,\ldots$, relative to $n$ cycles of $P_3$ about $P_2$, is given by

$$\mathcal{W}_n = \partial \bigcup_{e_2 \in [0,1), \theta_2 \in [0,2\pi]} U_n(e_2,\theta_2), \tag{8}$$

where $U_n$ consists of points that we refer to as $n$-stable,

$$U_n = \bigcup_{k \geq 1} I_k \tag{9}$$

Summarizing, the weak stability boundary of order $n$, denoted by $\mathcal{W}_n$ is the locus of all points $r^*(e_2,\theta_2)$ along the radial segment $l(\theta_2)$ for which there is a change of stability of the initial trajectory $\psi(t)$, that is, $r^*(e_2,\theta_2)$ is one of the endpoints of an interval $I_k = (r^*_{2k-1}, r^*_{2k})$ characterized by the fact that for all $r \in I_k$ the motion is $n$-stable and there exist $r' \notin I_k$, arbitrarily close to either $r^*_{2k-1}$ or $r^*_{2k}$ for which the the motion is $n$-unstable. Thus

$$\mathcal{W}_n = \{r^*(e_2,\theta_2) | \theta_2 \in [0,1), \theta_2 \in [0,2\pi]\}.$$

We can define a subset of the weak stability boundary of order $n$, $\mathcal{W}_n(e_2)$, obtained by fixing the eccentricity of the osculating ellipse,

$$\mathcal{W}_n(e_2) = \{r^*(e_2,\theta_2) \mid \theta_2 \in [0,2\pi]\}. \tag{10}$$

The sets $\mathcal{W}_n$, $U_n$ are graphically determined in [13,19] for many different parameter values. This takes a substantial amount of numerical work and the details can be found in these papers. For the sake of brevity just two figures are displayed from [19]. Figure 3 shows the weak stability boundary of order 1 for the case $e_2 = 0$ as the boundary of the set $U_1$ of 1-stable points. Multiple components are shown indicating a complicated structure. Figure 4 shows the sets $U_n$ with respective boundaries $\mathcal{W}_n$, $n = 1,\ldots,6$ for $e_2 = 0$. These sets become more sparse as $n$ increases.

Results in [19] show that the size of the sets $U_n$, and hence $\mathcal{W}_n$, reduce in size as $e_2 \to 1$. It is also seen that these sets exist for a certain range of the Jacobi constant.

Preliminary results obtained in [19] indicate that the weak stability boundary of order $n$ is related to the invariant manifold structure associated with the limit sets of the stable manifolds associated to the Lyapunov orbits near $L_1$,$L_2$ for a specific range of Jacobi constant. This is currently being studied by F. Topputo, M. Gidea and this author. Figure 5 shows trajectories with initial conditions on $\mathcal{W}_1(0)$ which in forward time move near to the stable manifold on the Lyapunov orbit about $L_1$.

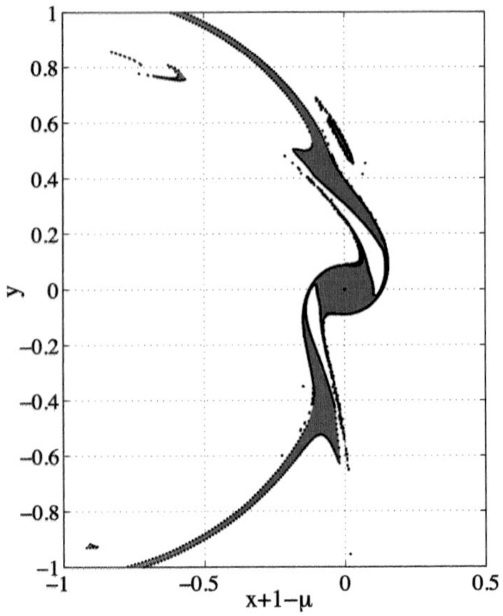


**Fig. 3** The open set $U_1$ of 1-stable points for the case $e_2 = 0$ with boundary $\mathcal{W}_1(0)$ centered at $P_2$. Multiple components of the boundary are shown

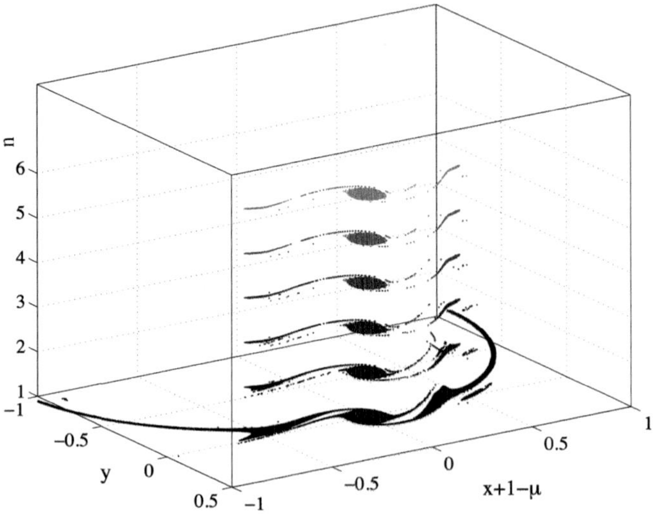


**Fig. 4** $n$-stable sets and $n$th order weak stability boundaries for $e_2 = 0$, $n = 1, \ldots, 6$

# 4 Chaos Associated with the Weak Stability Boundary and Parabolic Motion

The location of the weak stability boundary can be roughly approximated by the set,

$$W_E = \mathcal{J}(C) \cap \Sigma \cap \sigma, \tag{11}$$

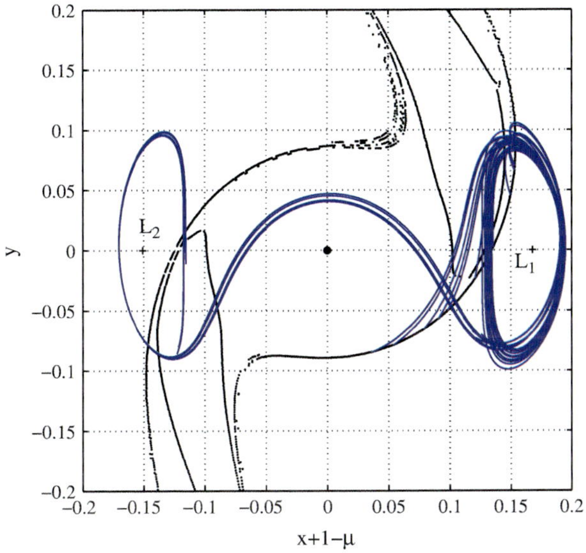


**Fig. 5** Trajectories with initial conditions on $\mathcal{W}_1(0)$ for $\theta_2 \in [-\pi/4, \pi/4]$ moving close to the stable manifold of the Lyapunov orbit about $L_1$

where

$$\mathcal{J}(C) = \{(x,y,\dot{x},\dot{y}) \in \mathbb{R}^4 | J = C,\ C^* \leq C < C_1\},$$
$$\Sigma = \{(x,y,\dot{x},\dot{y}) \in \mathbb{R}^4 | H_2 \leq 0\},$$
$$\sigma = \{(x,y,\dot{x},\dot{y}) \in \mathbb{R}^4 | \dot{r}_2 = 0\}.$$

The approximation given by (11) is derived in detail in [5]. It is based on the formulation of the algorithmic definition of the weak stability boundary. The value of $C^*$ is determined so that $W_E$ is well defined. It can be shown that (11) implicitly yields an expression of the form $r_2 = g(e_2,\theta_2,C)$ for a function $g$. This gives rise generally to an annular region for each fixed value of $C$.

The set $W_E$ does not yield significantly useful information on the structure of the weak stability boundary of order $n$, $\mathcal{W}_n$. This is clear due to the complexity of the $n$th order boundary as seen in the previous section. However, this set is convenient since one can obtain estimates of $C$ from the formula $r_2 = g(e_2,\theta_2),C$ for ranges of $r_2,e_2,\theta_2$. These estimates turn out to be useful when trying to prove that ballistic capture is associated to a hyperbolic invariant set and therefore a chaotic process, carried out in [5]. This is accomplished by using a theorem due Z. Xia proving that parabolic motion in the restricted three-body problem is a chaotic motion [20]. He did this by proving that for the Jacobi energy $C$ restricted to lie near a specific value, $C' = -2\sqrt{2}$, for the set of parabolic orbits in the restricted three-body problem which pass by $P_2$ sufficiently close, i.e. $r_2 << 1$, gives rise to a hyperbolic invariant set $H$ on a surface of section $S$ due to a transversal homoclinic point on $S$. The existence of $H$ implies by the Smale-Birkhoff theorem that the dynamics of motion of the trajectories passing through $S$ is chaotic. It is proven in [5] that $W_E$ does not intersect $H$ since, on $H$, it can be shown that it is difficult to match the Jacobi energy values for the set $W_E$, with $C'$. This can be solved by extending the energy of $H_2$ to be slightly positive, $H_2 \gtrsim 0$ as is described in [5]. This implies that we need to define a new set $W_H$ by extending $W_E$,

$$W_H = \{(x,y,\dot{x},\dot{y}) \in \mathbb{R}^4 \mid H_2 \gtrsim 0\}. \tag{12}$$

$W_H$ represents a slight hyperbolic extension of $W_E$ defining pseudo-ballistic capture. It is seen that $W_H$ has no constraints on $\dot{r}_2$ unlike $W_E$. $J$ can achieve the value of $J = C'$ on the set $W_H$ for $r_2 << 1$, $\mu << 1$.

It can be proven that $H$ intersects $W_H$. Thus, pseudo-ballistic capture is a chaotic process. Since $W_H$ lies arbitrarily close to $W_E$, then ballistic capture is also chaotic.

This result shows that the dynamics associated to ballistic capture is complicated and near to that of a hyperbolic invariant set on $S$. In the full phase space, then this dynamics would lie near to a network of hyperbolic invariant manifolds, for the Jacobi energy values required for the parabolic motion that moves close to $P_2$.

It is important to note that the above results were obtained for motion of $P_3$ that was approximately parabolic with respect to $P_1$. This required a Jacobi constant that was near the value of $C'$. In this case $P_3$ had enough velocity to closely fly by $P_2$ at approximately its periapsis with respect to $P_1$. Under these conditions, rigorous results can be proven on the existence of hyperbolic invariant manifolds near the set $W_E$. It is not clear if that would imply the same for $\mathcal{W}_n$, but it is reasonable it would.

Now, the parabolic motions for $P_3$ are relatively energetic as was just described. It is interesting to understand the existence of hyperbolic manifolds associated with $W_E$ when the Jacobi energy $C$ is larger than $C'$ and less than $C_1$. For $C$ sufficiently near $C_1$, $P_3$ will not fly by $P_2$ parabolically, but rather be able to move for extended periods of time near $P_2$. In this case, it would be interesting to understand the types of motions that could occur on the set $W_E$. If, in fact, $C \lesssim C_1$, then it is numerically known that a complex network of hyperbolic manifolds exists around $P_2$ due to the transverse intersection of invariant manifolds from the Lyapunov orbits about $L_1$ and $L_2$ in special cases that have been studied for values of $\mu$ and $C$ [15]. The existence of this network implies that the set $W_E$ for similar values of $C,\mu$ would give rise to a complicated dynamics. This was shown to be the case for a wide range of $C < C_1$ and for a specific case of $\mu = 0.012$ by Belbruno, Topputo, Gidea [8]. This was revealed on two-dimensional surfaces of section in $x,\dot{x}$ coordinates, for different values of $C$. The sections were chosen to lie along the line between $P_1$ and $P_2$, taken to the Earth and Moon, respectively. The dynamics was seen to largely consist of resonance motions of types $m{:}n$, where $P_3$ has a period of motion about the Earth that is $m/n$ times the period of the Moon, $m,n$ are positive integers. The initial conditions for the sections start with a 2:1 resonance state, so it is not too surprising to observe resonance motions. What is surprising are the richness of them and that $W_E$ can be approximately viewed on these sections. The set $W_E$ was identified on these sections for a range $C \in [C^{**},C_1]$, where $C^{**}$ is explicitly determined. One of the sections is shown in Fig. 6. Many different resonances are indicated. The dynamics is generally that of resonance and near resonance motions

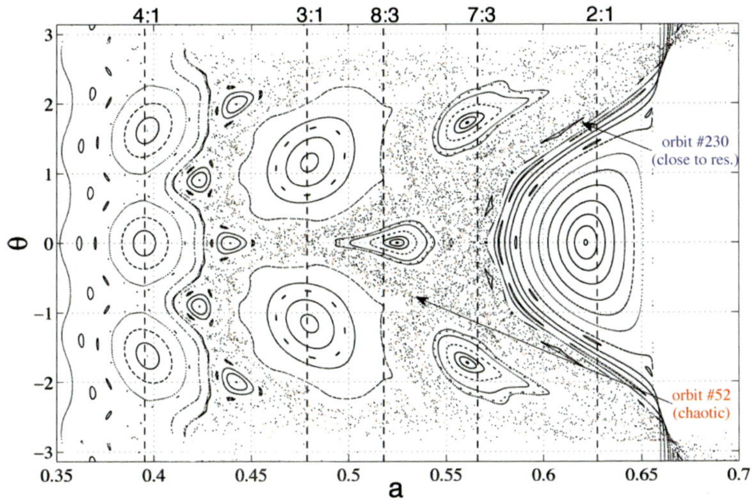


**Fig. 6** Surface of section for $C = 3.1817683$ showing resonance and chaotic motions. The Earth, Moon are located at (0,0), (1,0), respectively (not shown). (The orbits labeled by the numbers *52* and *230* refer to orbits in the chaotic region and boundary of a 2:1 resonance, respectively. The former yields unstable motion and the latter yields stable motion [8])

within large regions of chaotic motions. As $C$ changes, the dynamical picture on the sections can change significantly. The coordinates on this section are $a$,$\theta$, which are the osculating semi-major axis and true anomaly, respectively. As is discussed in [8], the set $B_E = \mathcal{J}(C) \cap \Sigma$ is, in some regards, more advantageous to study than $W_E$ since it contains more information on the section. However, both these sets only roughly indicate what part of phase space that the weak stability boundary of order $n$ exists and are limited in their usefulness understanding this region. The insight obtained on using the sections, for the range of $C$ considered where $W_E$ exists, is to give a sense of the resonance motions that exist.

## 5 Applications to Astronomy

We briefly discuss two applications of the weak stability boundary to astronomy. The first is on a theory on the origin of the Moon and the other is for minimal energy transfer of solid material between planetary systems. In each problem we will indicate where this boundary plays a role.

### *5.1 Formation of the Moon*

The current generally accepted theory for the formation of the Moon, called the "giant impact hypothesis" [10,14], describes how the early Earth was hit by a Mars-sized object($P_3$), about 0.1 mass of the Earth, and from that collision, the debris coalesced to form the Moon.

What was not addressed in this hypothesis was where the Mars-sized object(impactor) came from. A theory of the origin of the impactor and how it could have collided with the Earth was published in 2005 by Belbruno and Gott [4]. The general idea is that $P_3$ initially formed by material collecting near one of the equilateral Lagrange points $L_4$,$L_5$. By Theorem 1, these are stable locations, and material can collect near these locations without moving away. Then, as more material collects there, and grows by accretion in mass, small velocity displacements on the forming object cause it to move in horseshoe orbits [17] moving back and forth in approximately Earth's orbit. The horseshoe motion eventually bifurcates into another type of motion called breakout, where $P_3$ cycles about the Sun in a monotone fashion, repeatably closely flying by the Earth. This results in collision with the Earth with high probability.

The collision trajectories generated by this methodology satisfy the necessary conditions that would be required: First, the approach velocity with the Earth is relatively low, on the order of 0.2 km/s. This velocity is much lower that typical collision trajectories obtained by Monte Carlo methods, which would have velocities on the order of 6 km/s (30 times more). With such trajectories collision is very unlikely. Second, the collision itself with the Earth is a grazing impact.

This problem was initially modeled in [4] by the restricted three-body problem, defined by (1), giving the simplest modeling where the mass of $P_3$ is 0 instead of 0.1 Earth mass. (A more realistic model is eventually used in [4] but that is not necessary for our discussion.) The simplest way to generate collision orbits with the Earth from our methodology, is to put the impactor, $P_3$ at $L_4$, for example, with zero velocity. By Theorem 1, $P_3$ will remain there for all time. We randomly choose a direction at this location from which to apply velocity perturbations, due to tiny impacts, or near impacts, by material that is contributing to the accretion. For the sake of our discussion, we will assume that the direction is parallel to the $y$-axis. If a tiny velocity is applied, then the particle will have a new two-body energy energy with respect to the Sun that will hardly change. The particle will move back and forth in an orbit about the Sun keeping the same distance from the Sun as when it was at $L_4$; that is, the distance of the Earth from the Sun. So, it is moving back and forth approximately on the Earths orbit. This is a horseshoe orbit and one is shown in Fig. 7. In this figure, the Earth is fixed at $x = -1, y = 0$. In this figure, the loops occur since we are in an Earth fixed frame. Each loop corresponds to 1 year. As the velocity increases slightly, the horseshoe orbit will come closer and closer to the Earth, until it moves beyond the Earth. When this happens, the breakout state occurs where $P_3$ ceases to move back and forth on a horseshoe orbit, and instead, monotonically cycles around the Sun in an Earth-like orbit repeatably passing closely by the Earth until, or if, collision eventually occurs.

It is the transition from horseshoe motion to breakout that is of most interest since this is where collision trajectories can be found. It turns out that in the current example, that breakout occurred when the velocity increased to 0.011. This implies that for an initial velocity direction at $L_4$ parallel to the $y$-axis, i.e. the angle $\theta$ from the $x$-axis was $\pi/2$, then the velocity magnitude $V = 0.001$ is required to achieve breakout. In general, it is found that for a given value of the initial velocity direction $\theta$,

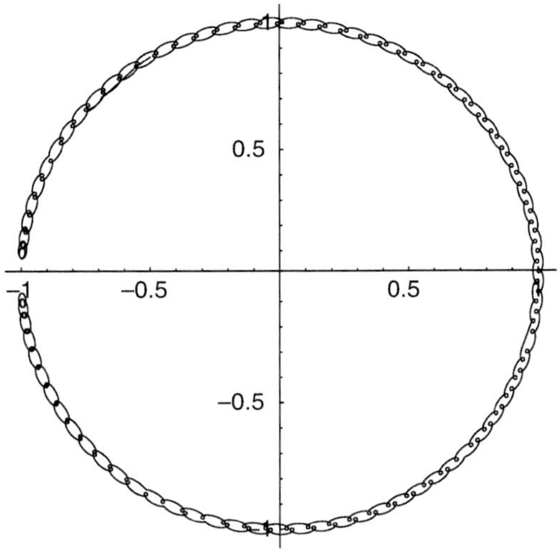


**Fig. 7** Horseshoe orbit, initial velocity $= 0.009$, $t \in [0,1000]$, $x$ vs. $y$, Sun centered

there exists a critical breakout velocity value. The set $B$ of critical breakout velocity values $V$ as a function of $\theta$,

$$B = \{\theta \in [0,2\pi] | V(\theta)\}$$

forms a transition boundary about $L_4$ in velocity space. This can be viewed as a weak stability boundary about $L_4$. One obtains a similar one about $L_5$.

Thus, the determination of the weak stability boundary $B$ about $L_4$ solves the problem on the existence of low energy collision trajectories with the Earth consistent with the giant impact hypothesis. This boundary region has only been roughly computed to solve this problem. It's precise determination has not been carried out and would be an interesting problem to study.

It is remarked that in the more realistic modeling of the motion of $P_3$, rather than increasing the velocity of $P_3$ at $L_4$, where the horseshoe orbits increase in the size of their oscillation, it is more realistic to apply random velocity perturbations at random times along the trajectory. This gives rise to a stochastic random walk dynamics superimposed on a deterministic motion defined by the differential equations. This is described in detail in [4]. This is also further explored in [2].

## 5.2 *Minimal Energy Transfer of Material Between Planetary Systems*

The problem of transfer of solid material between planets of different planetary systems is of interest for the possibility of the transfer of life bearing material within rocks that can be transferred from a planet, $P_2$, orbiting a star $P_1$, to a planet $P_2^*$ of another star $P_1^*$. We will refer to $S_1$ as the first planetary system consisting of $P_1$,$P_2$, and $S_1^*$ as the second planetary system consisting of $P_1^*$,$P_2^*$. We assume that the stars are in a open star cluster where the relative velocities of the stars is fairly low, about 1 km/s.

Previous work in this problem has relied on Monte Carlo methods to show that solid material, e.g. a rock, we call $P_3$, approximately of zero mass, escapes $S_1$ with a high velocity of about 6 km/s. Under this condition, the likelihood of $P_3$ flying into a planet $P_2^*$ of another star $P_1^*$ for the system $S_1^*$ is very small due to the small cross section impact area. However, using low velocity escape on the order of 0.1 km/s of approximately parabolic trajectories, $P_3$ should have more likelihood to be captured by another planetary system, $S_2$, by the mechanism of weak capture. This would make it more likely that it could end up on another planet $P_2^*$.

This idea of using low velocity parabolic escape from $S_1$ and weak capture by $S_1^*$ was studied by Belbruno, Moro-Martin, Malhotra [9]. It turns out that the probability of capture increased substantially as compared to previous studies. To accomplish this analysis, several weak stability boundaries were used. The first was about $P_2$ to give rise to the mechanism of parabolic escape from $S_1$, we discussed in Section 4. Next, when $P_3$ slowly escaped $S_1$ when it was sufficiently far away, a weak stability boundary was used for this weak escape. This boundary is about $P_1$ and the main perturbation consists of the resultant gravitational field from the other stars of the cluster including $P_1^*$. Finally, the third weak stability boundary for weak capture into $S_1^*$ is about $P_1^*$ where the main perturbation is again the resultant gravitational field of the remaining stars of the cluster, now including $P_1$.

This mechanism provides a way to obtain insights on the way solid material can transfer between planetary systems using minimal energy transfers. It is remarked that this mechanism can be veiwed as a 5-body problem between $P_1, P_2, P_1^*, P_2^*, P_3$.

**Acknowledgments** I would like to thank Telespazio and, in particular, Ettore Perozzi for the invitation to attend the workshop at Fucino in October 2007 on *Novel Spaceways for Scientific and Exploration Missions*. This work is partially supported by grants from NASA and the AISR program.

## References

1. Arnold, V.: Mathematical Methods in Classical Mechanics, Springer-Verlag (1983)
2. Belbruno, E.: Random walk in the three-body problem and applications, DCDS-S, **1**, n. 4, 519–540 (Dec. 2008)
3. Belbruno, E.: Fly Me to the Moon: An Insider's Guide to the New Science of Space Travel, Princeton University Press (2007)
4. Belbruno, E., Gott III, J.R.: Where did the moon come from? Astr. J., **129**, 1724–1745 (2005)
5. Belbruno, E.: Capture Dynamics and Chaotic Motions in Celestial Mechanics, Princeton University Press (2004)
6. Belbruno, E.: The dynamical mechanism of ballistic lunar capture transfers in the four-body problem from the perspective of invariant manifolds and hill's regions, Centre de Recerca Mathematica (CRM) Bellaterra(Barcelona) (www.crm.es), Preprint n. 270 (Dec. 1994)
7. Belbruno, E., Miller, J.: Sun-perturbed earth-to-moon transfers with ballistic capture, J. Guid. Control Dyn., **16**, 770–775 (1993)

8. Belbruno, E., Topputo, F., Gidea, M.: Resonance transitions associated to weak capture in the restricted three-body problem, Advances in Space Research, **42**, n. 2, 1330–1352 (2008)
9. Belbruno, E., Moro-Martin, A., Malhotra, R.: Minimal energy transfer of solid material between planetary systems, ArXiv:0808.3268v2(astro-ph), 1–30 (Sep. 3, 2008)
10. Cameron, A.G.W., Ward, W.R.: The Origin of the Moon, Lunar and Planetary Science VII, Lunar Planetary Inst., 120–122 (1976)
11. Circi, C., Teofilatto, P.: On the dynamics of weak stability boundary lunar transfers, Cel. Mech. Dyn. Ast., **79**, 41–72 (2001)
12. Conley, C.: Low energy transit orbits in the restricted three-body problem, SIAM J. Appl. Math., **16**, 732–746 (1968)
13. Garcia, F., Gomez, G.: A note on the weak stability boundary, Cel. Mech. Dyn. Sys., **97**, 87–100 (2007)
14. Hartmann, W.K., Davis, D.R.: Satellite-sized planetesimals and lunar origin, Icarus, **24**, 504–515 (1975)
15. Koon, W.S., Lo, M., Marsden, J.E., Ross, S.: Low energy transfer to the moon, Cel. Mech. Dyn. Astr., **81**, 63–73 (2001)
16. Koon, W.S., Lo, M., Marsden, J.E., Ross, S.: Heteroclinic connections between periodic orbits and resonance transitions in celestial mechanics, Chaos, **10**, 427–469 (2000)
17. Llibre, J., Olle, M.: Horseshoe periodic orbits in the restricted three-body problem, "New Advances in Celestial Mechanics: HAMYSYS 2001, (eds J. Delgado, E.A. Lacomba, J. Llibre and E. Perez-Chavela), Kluwer Academic Publishers, 137–152 (2004)
18. Siegel, C.L., Moser, J.K.: Lectures on Celestial Mechanics, Springer-Verlag (1971)
19. Topputo, F., Belbruno, E.: Computation of weak stabilities: Sun-Jupiter case, Cel. Mech. Dyn. Astr., **105**(1–3) (2009) yet to be published
20. Xia, Z.: Melnikov method and transversal homoclinic points in the restricted three-body problem, J. Differ. Equ., **96**, 170–184 (1992)

# On the Accessibility of the Moon

**Ettore Perozzi, Riccardo Marson, Paolo Teofilatto, Christian Circi, and Alessio Di Salvo**

**Abstract** The large mass fraction of the Moon with respect to the Earth implies an extended sphere of influence which can be exploited in planning exploration missions either directed to our satellite or to other solar system bodies. The dynamical systems approach to mission design has shown the existence of novel trajectories in the Earth-Moon system, which can respond to widely different exploration goals such as low-energy lunar orbit insertion, reaching Mars from the Moon or bringing lunar resources to Earth. Within this framework the general topic of the accessibility of our satellite is discussed and examples of actual mission profiles are given.

## 1 Introduction

The Moon is unique among the natural satellites of the Solar System. The mass ratio with respect to the primary is unusually high (1/81, whereas Titan/Saturn = 1/4207) while the distance from Earth is large enough to allow consistent solar perturbations which result in periodic and secular variations of its orbital parameters. Semimajor axis, eccentricity and inclination exhibit complex patterns while the angular parameters, namely the nodal and apsidal lines, follow precessional (the former) and prograde (the latter) motions. Finally, the orbital plane of the Moon is inclined about 5° on the ecliptic; this, coupled with a low value of the obliquity accounts for reduced polar circles.

The study of the motion of the Moon (main lunar problem) belongs to the general three-body problem, which is well known to be non-integrable: computing accurate ephemeris of the Moon has been historically challenging for celestial mechanics. It is reported that Isaac Newton himself found the lunar problem so difficult that "it made his head ache and kept him awake so often that he would think of it no more" [33].

Modern digital computing and the use of high accuracy numerical methods for orbit integration have solved the problem as far as short-term precise ephemeris

---

E. Perozzi (✉)
Telespazio, Via Tiburtina 965, 00156, Roma, Italy
e-mail: ettore.perozzi@telespazio.com

E. Perozzi, S. Ferraz-Mello (eds.), *Space Manifold Dynamics*,
DOI 10.1007/978-1-4419-0348-8_5, 

are concerned. The existence of periodic orbits close to that of the Moon [38] has pointed out that the ancient eclipse prediction cycle Saros (18-year and 10 or 11 days depending on the number of leap years included) should be considered as the natural averaging period of time by which solar perturbations can be most effectively removed. The importance of periodic orbits for understanding the N-body problem was first pointed out by the Henry Poincaré in his treatise "Les Méthodes Nouvelles de la Mechanique Celeste" [31] toward the end of the XIX.th century. Following Poincaré's footsteps, the so-called dynamical systems approach and its astrodynamics applications have allowed to investigate the structure of the Earth-Moon space in the three and four-body problems, i.e. involving the Earth, the Moon, the Sun and a spacecraft (e.g. [27]).

The discovery of periodic and quasi-periodic orbits in the vicinity of the Moon dates back to the Apollo mission studies [16]. Large amplitude librations around the collinear lagrangian points as seen from Earth result in elliptically-shaped paths encircling the Moon (hence the name "halo orbits"). Placing a telecommunication satellite around the translunar libration point $LL_2$ (Fig. 1) could prevent the telecommunication breakdown when a spacecraft flies on the far side of our satellite [17]. Refined modelling of the motion around the collinear lagrangian points of both the Earth-Moon and the Earth-Sun systems (e.g. [32,20]) has made the use of this kind of orbits well established for solar and astronomy missions. The seminal work by Conley [15] provided the connection among these orbits and the vicinity of the Earth, thus paving the way for the computation of low-energy transfer trajectories (e.g. [18]) and the related stable/unstable manifold theory [19]. A peculiar feature of this approach is the existence of trajectories allowing a spacecraft to achieve selenocentric orbit without performing an injection manoeuvre (ballistic capture). In this respect the so-called weak stability boundary (WSB) trajectories [2] exploit

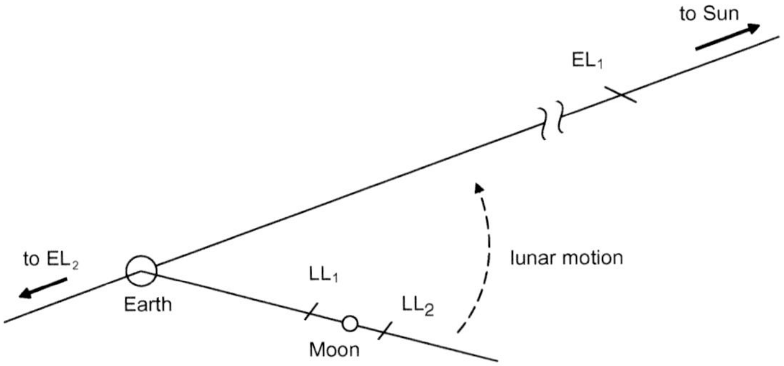


**Fig. 1** In the planar case the line encompassing the Moon and the collinear lagrangian points $LL_1$ and $LL_2$ revolves around the Earth at constant speed, periodically aligning with the Sun-Earth direction where the collinear lagrangian points of the Earth-Sun system $EL_1$ and $EL_2$ are located. This leads to the complex dynamical patterns exploited by space manifold dynamics. The figure is not to scale

the existence of "fuzzy boundaries" around the collinear lagrangian points generated by lunar and solar perturbations for redirecting the motion of a spacecraft in order to achieve ballistic capture around the Moon [1]. The possibility of connecting the surroundings of the lagrangian points of both, the Earth-Moon and the Earth-Sun systems (Fig. 1) has been also investigated in realistic cases (e.g the Genesis mission, [23]), as well as the advantages for accessing the lunar surface [25] and the interplanetary space [26,5]. We will refer to this kind of trajectories with the term "space manifold dynamics" (SMD).

Thus the dynamical characteristics of the Moon are of considerable help to spaceflight mechanics and a discussion of the accessibility of our satellite within the framework of the worldwide renewed interest for lunar exploration seems desirable. Future developments such as the construction of a Moon Base [14] and/or of lunar permanent infrastructures (e.g. a condominium of observatories, [39]) foresee many different servicing missions, each one characterized by different constraints: manned spacecrafts, cargo transportation, telecommunication satellites, etc. In what follows the efficiency of using SMD trajectories as compared to classical Hohmann-like mission profile will be analysed form a strictly dynamical point of view and the results discussed in the broader context of the accessibility of the Moon for exploration.

## 2 H-plot

The concept of "accessibility" of a celestial body was introduced at the beginning of the XX$^{\text{th}}$ century by Walter Hohmann in his pioneering work on the human exploration of Venus and Mars [28]. Using simple keplerian approximation Hohmann showed that the change of velocity ($\Delta$V) needed for orbital transfers is a useful measure of accessibility. When dealing with circular coplanar orbits this value can be straightforwardly computed from keplerian motion as the sum of two manoeuvres (Fig. 2). The first one injects the spacecraft into a transfer ellipse having the same apocentre distance as the target orbit; the second manoeuvre (also known as "circularization") is applied upon arrival at the apocentre of the transfer orbit, and its magnitude allows the spacecraft to achieve the orbital velocity of the target body. Hohmann transfers are also minimum-energy trajectories if the two orbits are not too distant, i.e. when the radius of the target orbit to that of the departure orbit does not exceeds 15.58. The validity of Hohmann-like transfers has been also extended to target orbits characterized by non-zero inclinations and eccentricities.

A graphical representation of the Hohmann transfer strategy (hereinafter called H-plot, Fig. 3) has been used for carrying out meaningful comparisons among different Earth-Moon transfer strategies [30]. In the diagram of Fig. 3 the $\Delta$V requirements needed for performing two different kinds of Hohmann transfers are shown. The former involves a spacecraft leaving a 500 km circular orbit around the Earth toward target orbits of ever increasing size, while the latter is concerned with orbital transfers from the distance of the Moon outwards. In both cases the lower thin curve refers to the magnitude of the manoeuvre which injects the spacecraft into

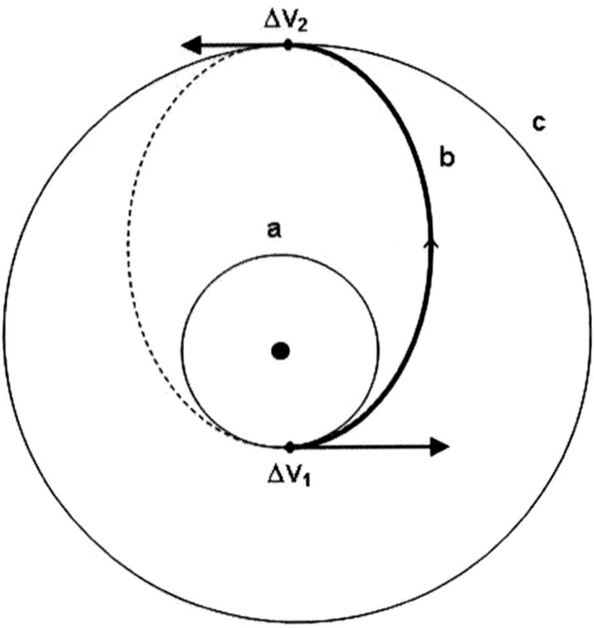


**Fig. 2** The Hohmann transfer strategy between two circular coplanar orbits (a, c) foresees two manoeuvres. The first one, of magnitude $\Delta V_1$, injects the spacecraft into an elliptic path (b) with apocenter at the target distance, while the second ($\Delta V_2$) is applied upon reaching it, of the exact amount for circularization. Relying on simple keplerian motion, the actual value of the $\Delta V$ depends on the choice of the central body (Sun, Earth, planet, etc)

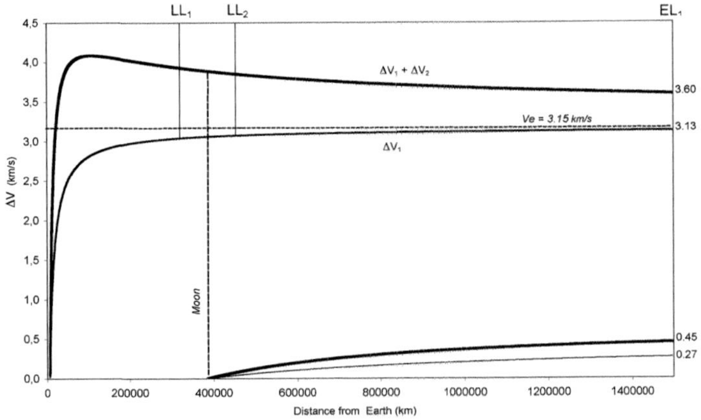


**Fig. 3** The H-plot diagram described in the text. The upper curves tend both to the Earth escape velocity *Ve*. The lower pair are representative of performing outbound transfers starting from an initial orbit of radius equal to the distance of the Moon. The location of the collinear lagrangian points $LL_1$ and $LL_2$ is remarked and the diagram is bounded on the right by the Earth-Sun lagrangian point $EL_1$, where some reference values are also reported

transfer ellipses of increasing apogee distance (i.e. the Hohmann $\Delta V_1$ of Fig. 2) while the upper curve includes also the circularization manoeuvre ($\Delta V_1 + \Delta V_2$).

The computations have been carried out until reaching the border of the gravitational domain of the Earth in the sunward direction, corresponding to the collinear lagrangian point $EL_1$ located at approximately 1.5 million km from the Earth.

A number of considerations can be drawn from the plots of Fig. 3. The behaviour of the upper curve clearly shows how deep is the gravitational well surrounding the Earth at distances below 100,000 km. Note that its maximum occurs at a distance corresponding to the afore mentioned limiting value for a Hohmann transfer to be optimal. Proceeding outwards the total $\Delta V$ slowly decreases tending monotonically to the escape Earth velocity from a 500 km altitude circular orbit. This means that it is in principle more demanding in terms of $\Delta V$ to insert a satellite into a circular 100,000 km altitude orbit, than to achieve an orbit having the same distance as the Moon and beyond, as already pointed out by Keaton [22]. In particular it can be shown that the overall energy budget needed for achieving geostationary orbit from LEO is not too far from the value characterizing a lunar rendezvous mission, and largely sufficient for interplanetary escape [30].

In general one can conclude that most of the region of space extending from an altitude of about 50,000 km up to $EL_1$ is comparable in terms of accessibility, the differences being less than 0.5 km/s. A similar conclusion can be drawn focusing on the lower two curves, which represent Hohmann transfers starting from a circular orbit having the same distance as the Moon.

## 3 SMD Transfers

The H-plot of Fig. 3 describes the keplerian motion of a spacecraft and it is used to give an overall picture of the accessibility of the near Earth space. Nevertheless it provides a convenient framework for understanding some basic characteristics of SMD transfer strategies, which exploit also the gravitational pull of the Moon and the Sun (Fig. 4).

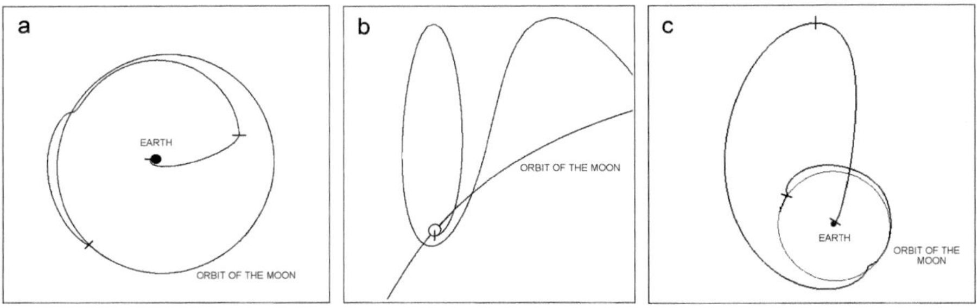


**Fig. 4** **a**), **c**) SMD transfers leading to ballistic capture, i.e when the energy of the spacecraft becomes negative with respect to the Moon without the need of firing the propulsion system. The capture orbit is in general highly eccentric (**b**), thus an additional manoeuvre is needed if one wishes to reach a typical LLO (in the center plot a circular polar 100 km target orbit is drawn). Tick marks show the location of orbital manoeuvres

The trajectory of Fig. 4a is obtained by targeting the stable manifold which brings a spacecraft to the Moon by transiting inside the $LL_1$ region (e.g. [29]). Upon leaving the Earth a "bridging" elliptic path is needed before performing the manifold insertion manoeuvre leading to the lunar ballistic capture shown in Fig. 4b. Figure 4c shows the so called "WSB external transfer" [2]: the spacecraft is initially placed on an highly eccentric orbit reaching up the $EL_1$ distance; perturbations by the Sun are then exploited to change the spacecraft trajectory into one bringing it back to the Moon and undergo a ballistic capture (similar to that shown in Fig. 4b).

From a strictly dynamical point of view a qualitative description of using SMD transfers with respect to Hohmann-like trajectories can be done by comparing the corresponding sequence of events. A Hohmann transfer is a two-impulse strategy where the manoeuvres are executed at fixed positions in the orbit (pericente and apocentre) and the total $\Delta V$ budget depends upon the balancing between them. The H-plot of Fig. 3 clearly shows that as the target distance increases, the $\Delta V_1$ increases accordingly because transfer ellipses of larger semimajor axis are needed; on the contrary the $\Delta V_2$ follows an opposite trend as receding from the centre of motion implies a steady decrease in the magnitude of the orbital velocities. Thus the peculiar shape of the upper curve of Fig. 3 is obtained.

Using SMD, more complex three- and four body systems are involved. The location and magnitude of the manoeuvres are not fixed but may vary according to the specific strategy adopted. As an example, in the trajectory shown in Fig. 4a the location of the second manoeuvre may vary considerably depending where the "bridging" arc meets the stable manifold leading to the Moon. The transfer of Fig. 4c uses solar perturbation for decreasing as much as possible the second manoeuvre whose aim is to raise the perigee of the orbit until it becomes nearly tangent to the lunar orbit, thus allowing a low-velocity encounter leading to lunar ballistic capture. In passing, the connection between manifolds, periodic orbits and temporary satellite capture (TSC) of short-period comets by Jupiter (e.g. [9]) has been recently investigated [21], bearing strong similarities with SMD ballistic capture events.

## 4 A Lunar Satellite Constellation

The feasibility of WSB transfers to the Moon has been demonstrated in a real case when the Japanese small Hiten spacecraft was successfully inserted into selenocentric orbit [3] after travelling on a trajectory similar to that shown in Fig. 4c. Although having at disposal a $\Delta V$ budget less than the one required for a lunar Hohmann transfer (Hiten was originally designed as an Earth orbiting mission) the gravitational perturbations by the Sun allowed enough fuel savings to send the spacecraft in orbit around the Moon. Further investigations [4,12] have shown that this kind of transfer has also a great flexibility in terms of launch windows because when the geometries shown in Fig. 5 are satisfied the trajectory experiences a solar gravitational help at apogee and a solar brake at perigee. The former allows the spacecraft to reach the lunar orbit, the latter allows the ballistic capture by transiting inside the

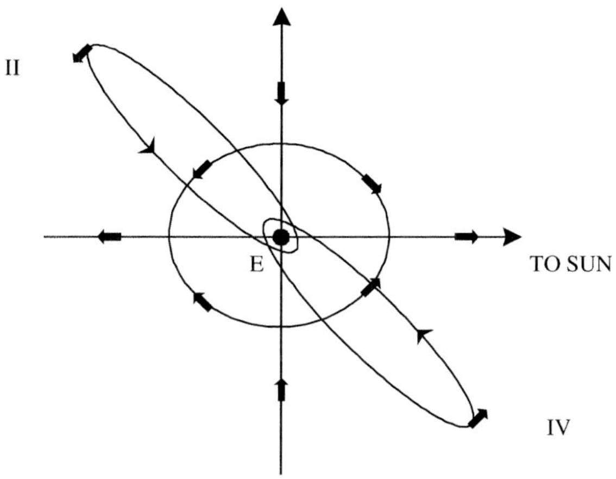


**Fig. 5** The sun gravity gradient field line directions: a trajectory with apogee in the second and fourth quadrant experiences gravitational perturbations which increase the orbital energy. The opposite happens in the first and third quadrants

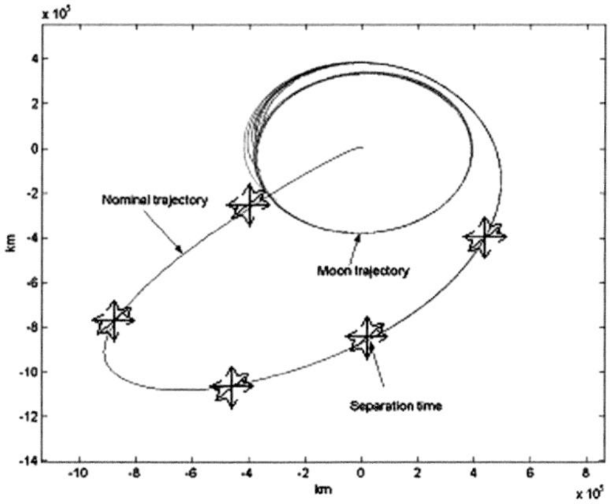


**Fig. 6** A series of separation burns along a nominal trajectory lead to different lunar encounter geometries

tiny neck open in the zero velocity curves surrounding the exterior Lagrangian point of the Earth-Moon system.

Upon reaching the Moon the chaotic nature of this transfer allows to achieve very different orbital configurations by applying small perturbations to a nominal transfer trajectory. It is then possible develop a strategy for increasing the payload weight in the orbit acquisition and the deployment of a constellation of lunar satellites. As shown in Fig. 6 the deployment of each satellite is performed long before the gravitational capture occurs and involves separation impulses of magnitudes extremely small (as small as 20 cm/s). The slightly different encounter trajectories will then lead to the widely different circumlunar orbits (from equatorial to polar) required for fulfilling mission constraints such as the continuous coverage of the lunar surface [13].

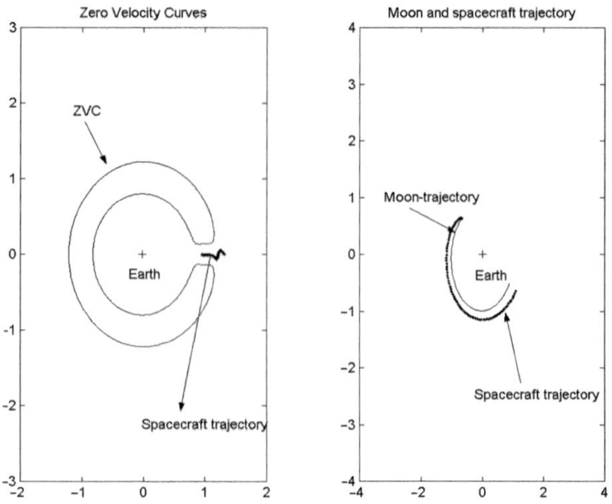


**Fig. 7** The escape trajectory from the Moon into interplanetary space described in the text is shown in two difference reference frames. In the *left* plot an Earth-centered coordinate system rotating with the same angular velocity of the Moon is used, which allows to draw the corresponding zero velocity curves (ZVC). On the *right* plot the trajectories of the Moon and of the spacecraft are drawn in an inertial Earth-centered reference frame

# 5 Mars and Geostationary Missions

The advantage of using the Moon as initial launch base is interesting for direct Moon-Mars transfers. The launcher ascent trajectory from the lunar surface to a Low-Moon-Orbit, due the low Moon gravity and the absence of atmosphere, gives high performance with respect to the Earth case. A simulation using a minimum energy escape trajectory from the Earth-Moon system (Fig. 7) coupled with electric propulsion for the interplanetary transfer phase has been carried out. The results show that a total launch vehicle mass of 459,146 kg (launch vehicle plus payload mass) is able to deliver a 64,639 kg payload into a synchronous Mars orbit [10].

Using direct Earth-Mars transfer, to obtained the same payload for synchronous Mars orbit, are necessary 525000 kg as initial mass in a circular parking orbit at 1200 Km altitude [36,37]. Transfer to Mars via the Moon is very advantageous with

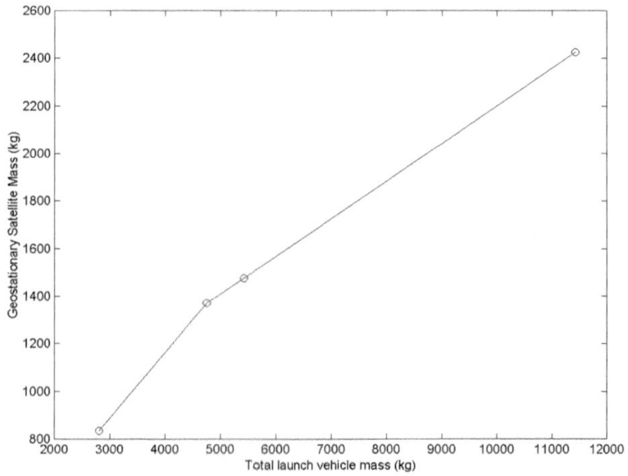


**Fig. 8** Performances for Moon-GEO transfer trajectory

respect to classical direct Earth-Mars missions and gives a good opportunity for future Mars exploration.

Similar advantages are found when choosing the Moon as a base for launching Earth geostationary satellites. Also in this case using a minimum escape energy trajectory and electric propulsion high performances are obtained. Figure 8 shows the satellite mass on GEO orbit vs. the launcher mass on the lunar surface [11].

## 6 Concluding Remarks

Comparing the efficiency of lunar transfer strategies is not easy because of the different underlying assumptions (dynamical model, propulsion system, choice of the initial and final orbits). From the published literature (e.g. [7,6,24]) emerges a clear correlation between the $\Delta V$ budget and the duration of the transfer. Decreasing fuel consumption (e.g. by applying numerical optimization techniques to SMD trajectories) means in general longer transfer times, even if a precise relationship between these two quantities has not been found. With respect to a Hohmann transfer, which lasts a few days, the SMD trajectory shown in Fig. 4a takes a couple of weeks to reach the Moon, with comparable $\Delta V$ performances. In this respect it is worthwhile mentioning that Bollt and Meiss [8] found a chaotic transfer to the Moon which allows a 38% advantage with respect to a 6.6 days Hohmann transfer: however the spacecraft spends 2 years looping around the Earth before being captured by the Moon.

External WSB transfers (Fig. 4c) seem more effective, although with a minimum transfer time of about 3 months, due to the unavoidable step of reaching $EL_1$ before going back. The target orbital configuration around the Moon is also a deciding parameter when a ballistic capture is involved (Fig. 4b). The additional manoeuvre required to stabilize the orbit lowering the aposelenium has in fact a significant magnitude if low-altitude circular orbits or lunar landings are foreseen [30]. This consideration, together with the H-plot of Fig. 3, provides the dynamical ground for the proposed long-term exploration scenario where $LL_1$ plays an important role as a gateway for supporting different kind of missions [35]: a terminal for uploading lunar resources (e.g. fuel), a permanent hub for manned missions, interplanetary travel and the servicing of astronomical observatories and scientific missions operating at high-altitude Earth orbits or at the Earth-Sun lagrangian points $EL_1$ and $EL_2$.

As far as propulsion is concerned, traditional high-thrust low specific impulse systems allow to perform almost impulsive manoeuvres, while low-thrust electric engines bring significant mass savings due to a higher specific impulse. At present the latter type of propulsion coupled with SMD transfers represents the most efficient way for reaching the Moon, because low-thrust navigation suits well the chaotic nature of SMD. The drawback is that transfer times may become even longer because of the slow spiralling outwards which characterizes the trajectory of such a spacecraft upon leaving LEO. Thus on a long-term perspective, the large space transportation requirements needed to sustain a Lunar base call for the development of high thrust, high specific impulse propulsion systems. As an example in the case of the proposed nuclear fuelled Americium 242 engine [34], the energy delivered by 1 g of Am is equivalent to 1 ton of chemical fuel.

An alternative way of looking at the accessibility of the Moon is by taking into account mission constraints and opportunities which are not solely related to celestial mechanics. If one considers the radiation hazard issue, fast transfers lasting less than 1 month are mandatory for manned missions. Intermediate and long duration SMD trajectories ranging from some months to more than one year are preferred when a large mass capability is required for unmanned cargo missions. The possibility of spending a large fraction of the transfer time at large distances from the Earth could be appealing for scientific disciplines studying the Sun-Earth interaction and the Earth magnetospheric environment. Payloads dedicated to space weather could then greatly improve the scientific return of a mission aimed to the Moon. Finally, operations may also profit of the many transfer options now available. The ballistic capture of a spacecraft by the Moon removes the lunar orbit insertion manoeuvre as a single-point-failure event: an engine malfunction will then have less dramatic consequences because enough time is left for attempting rescue operations.

## References


1. Belbruno, E.: Capture dynamics and chaotic motions in celestial mechanics. Princeton University Press, Princeton (2004).
2. Belbruno, E.A. and Carrico, J.P.: *Construction of Weak Stability Boundary ballistic Lunar transfer trajectories*. AIAA 2000-4142 (2000).
3. Belbruno, E.A. and Miller, J.: *A ballistic lunar capture trajectory for the Japanese spacecraft Hiten*. JPL report IOM 312/90.4-1317 (1990).
4. Bellò Mora, M., Graziani, F., Teofilatto, P., Circi, C., Porfilio, M., Hechler, M.: *A systematic analysis on WSB transfers to the Moon*, Paper IAF-00-A.6.03 (2000).
5. Bellò Mora, M., Castillo, A., Gonzalez, J.A., Janin, G., Graziani, F., Teofilatto, P., Circi, C.: *Use of weak stability boundary trajectories for planetary capture*, Paper No. IAF-03-A.P.31 (2003)
6. Bernelli Zazzera, F., Topputo, F., Massari, M.: *Assessment of mission design including utilization of libration points and weak stability boundaries* AO 4532 03/4130 Final Report, ESTEC (2004)
7. Biesbroek, R. and Janin, G.: *Ways to the Moon?*ESA Bull. 103 (2000).
8. Bollt, E.M., and D. Meiss, J.D.: Targeting *chaotic orbits to the Moon through recurrence*. Phys Lett A 204 (1995)
9. Carusi, A., Kresàk, L., Perozzi, E., Valsecchi, G.B.: *Long-Term Evolution of Short-Period Comets*. Adam Hilger Ltd. Bristol and Boston (1985).
10. Circi, C.: *Lunar Base for Mars Missions*, J Guid Contr Dynam 28, 2 372-374 (2005).
11. Circi, C. and Graziani, F.: *Basi lunari per satelliti geostazionari*. Aerotecnica Missili e Spazio, 86, 1 (2007).
12. Circi, C. and Teofilatto, P.: *On the dynamics of weak stability boundary lunar transfers,* Celest Mech Dyn Astr, 79(1), 41-72 (2000).
13. Circi, C. and Teofilatto, P.: *Weak stability boundary trajectories for the deployment of lunar spacecraft constellations,* Celest Mech Dyn Astr, 95, 371-390 (2006).
14. Compagnone, F. and Perozzi, E. (Editors): *Moon Base: a Challenge for Humanity*, Donzelli Editore, Roma (2007).

15. Conley, C.C. : *Low Energy Transit Orbit in the Restricted Three-Body Problem.* SIAM J Appl Math, 16, 4 (1968).
16. Farquhar, R.W.: *The control and use of libration-point satellites.* NASA TR R-346 (1970).
17. Farquhar, R.W. and Kamel, A.A.: *Quasi-Periodic Orbits about the Translunar Libration Point.* Celest Mech 7, 458 (1973)..
18. Gomez, G., Jorba, A., Masdemont, J., R., Simo, C.: *Study of the transfer from the Earth to a halo orbit around the equilibrium point* $L_1$, Celest Mech 4, 56 (1993)
19. Gómez, G., Koon, W.S., Lo, M.W., Marsden, J.E., Masdemont, J.J., Ross, S.D.: *Connecting Orbits and Invariant Manifolds in the Spatial Restricted Three Body Problem*, Nonlinearity, 17, 1571–1606 (2004)
20. Howell, K. C.: *Three dimensional periodic halo orbits.* Celest Mech 32 (1984).
21. Howell, K. C., Marchand, B.G., Lo, M.W.: *Temporary Capture of Jupiter Family Comets from the Perspective of Dynamical Systems.* J Astronaut Sci 49, 4 (2001).
22. Keaton, P.W.: , *A Moon Base / Mars Base Transportation Depot.* In 'Lunar Bases and Space Activities of the 21.st Century', WW. Mendell Editor, LPI (1985).
23. Koon, W.S., Lo, M.W. Marsden, J.E., Ross, S.D.: *The Genesis trajectory and heteroclinic connections,* AAS/AIAA Astrodynamics Specialist Conference, Girdwood, Alaska, AAS99-451 (1999).
24. Leiva, A.M. and Briozzo, C.B.: *Extension of fast periodic transfer orbits from the Earth–Moon RTBP to the Sun–Earth–Moon Quasi-Bicircular Problem.* Celest Mech Dyn Astr 101 (2008)
25. Lo, M.W. and Chung, M. J.: *Lunar Sample Return via the Interplanetary Superhighway.* Paper AIAA 2002-4718, (2002).
26. Lo, M.W. and Ross, S.D.: *The lunar L1 gateway: portal to the stars and beyond.* In proc. AIAA Space 2001 Conference, Albuquerque (2001)
27. Marsden, J.E. and Ross, S.D.: *New Methods In Celestial Mechanics and Mission Design.* Bull Amer Math Soc (New Series) 43, 1 (2005)
28. McLaughlin, W.I.: *Walter Hohmann's Roads In Space.* J SMA, 2, 1–14 (2001).
29. Parker, J.S.: *Low Energy Ballistic Lunar Transfers.* Ph.D. Thesis, Department of Aerospace Engineering Sciences, University of Colorado at Boulder (2007).
30. Perozzi, E, and Di Salvo, A.: *Novel spaceways for reaching the Moon: an assessment for exploration.* Celest Mech Dyn Astr 102, 207-218 (2008).
31. Poincaré, H.:*Les Méthodes Nouvelles de la Mechanique Celeste*, Gauthier Villars, Paris (1893)
32. Richardson, D.L.: A *Note on a Lagrangian Formulation for Motion around the Collinear Points.* Celest Mech 22 (1980).
33. Roy, A.E. *Orbital Motion*, Adam Hilger, Bristol (1982).
34. Rubbia, C.: *Neutrons in a Highly Diffusive Medium: a New Propulsion Tool for Deep Space Exploration?* In ' Space Exploration and Resources Exploitation', ESA WPP-151 (1999).
35. Thronson, H.A., Lester, D.F., Dissel, A.F. Folta, D.C., Stevens, J., Budinoff, J.: *Does the NASA constellation architecture offer opportunities to achieve Space science goals in space?* IAC-08-A5.3.6 (2008).
36. Vadali, S.R., Nah, R., Braden, E., Johnson Jr, I.L.: *Fuel-Optimal Planar Earth-Mars Trajectories Using Low-Thrust Exhaust-Modulated Propulsion,* J Guid Contr Dynam 23, 3 (2000)
37. Vadali, S.R., Nah, R., Braden, E.: *Fuel-Optimal, Low-Thrust, Three-Dimensional Earth-Mars Trajectories,* J Guid Contr Dynam 24, 6 (2001).
38. Valsecchi, G.B., Perozzi, E., Roy, A.E., Steves, B.A.: *Periodic Orbits close to that of the Moon.* Astron. Astroph. 271 (1993).
39. Van Susante, P. J. and Heiss, K. P.: 2005. *A Condominium of Observatories on the Moon.* J Wash Acad Sci, Summer issue (2005).

# Optimal Low-Thrust Trajectories to the Interior Earth-Moon Lagrange Point

**Christopher Martin, Bruce A. Conway, and Pablo Ibánez**

**Abstract** Minimum-time and hence minimum-fuel trajectories are found for a spacecraft using continuous low-thrust propulsion to leave low-Earth orbit and enter a specified periodic orbit about the interior Earth-Moon Lagrange point. The periodic orbit is generated with a new method that finds a periodic orbit as a solution to a numerical optimization problem, using an analytic approximation for the orbit as an initial guess. The numerical optimization method is then employed again to determine the low-thrust trajectory from low-Earth orbit to the specified periodic orbit. The optimizer chooses the thrust pointing angle time history and the point of arrival into the periodic orbit, in order to minimize the total flight time. The arrival position and velocity matching conditions are obtained from a parameterization of the orbit using cubic splines since no analytic description of the orbit exists.

## 1 Introduction

The three body system has been of interest to mathematicians and scientists for well over a century dating back to Poincaré [1]. Much of the interest in recent years has focused on using the interesting dynamics present around libration points to create trajectories that can travel vast distances around the solar system for almost no fuel expenditure traversing the so-called "Interplanetary Super Highway" (IPS) [2]. It has also been proposed to use Lagrange points as staging bases for more ambitious missions [3].

Lagrange points are equilibrium points of the three body system that describes the motion of a massless particle in the presence of two massive primaries in a reference frame that rotates with the primaries. There are five Lagrange points (labelled $L_1, \ldots, L_5$), the three collinear points along the line of the two primaries and the two equilateral points that form a equilateral triangle with the two primaries. It

C. Martin (✉)
Department of Aerospace Engineering, University of Illinois, 61801, IL, USA
e-mail: cmartin6@uiuc.edu

E. Perozzi, S. Ferraz-Mello (eds.), *Space Manifold Dynamics*,
DOI 10.1007/978-1-4419-0348-8_6, 

is the collinear points that are of the most interest and in particular the $L_1$ point between the two primaries and the $L_2$ point on the far side of the smaller primary. Since Poincaré there has been much work on finding periodic solutions to the three body problem. Early work was confined to analytic studies, which are restricted to approximations as there exists no closed form analytical solution to the three body system equations of motion.

Most of the analytical studies focus on finding families of periodic orbits. Of particular interest for this study is a method for accurately calculating individual periodic orbits. Richardson [4] constructed 1st and 3rd order analytic approximations by linearizing the equations of motion about the $L_1$ and $L_2$ points. Higher order approximations were computed by Gomez and Mancote [5]. Various others methods have been employed; generating functions [6], Fourier analysis [7] and multiple shooting [8].

In this work the orbit generation problem is reposed as an optimal control problem and direct transcription is used to compute individual trajectories. In the method of direct transcription the continuous trajectory is discretized and the continuous optimal control problem is converted into a non-linear programming (NLP) problem. The system dynamic equations are then enforced as constraints. The particular direct transcription method used here, direct collocation, is normally quite robust. That is, it will converge from a poor initial guess of the optimal solution.

The high cost of putting mass into orbit has encouraged the use of low-thrust propulsion, taking advantage of its high specific impulse to realize savings in propellant mass. However low-thrust engines impart very small thrust. For example the recently launched Dawn mission is powered by Deep Space 1 heritage xenon-ion thrusters producing only 90 mN of thrust [9, 10]. Therefore the engine may need to be operating for a significant portion of the trajectory to impart the $\Delta v$ required. This creates new challenges for optimization as there is no analytic solution for the motion when low-thrust engines are used. Trajectory optimization problems can be formulated as optimal control problems. The dynamic system is described by a system of differential equations, the goal is then to minimize a cost functional subject to path constraints on the states and controls.

The optimal control for this dynamical system can be computed using direct or indirect methods. Indirect methods introduce adjoint variables and use the calculus of variations or the maximum principle to determine necessary conditions that must be satisfied by an optimal solution [11]. The determination of the controls from the necessary conditions generally results in a two point boundary value problem (TPBVP). Indirect methods allow very accurate computation of the optimal solution however the region of convergence can be small and therefore an accurate initial guess of the states, adjoints and controls is required. Direct methods, in which the optimal control problem is converted into a nonlinear programming problem (NLP) also require an initial guess but are more robust. They do not explicitly use the necessary conditions and therefore do not require the addition of adjoint variables.

## 2 Governing Equations

The differential equations of the circular restricted three body problem (CR3BP) describe the motion of a point mass $P_3$ with mass $m_3$ under the gravitational influence of two massive primaries $P_1$ and $P_2$ with masses $m_1$ and $m_2$ respectively, where $m_1 > m_2 \gg m_3 \approx 0$. Therefore $P_3$ exerts negligible influence on the primaries. The motion is considered in a non-inertial frame that moves with the two primaries as they rotate about the system barycenter at constant radius (Fig. 1).

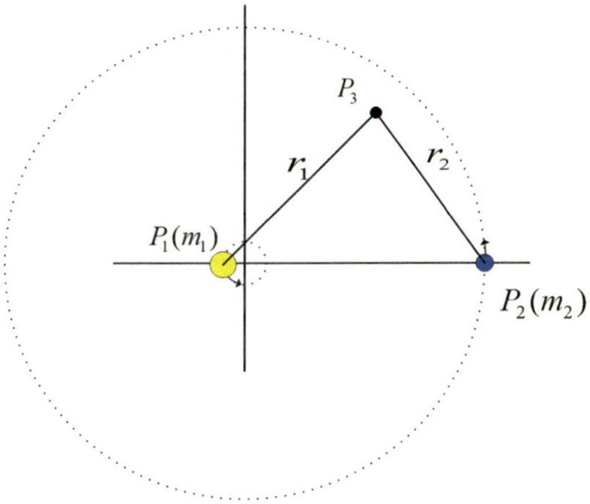


**Fig. 1** Circular restricted 3 body system geometry

It is convenient to normalize the system. The constant separation of the two primaries $P_1$ and $P_2$ is chosen to be the length unit, the combined mass of the two primaries $m_1 + m_2$ becomes the mass unit, the time unit is then selected to make the orbital period of the two primaries about the system barycenter equal to $2\pi$ time units. To further simplify the universal gravitational constant $G$ becomes unity, therefore the mean motion $n$ of the primaries is also equal to 1. The system can now be solely described by a single parameter, the mass ratio $\mu$.

$$\mu = \frac{m_2}{m_1 + m_2} \tag{1}$$

By convention $m_1 \geq m_2$, therefore $\mu \in [0,0.5]$. The normalized masses of the primaries are then $P_1 = (1 - \mu)$ and $P_2 = \mu$ and they orbit the system barycenter at radii $\mu$ and $(1 - \mu)$ length units respectively.

The circular restricted three body system equations of motion are:

$$\begin{aligned}\dot{x} &= v_x \\ \dot{y} &= v_y \\ \dot{z} &= v_z\end{aligned}$$

$$\dot{v}_x = 2v_y + x - \frac{(1-\mu)(x+\mu)}{r_1^3} - \frac{\mu(x-1+\mu)}{r_2^3} \quad (2)$$
$$\dot{v}_y = -2v_x + y - \frac{(1-\mu)y}{r_1^3} - \frac{\mu y}{r_2^3}$$
$$\dot{v}_z = \frac{-(1-\mu)z}{r_1^3} - \frac{\mu z}{r_2^3}$$

where,

$$\mathrm{r}_1 = \sqrt{(x+\mu)^2 + y^2 + z^2}$$
$$\mathrm{r}_2 = \sqrt{(x-1+\mu)^2 + y^2 + z^2}$$

The optimal low thrust transfer study will be restricted to the planar case of the CR3BP. The idealized low-thrust engine used in this study provides a constant thrust acceleration of magnitude $T_a$. The thrust pointing angle $u$ is defined with respect to the instantaneous velocity vector $\mathbf{v}$ which makes an angle $\gamma$ (the flight path angle) with the x axis as shown in Fig. 2. The equations of motion (in 1st order form) then become:

$$\dot{x} = v_x$$
$$\dot{y} = v_y$$
$$\dot{v}_x = 2v_y + x - \frac{(1-\mu)(x+\mu)}{r_1^3} - \frac{\mu(x-1+\mu)}{r_2^3} + T_a \cos(\gamma + u) \quad (3)$$
$$\dot{v}_y = -2v_x + y - \frac{(1-\mu)y}{r_1^3} - \frac{\mu y}{r_2^3} + T_a \sin(\gamma + u)$$

where,

$$\gamma = \tan^{-1}\left(\frac{v_y}{v_x}\right) \quad (4)$$

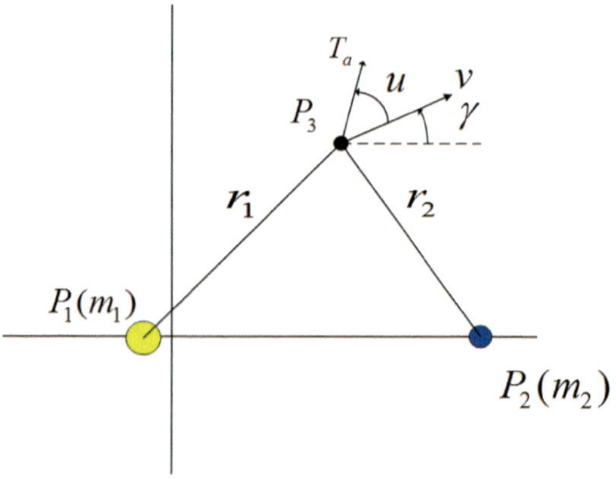


**Fig. 2** Locating the low-thrust spacecraft $P_3$ in the three body system

## 3 Numerical Optimization Method

Consider the optimal control problem with system governing equations:

$$\dot{\mathbf{x}}(t) = f(\mathbf{x}(t),\mathbf{u}(t),t), t_0 \leq t \leq t_f \tag{5}$$

and objective function:

$$J = \Phi(\mathbf{x}(t_f),t_f) + \int_{t_0}^{t_f} L(\mathbf{x}(t),\mathbf{u}(t),t)dt \tag{6}$$

subject to the path constraints:

$$\mathbf{g}(\mathbf{x},\mathbf{u},t) \leq 0, t_0 \leq t \leq t_f \tag{7}$$

and terminal constraints:

$$\Psi(\mathbf{x}(t_f),t_f) = 0 \tag{8}$$

The method used to solve optimal control problems in this work is the method of direct collocation with non-linear programming [12]. In this method the trajectory is approximated by a piecewise polynomial defined by the values of the state and control variable at discrete nodes. The trajectory of each state and control are discretized; an example for the state variable history is shown in the cartoon of Fig. 3.

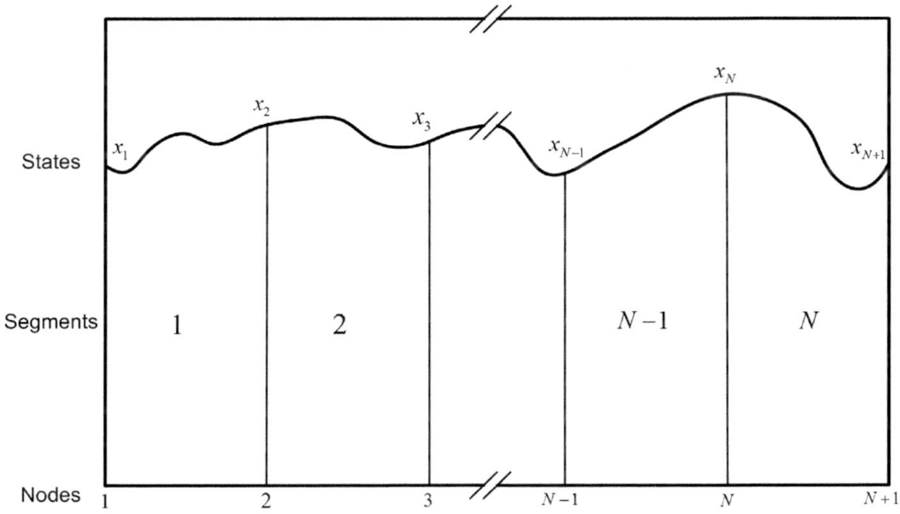


**Fig. 3** Illustration of the discretization of the continuous system

The trajectory is now wholly defined by the state variables, $\mathbf{x}$, and control variables, $\mathbf{u}$, defined at (N+1) nodes and possibly at some interior points. For each interval the time history of each state is approximated over the interval by a polynomial. Herman and Conway [12] found the accuracy and computational efficiency of direct collocation using 5th degree polynomial trajectory approximations superior to approximations using polynomials of lower degree. Six conditions are required to uniquely define the approximating quintic polynomial over the segment. The values of the states at the segment boundary nodes $(i,i+1)$ are defined and the derivatives can be calculated from the system equation $\dot{\mathbf{x}} = \mathbf{f}(\mathbf{x},\mathbf{u},t)$ providing four conditions $(\mathbf{x}_i,\mathbf{f}_i,\mathbf{x}_{i+1},\mathbf{f}_{i+1})$. That is, the approximating polynomial must evaluate to $\mathbf{x}_i$ at the left side node and its slope must correspond to $\mathbf{f}_i = \mathbf{f}(\mathbf{x}_i)$ there. At the right side node the polynomial must evaluate to $\mathbf{x}_{i+1}$ and its slope must be $\mathbf{f}_{i+1}$. The remaining two conditions are defined by adding a node at the center of the segment with corresponding state $\mathbf{x}_c$ and time rate of change $\dot{\mathbf{x}}_c = \mathbf{f}(\mathbf{x}_c,\mathbf{u}_c,t_c)$, where $\mathbf{u}_c$, the control at the center point, and $\mathbf{x}_c$ are free parameters.

These six conditions determine a quintic polynomial satisfying the governing equations at the left and right nodes and at the center of the segment. An additional constraint is imposed that the derivative of the quintic polynomial be equal to the derivative calculated from the system equation (2) at two interior points.

The system equations are satisfied by enforcing the derivative of the quintic polynomial to be equal to the derivative calculated from the system equation (5) at two interior collocation points. The collocation points $(t_1,t_3)$ are selected to minimize the error in the polynomial estimation of the state [12]. For a 5th degree Gauss-Lobatto polynomial the collocation points are located at

$$\begin{aligned} t_1 &= t_c - \sqrt{\frac{3}{7}\frac{1}{2}}\Delta t_i \\ t_3 &= t_c + \sqrt{\frac{3}{7}\frac{1}{2}}\Delta t_i \end{aligned} \tag{9}$$

where $\Delta t_i$ is the width of the $i$th time segment.

The states at the collocation points in segment $i$ are obtained from the polynomial evaluated at $t_1$ and $t_3$:

$$\mathbf{x}_1 = \frac{1}{686}\left\{\begin{aligned} &\left(39\sqrt{21}+231\right)\mathbf{x}_i + 224 + \left(-39\sqrt{21}+231\right)\mathbf{x}_{i+1} \\ &+\Delta t_i\left[\left(3\sqrt{21}+21\right)\mathbf{f}_i - 16\sqrt{21}\mathbf{f}_c + \left(3\sqrt{21}-21\right)\mathbf{f}_{i+1}\right] \end{aligned}\right\} \tag{10}$$

$$\mathbf{x}_3 = \frac{1}{686}\left\{\begin{aligned} &\left(-39\sqrt{21}+231\right)\mathbf{x}_i + 224 + \left(39\sqrt{21}+231\right)\mathbf{x}_{i+1} \\ &+\Delta t_i\left[\left(-3\sqrt{21}+21\right)\mathbf{f}_i + 16\sqrt{21}\mathbf{f}_c + \left(-3\sqrt{21}-21\right)\mathbf{f}_{i+1}\right] \end{aligned}\right\} \tag{11}$$

where $\mathbf{f}_i = \mathbf{f}(\mathbf{x}_i,\mathbf{u}_i,t_i)$, $\mathbf{f}_c = \mathbf{f}(\mathbf{x}_c,\mathbf{u}_c,t_c)$ and $\mathbf{f}_{i+1} = \mathbf{f}(\mathbf{x}_{i+1},\mathbf{u}_{i+1},t_{i+1})$.

The system constraints are then:

$$\mathbf{C}_{5,1}(\mathbf{x}_i,\mathbf{x}_{i+1}) = \frac{1}{360}\left\{\begin{array}{l}\left(32\sqrt{21}+180\right)\mathbf{x}_i - 64\sqrt{21}\mathbf{x}_c + \left(32\sqrt{21}-180\right)\mathbf{x}_{i+1} \\ \Delta t_i\left[\left(9+\sqrt{21}\right)\mathbf{f}_i + 98\mathbf{f}_1 + 64\mathbf{f}_c + \left(9-\sqrt{21}\right)\mathbf{f}_{i+1}\right]\end{array}\right\} = 0 \quad (12)$$

$$\mathbf{C}_{5,3}(\mathbf{x}_i,\mathbf{x}_{i+1}) = \frac{1}{360}\left\{\begin{array}{l}\left(-32\sqrt{21}+180\right)\mathbf{x}_i + 64\sqrt{21}\mathbf{x}_c + \left(-32\sqrt{21}-180\right)\mathbf{x}_{i+1} \\ \Delta t_i\left[\left(9-\sqrt{21}\right)\mathbf{f}_i + 64\mathbf{f}_c + 98\mathbf{f}_3 + \left(9+\sqrt{21}\right)\mathbf{f}_{i+1}\right]\end{array}\right\} = 0 \quad (13)$$

where $\mathbf{f}_1 = \mathbf{f}(\mathbf{x}_1,\mathbf{u}_1,t_1)$, $\mathbf{f}_3 = \mathbf{f}(\mathbf{x}_3,\mathbf{u}_3,t_3)$. Note that the values of the controls at collocation points $t_1$ and $t_3$ must be specified. This then adds two more control parameters per segment per control.

The state parameters $\mathbf{x}$, control parameters $\mathbf{u}$, and the event variables $\mathbf{E}$ are collected into a single vector $\mathbf{P}$. Event variables are extra parameters required to describe the trajectory such as time of flight, engine burn times, departure date, etc...

$$\mathbf{P} = [\mathbf{Z},\mathbf{E}] \quad (14)$$

where:

$$\mathbf{Z} = [\mathbf{x}_1,\ldots\mathbf{x}_{N+1},\mathbf{u}_1,\ldots,\mathbf{u}_{N+1}]$$

The problem is then to minimize $\Phi(P)$ subject to:

$$\mathbf{b}_L \le \left\{\begin{array}{c}\mathbf{P} \\ A\mathbf{P} \\ \mathbf{C}(\mathbf{P})\end{array}\right\} \le \mathbf{b}_U \quad (15)$$

where $A\mathbf{P}$ is a vector of linear constraints defined by the matrix $A$, and $\mathbf{C}(\mathbf{P})$ is a vector of non-linear constraints virtually all of which are the implicit integration constraints (12) and (13).

## 4 Generation of Periodic Orbits

Periodic orbits about Lagrange points are of interest to scientists because the orbit characteristics are attractive for scientific missions. Missions with varying goals such as collecting "solar wind" (Genesis [13]), high precision imaging (James Webb Space Telescope [14]), and measuring the background cosmic microwave radiation (Wilkinson Microwave Anisotropy Probe [15]) have used (or will use) Lagrange point orbits. Periodic orbits are also the connections for the invariant manifolds that make up the Interplanetary Super-Highway [2].

Periodic solutions to the CR3BP have been sought since Poincaré. Since computational power was not readily available early work focused on analytic approximations. Richardson [4] constructed 1st and 3rd order analytic approximations by linearizing the equations of motion about the $L_1$ and $L_2$ points. Gomez and Marcote [5] extended the analysis to higher orders. More accurate solutions have been obtained through computational means, however, all numerical methods have their disadvantages. Shooting methods such as those used by Howell and Pernicka [8] are highly sensitive to the quality of the initial guess. Generating functions as used by Scheeres and Guibout [6] have a limited (spatial) range of applicability. In this work a new more robust method for generating periodic orbits is found.

Periodic orbits are trajectories where $\tilde{x}(t_0) = \tilde{x}(t_0 + NT)$ for some integer $N$ and orbital period $T$. Therefore, the search for periodic solutions to the circular restricted three body problem of specified amplitude or period can be described as an optimal control problem [16] with the objective of minimizing the function:

$$J = \left|\tilde{\mathbf{x}}(t_o) - \tilde{\mathbf{x}}(t_0 + NT)\right| \tag{16}$$

subject to the system equations (2). The problem can be transposed into a problem to be solved with direct collocation. The native MATLAB NLP solver *fmincon* was used in this study.

Periodic orbits were sought about the interior Lagrange point $L_1$ in the Earth-Moon system although the method would apply equally well to other Lagrange points and/or different three body systems. In normalized units the Earth-Moon system is fully defined by the mass ratio $\mu = m_{\text{moon}}/(m_{\text{earth}} + m_{\text{moon}}) = 0.0122$. The full three-dimensional solution to equations (2) was sought. The time histories of the states were discretized into $N$ segments (as shown in Fig. 3). The values of state variables at the boundaries and at the centers of each segment and the final time $t_f$ then constitute the NLP parameter vector (**P**). This vector is typically 100's or 1000's of elements in length.

The 5th degree Gauss-Lobatto collocation constraints (equations (12,13)) were employed to ensure the trajectory satisfies the system equations of motion. To ensure that the resulting trajectory has the required amplitude, the position $r_1 = [x_1, y_1, z_1]$ at the initial node is specified. Since any periodic orbit must return to its initial position after one orbit period, the position state at the final node is also fixed, i.e.

$$r_{N+1} = [x_{N+1}, y_{N+1}, z_{N+1}] = [x_1, y_1, z_1] = r_1 \tag{17}$$

For a periodic solution the velocities at the initial and final nodes must also be equal. This is achieved by minimizing the objective (penalty) function:

$$\Phi(\mathbf{x}) = \sqrt{(v_{x_i} - v_{x_f})^2 + (v_{y_i} - v_{y_f})^2 + (v_{z_i} - v_{z_f})^2} \tag{18}$$

This function has a global minimum $\Phi(x) = 0$ when $v_{x_i} = v_{x_f}$, $v_{y_i} = v_{y_f}$ and $v_{z_i} = v_{z_f}$, (i.e. when the trajectory is a periodic orbit). Thus, the pure penalty function of equation (16) is not required. Solutions were obtained with the equality constraint (17) for the positions and the penalty function constraint (18) for the velocities.

The NLP solver requires an initial guess of the optimal solution. For this case Richardson's 1st order analytical approximation [4] is used, in which

$$\begin{aligned} x &= A_x \cos(\lambda t + \Phi) \\ y &= kA_x \sin(\lambda t + \Phi) \\ z &= A_z \sin(\lambda t + \psi) \end{aligned} \tag{19}$$

where $\Phi$ and $\psi$ are phase angles and,

$$k = \frac{2\lambda}{\lambda^2 + 1 - c_2} \tag{20}$$

The linearized frequency $\lambda$ is found from the solution to:

$$\lambda^4 + (c_2 - 2)\lambda^2 - (c_2 - 1)(1 + 2c_2) = 0 \tag{21}$$

Constants $c_1$ and $c_2$ defined by:

$$c_n = \frac{1}{\gamma_L^3}\left[(\pm 1)^n \mu + (-1)^n \frac{(1-\mu)\gamma_L^{n+1}}{(1 \mp \gamma_L)^{n+1}}\right] \tag{22}$$

with

$$\gamma_L = n_1^{\frac{2}{3}} \tag{23}$$

Higher order approximations are available [4, 5], however, this simple initial guess proved to be sufficient and demonstrated the robustness of the solution technique. For calculations of families of periodic orbits homotopy was employed, i.e. the solution for a periodic orbit of a certain amplitude is used as an initial guess in the calculation of a periodic orbit with a similar amplitude and this process is continued.

## *4.1 Results*

Periodic orbits of a given amplitude are obtained by specifying the initial position. For an orbit around $L_1$ with $x$ amplitude $A_x$ and $z$ amplitude $A_z$, initial position is $r_1 = [L_1 + A_x, 0, A_z]$. Figure 4 shows a periodic orbit with $A_x = 1.0 \times 10^4$ km and $A_z = 5 \times 10^3$ km as well as the initial guess from the 1st order Richardson approximation (equation (20)) for a periodic orbit with amplitudes $(A_x, A_z)$.

Families of orbits were also constructed. Figure 5 shows a family of periodic orbits about $L_1$ with $A_x = 1.2 \times 10^4$ km and $A_z$ varying between $\pm 1 \times 10^4$. Figure 6 shows a family of Lyapunov (planar) orbits about $L_1$.

Periodic orbits of a given period are obtained by removing the constraints on the initial position and adding a constraint on the final time $t_f$, all other constraints and the objective function remaining the same. The initial guess was a 1st order Richardson approximation for randomly selected amplitudes $A_x$ and $A_z$. Figure 7

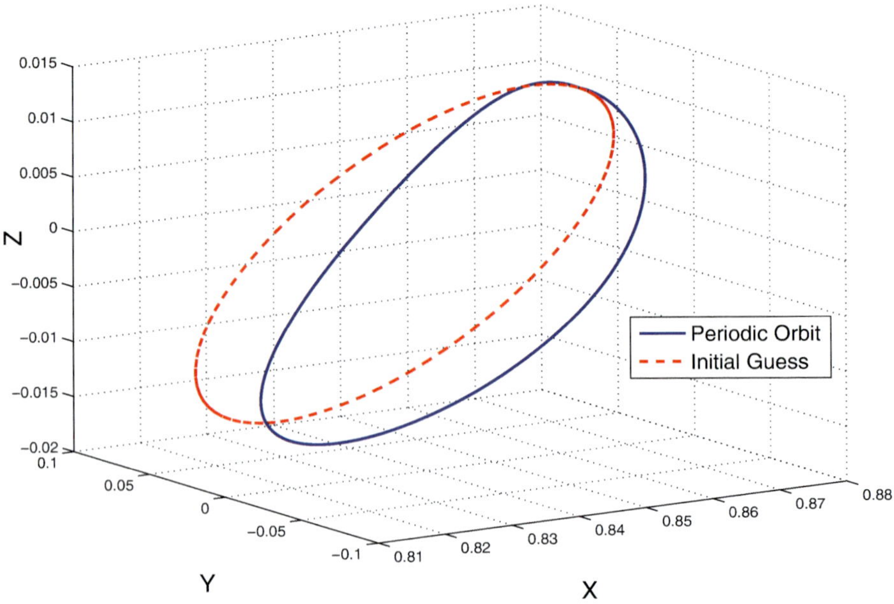


**Fig. 4** Single periodic orbit about Earth-Moon $L_1$

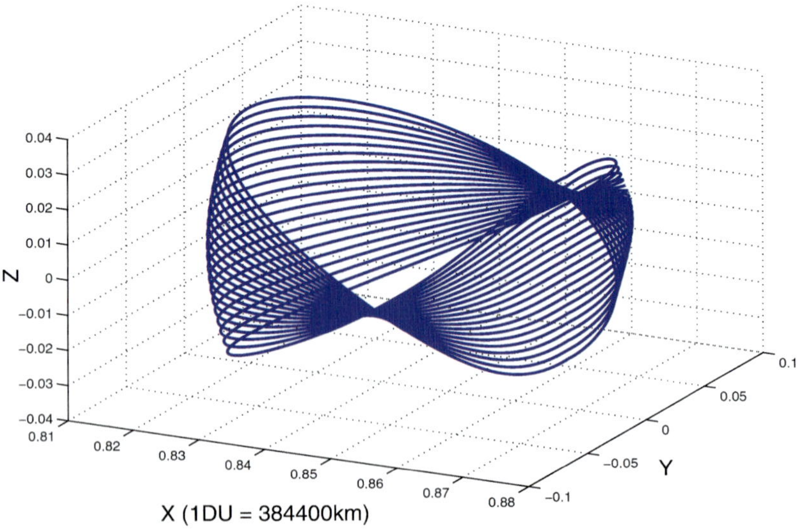


**Fig. 5** Family of periodic orbits with $A_x = 1.2 \times 10^4$ km

shows a periodic orbit with period 3.087 TU and the initial guess of the periodic orbit given to the numerical optimizer.

Specifying amplitude or period are just two examples of the constraints that can be used with this method to create periodic orbits with desired characteristics. It

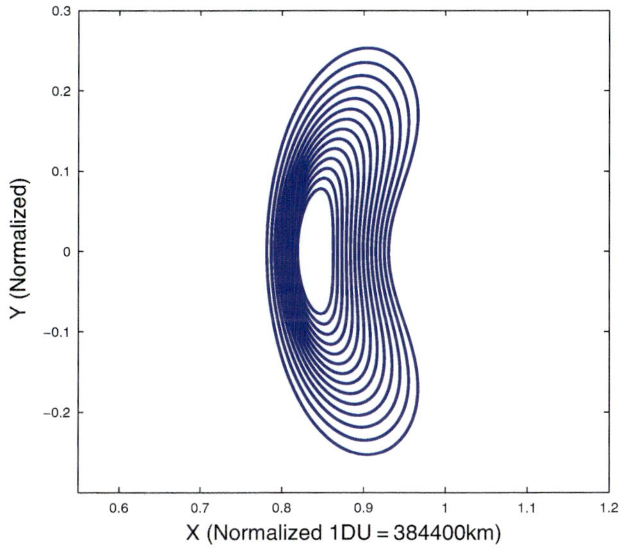


**Fig. 6** Family of Lyapunov periodic orbits about Earth-Moon Lagrange point $L_1$

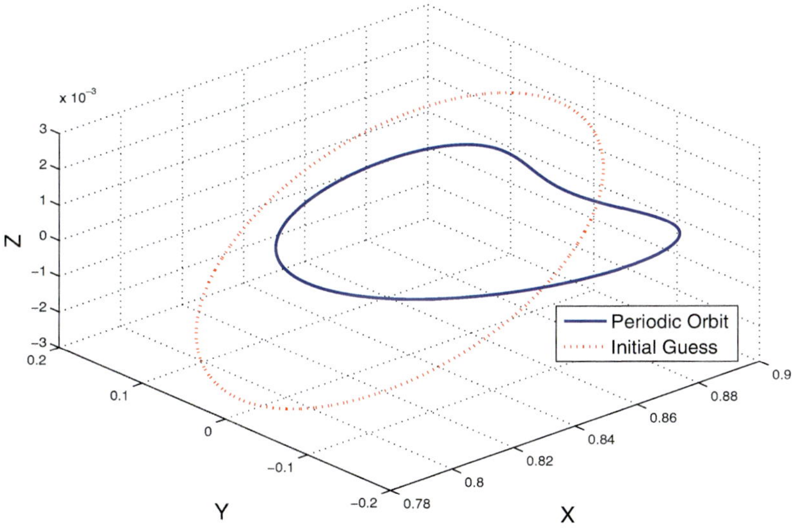


**Fig. 7** Single periodic orbit with period 3.087 TU about Earth-Moon Lagrange point $L_1$

would be possible to search for an orbit with amplitude and period in given ranges. The method could apply equally well to the generation of periodic orbits about the other Lagrange points ($L_2, \ldots, L_5$) and for different three body systems such as Earth-Sun-Spacecraft, Sun-Jupiter-Spacecraft etc.

## *4.2 Comparison with Shooting Method*

The method based on direct collocation for generating (quasi) periodic orbits derived here is compared to a shooting method, the conventional way of generating such trajectories.

The periodic orbits desired are symmetric about the x–z plane; the shooting method takes advantage of this symmetry. The symmetry implies that the components of the velocity in the x and z directions must be zero as the orbit crosses the x–z plane (y = 0). A starting point is chosen on the x–z plane based on the desired periodic orbit amplitude. Then an initial guess is made for the y component of velocity ($v_{yi}$), the x and z components are zero ($v_{xi} = v_{zi} = 0$). The system equations are integrated forward in time until the trajectory crosses the x–z plane. If the velocity components in the x and z directions are not zero, the initial velocity is adjusted and the integration repeated. This process continues until an orbit is found that crosses the x–z plane at two points (including the starting point) with the velocity solely in the y direction. Solutions found for various initial guesses of initial velocity ($v_{yi}$) and for orbits with x and z amplitudes $A_x = 20{,}000$ km and $A_z = 0$ km respectively are shown in Fig. 8.

The orbit with $v_y(0) = -0.2$(DU/TU) is one that orbits the moon. The desired periodic orbit has an initial y velocity $v_y(0) = -0.34$(DU/TU). As the figure shows, the shooting method does not converge to the desired solution for poor

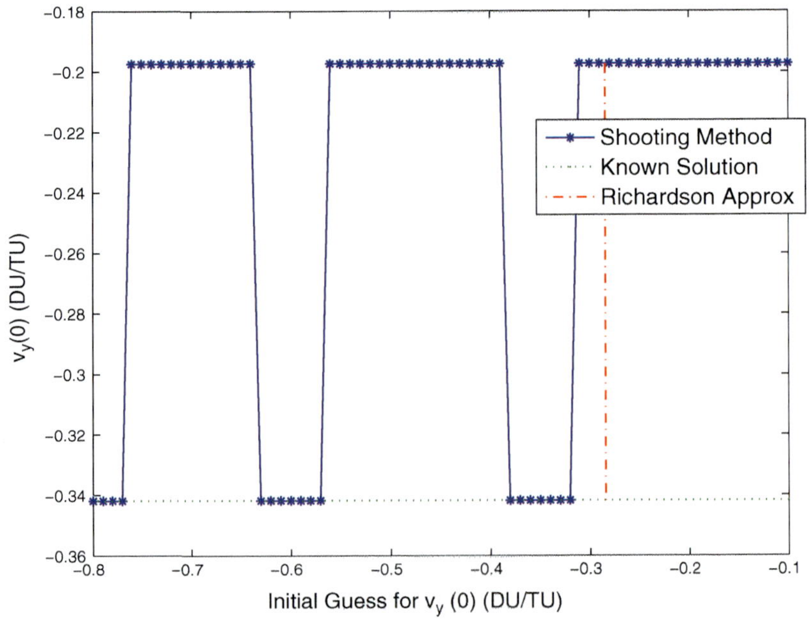


**Fig. 8** Converged solution for $v_y(0)$ for varying initial guess for $v_y(0)$ for periodic orbit with $A_x = 20{,}000$ km and $A_z = 0$ km

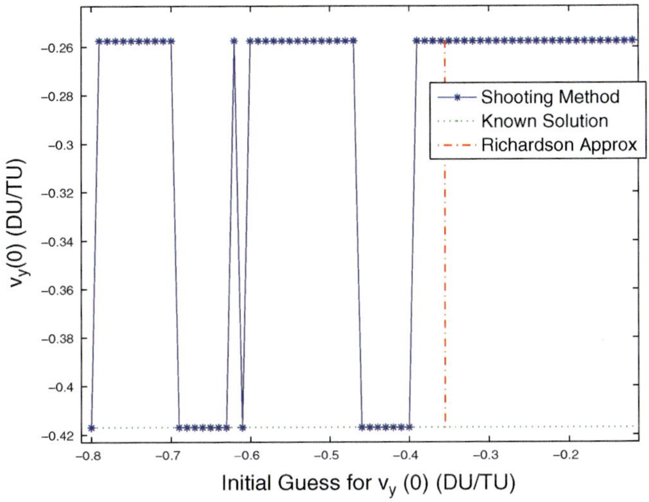


**Fig. 9** Converged solution for $v_y(0)$ for varying initial guess for $v_y(0)$ for periodic orbit with $A_x = 25{,}000\,\text{km}$ and $A_z = 0\,\text{km}$

initial guesses (including the Richardson approximation used as an initial guess for the collocation method). The convergence for extremely poor initial guesses e.g. $v_y(0) = 0.6(\text{DU/TU})$ or $v_y(0) = 0.8(\text{DU/TU})$) is due to a fortuitous starting point where the dervivative is such that the next estimation for $v_y(0)$, calculated by Newton's method, is in the region of convergence. The shooting method however is much less computationally expensive than the collocation solution (w/ 40 segments), which took 60 s. Solutions using the shooting method never required more than 0.032 s. The results are qualitatively similar for larger x amplitudes; Fig. 9, 10 and 11 show the results for larger x amplitudes (25,000,30,000 and 35,000 km respectively). The shooting method converges to the desired periodic orbit only when the initial guess is very close to the true solution.

The Richardson 1st order approximation for the orbit is of the form:

$$
\begin{aligned}
lx(t) &= L_1 + A_x \cos(\lambda t) \\
y(t) &= -kA_x \sin(\lambda t) \\
z(t) &= A_z \cos(\lambda t) \\
v_x(t) &= -\lambda A_x \sin(\lambda t) \\
v_y(t) &= -\lambda k A_x \cos(\lambda t) \\
v_z(t) &= -\lambda A_z \sin(\lambda t) \\
T &= \frac{2\pi}{\lambda}
\end{aligned}
\tag{24}
$$

where the amplitude ratio k and the frequency $\lambda$ are 2.922 and 1.869 for orbits about the Earth-Moon L1 point. The sensitivity of the collocation solution to variations of

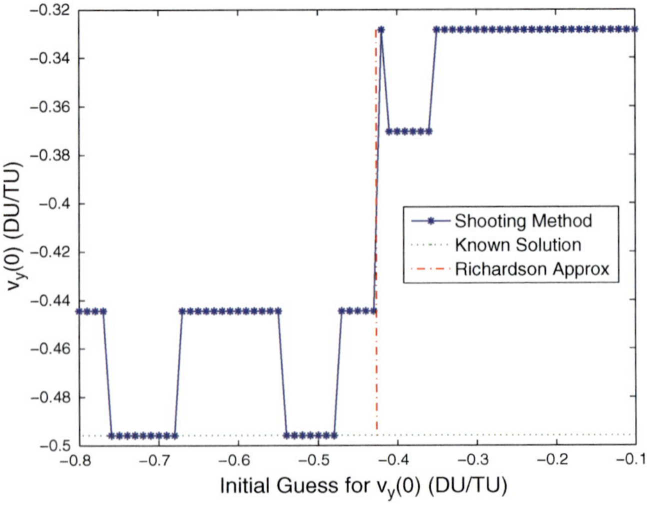


**Fig. 10** Converged solution for $v_y(0)$ for varying initial guess for $v_y(0)$ for periodic orbit with $A_x = 3000\,\text{km}$ and $A_z = 0\,\text{km}$

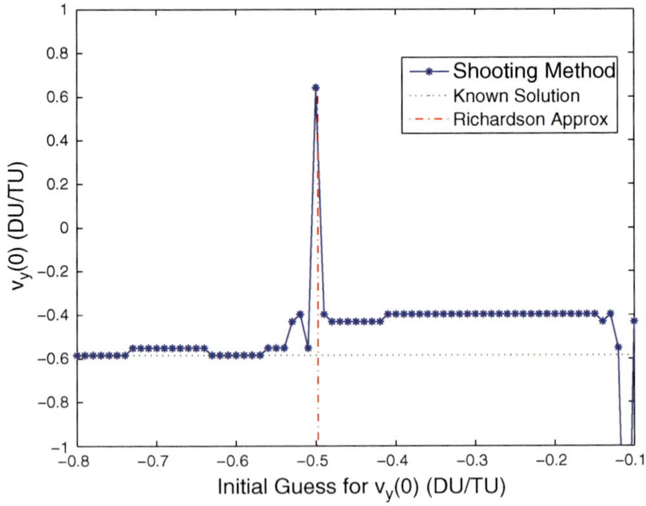


**Fig. 11** Converged solution for $v_y(0)$ for varying initial guess for $v_y(0)$ for periodic orbit with $A_x = 35000\,\text{km}$ and $A_z = 0\,\text{km}$

these constants is shown in Fig. 12. The figure shows that the optimizer finds the true solution $v_y(0) = -0.34DU/TU$ within the iteration limit (1000 iterations) for most combinations of $k$ and $\lambda$, having difficulty principally when $k$ and $\lambda$ are assumed much smaller than their true values. It is possible that the optimizer would converge to the correct solution if further iterations were performed.

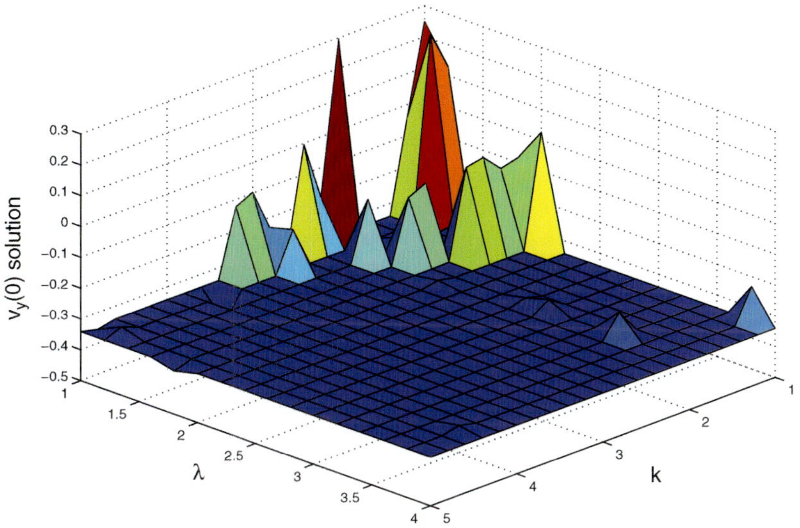


**Fig. 12** Converged solution for $v_y(0)$ using direct collocation for varying $k$ and $\lambda$ . $A_x =$ 20,000 km (true solution $v_y(0) - 0.34$DU/TU )

## 5 Low Thrust Transfer to a Periodic Orbit

The purpose of this study was to find a trajectory that leaves a circular Earth orbit and enters a target periodic orbit about the interior Earth-Moon Lagrange point in minimum-time using a low-thrust engine.

For this problem the virtual spacecraft is confined to the x–y plane in the CR3BP. This was done in order to limit the number of NLP parameters required; the full 3D problem would require an additional 4 state variables and 4 control variables per segment (2 extra states $(z, v_z)$ at each node and 1 extra control at each node as well as each collocation point). Note that the planar restricted problem solution is also a solution to the full 3D problem as $z = 0$ is a valid solution of the equations of motion (equation (2)). The virtual spacecraft has a constant thrust acceleration of $1 \times 10^{-3} g$ and starts from a circular Earth orbit at 3 Earth radii (19,134 km). It is assumed that there are no coasting arcs, i.e. the engine is always on and providing maximum thrust. With this assumption minimizing time of flight is equivalent to minimizing fuel consumption.

The spacecraft begins in a circular orbit at radius $3R_{\text{earth}}$. The initial position was chosen to be on the x axis with the velocity in the positive y direction. The initial states were fixed to reduce the number of degrees of freedom in the system, allowing the optimizer to more easily find a solution. There is a minimal loss in generality as the time taken to complete a revolution at $3R_{\text{earth}}$ is small compared to the total transfer time. Therefore the small acceleration provided by the low-thrust engine would not be able to change the velocity significantly in either magnitude or direction in the time taken to complete one orbit.

If the insertion point on the target periodic orbit is chosen *a priori* then final position and velocity are known and appropriate constraints can be placed on the states at the final node and an optimal trajectory can be calculated readily. However, if the goal is simply to finish on the periodic orbit, this insertion point can almost certainly be improved upon. This constraint, i.e. simply arriving on the periodic orbit, is problematic as there exists no analytical solution to the periodic orbit generation problem, hence there exists no terminal constraint of the form of equation (8).

The solution developed in this work is to express the target periodic orbit in terms of parameters available to the optimizer. To achieve this an additional parameter ($s$), hereafter referred to as the periodic orbit insertion point parameter, is introduced that represents the normalized position around the periodic orbit from some arbitrary starting location (where $0 \leq s = \frac{t}{T} \leq 1$). The states are then parameterized in terms of this new parameter.

A target periodic orbit with $A_x = 30000$ km was generated by the procedure in Section 4. This procedure generates a discretized periodic orbit. For each state $(x,y,v_x,v_y)$ a curve was fitted to create functions of the states in terms of the periodic orbit insertion parameter ($s$). A cubic spline interpolant was fitted using the MATLAB curve fitting toolbox program *cftool*. These fitted curves are continuous and differentiable functions for the terminal states in terms of $s$, the periodic orbit insertion point parameter. Examples of the fitted curves $x_f = f_1(s), y_f = f_2(s), v_{x_f} = f_3(s), v_{y_f} = f_4(s)$ for a target periodic orbit with amplitude $A_x = 30000$ km are shown in Figs. 13, 14, 15, 16.

The NLP parameters for this problem are the positions and velocities in Cartesian coordinates $(x,y,v_x,v_y)$ at each node and at the center points of the segments and the

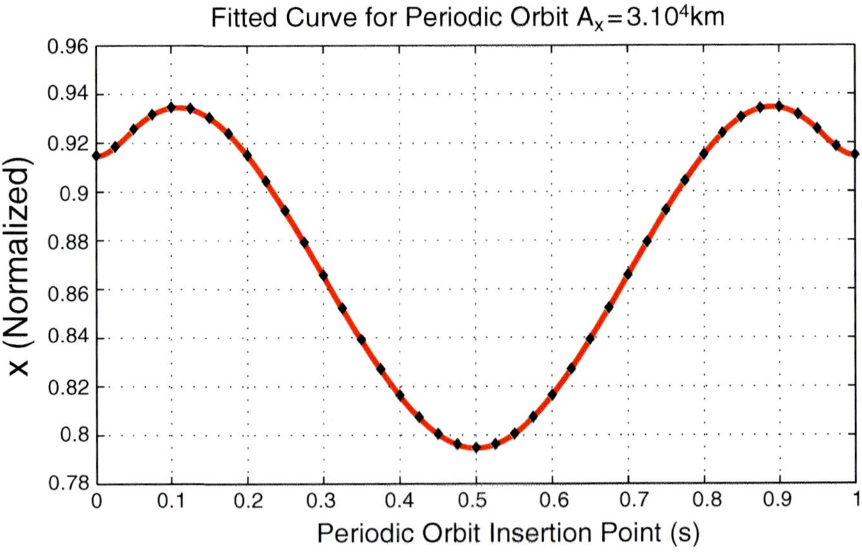


**Fig. 13** Periodic orbit $x$ position curve fit $x_f = f_1(s)$

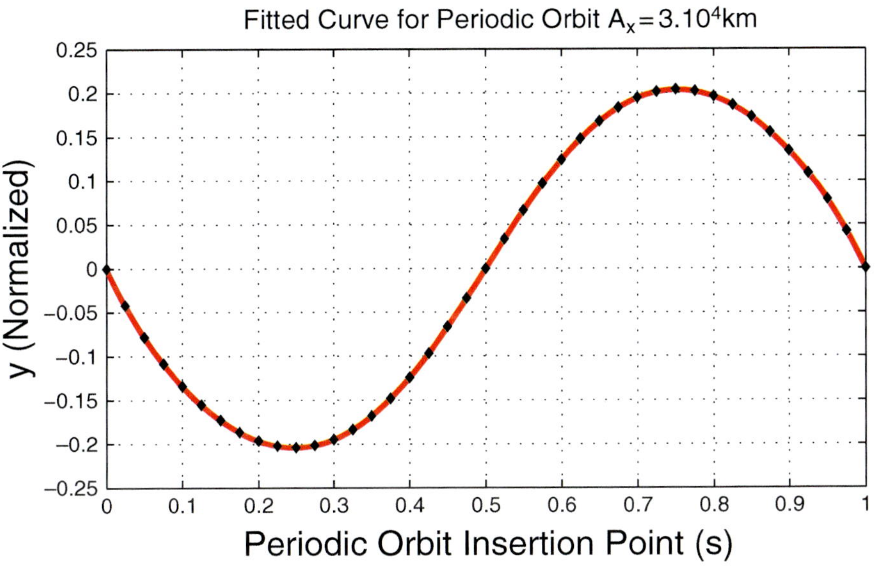


**Fig. 14** Periodic orbit $y$ position curve fit $y_f = f_2(s)$

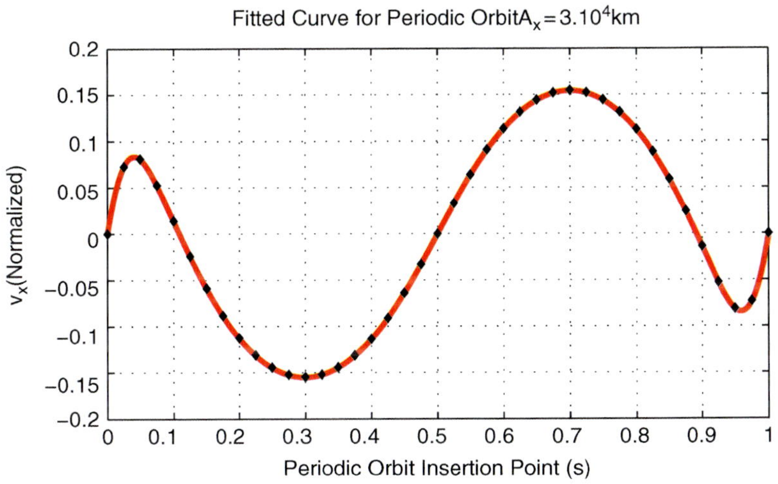


**Fig. 15** Periodic orbit $v_x$ position curve fit $v_{x_f} = f_3(s)$

control thrust pointing angle $u$ at each node, at the center point of each segment and at each collocation point. The thrust pointing angle $u$ is measured from the tangential direction, i.e. the direction of the velocity vector $v$. That is, if $u$ were always zero the thrust would always be in the direction of the velocity vector (Fig. 2). Two additional event (or decision) parameters are included, the periodic orbit insertion parameter $s$

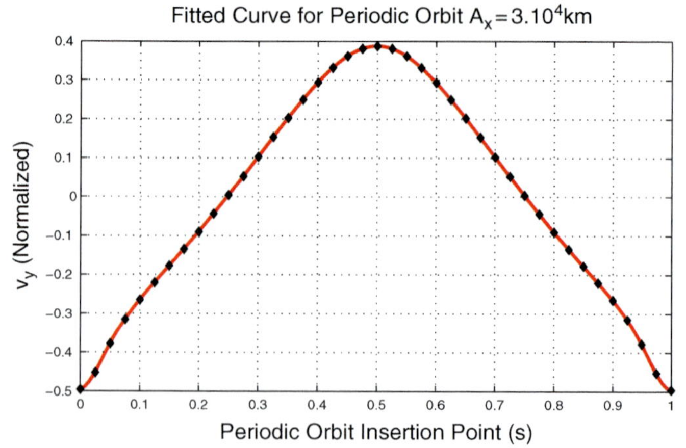


**Fig. 16** Periodic orbit $v_y$ position curve fit $v_{y_f} = f_4(s)$

and the final time $t_f$, the latter also being the objective function. Thus the parameter vector $P$ is,

$$\mathbf{P} = [x_1, y_1, v_{x_1}, v_{y_1}, \ldots, x_{N+1}, y_{N+1}, v_{x_{N+1}}, v_{y_{N+1}}, u_1, \ldots, u_{N+1}, s, t_f] \tag{25}$$

The quality of the initial guess significantly influences the ability of the NLP solver to obtain a convergent solution. It was chosen to have a feasible trajectory as an initial guess, that is a trajectory that satisfies the initial and final constraints as well as the Gauss-Lobatto collocation constraints. This increases the likelihood that the optimizer will converge to a satisfactory solution. However, a valid trajectory that inserts into the target periodic orbit is not known a priori and must be computed. The initial guess trajectory was created using a three step process.

1) The equations of motion (3) were integrated forward with the assumption that the thrust is always aligned with the instantaneous velocity vector, that is a thrust pointing angle of zero. The time of flight was selected so that at the end of the integration the spacecraft is in the vicinity of the target periodic orbit; an example trajectory is shown in Fig. 17.
2) The next step was to create a valid trajectory that finished at a specified position on the periodic orbit but with the velocity unconstrained. A direct collocation optimization was used with the initial states and Gauss-Lobatto constraints (12,13). There were no constraints applied to the final states. The objective function was the difference between the x and y states at the final node $(x_f, y_f)$ and the specified final position $(x_s, y_s)$.

$$\phi(\tilde{\mathbf{x}}) = \sqrt{(x_f - x_s)^2 + (y_f - y_s)^2} \tag{26}$$

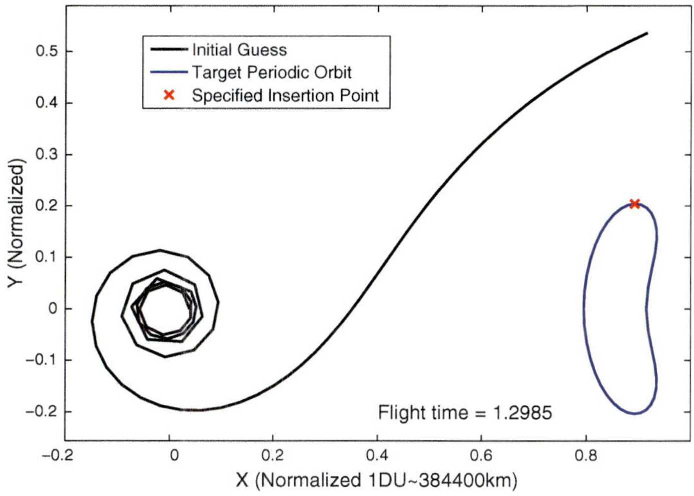


**Fig. 17** Trajectory assuming tangential thrust

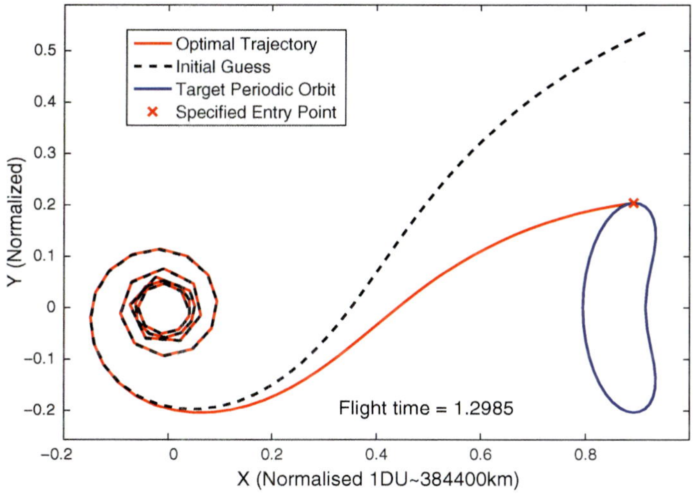


**Fig. 18** Trajectory matching position

The trajectory obtained with tangential thrust is used as the initial guess. A solution with an objective function value of zero, i.e. reaching precisely the specified entry point, was obtained, as shown in Fig. 18.

3) The final step is to find a trajectory that reached the same position but that also ha d the velocity required to complete the periodic orbit. A similar procedure is used with the x and y states at the final node $(x_f, y_f)$ now constrained to the specified final position $(x_s, y_s)$. The objective then becomes minimization of the

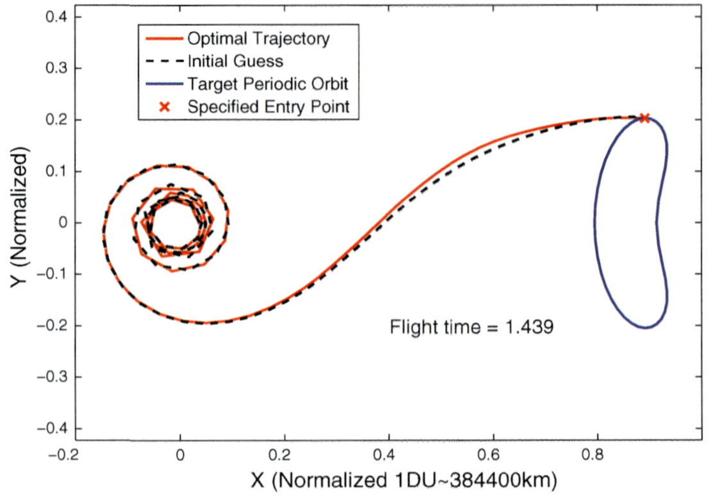


**Fig. 19** Trajectory matching velocity

difference in the velocities at the final node $(v_{x_f}, v_{y_f})$ and the known velocity required at the specified end point $(v_{x_s}, v_{y_s})$ . The objective function is then:

$$\phi(\tilde{\mathbf{x}}) = \sqrt{(v_{x_f} - v_{x_s})^2 + (v_{y_f} - v_{y_s})^2} \tag{27}$$

The trajectory arriving at $(x_s, y_s)$ is used as an initial guess. The resulting trajectory is shown in Fig. 19.

The resulting trajectory is a feasible insertion trajectory that can be used as an initial guess for the problem of finding a time-minimizing insertion trajectory as described in Section 5.

## *5.1 Results*

The numerical optimization procedure previously described was applied using the NLP parameters (25) and constraints (12,13) to generate optimal insertion trajectories for periodic orbits. The periodic orbit insertion parameter $s$ is now included as an NLP parameter. The numerical method sought a trajectory that leaves Earth orbit and arrives at a point on the periodic orbit with the required velocity to enter the periodic orbit such that the transfer is completed in minimum time.

The initial target orbit was a Lyapunov orbit about the Earth-Moon $L_1$ Lagrange point with amplitude $A_x = 3.0 \times 10^4$ km . The optimal trajectory computed and the successive initial guesses required are shown in Fig. 20. The thrust pointing angle time history for the optimal trajectory is shown in Fig. 21. The initial value of $s$ was 0.75 and the optimal value was found to be 0.6357.

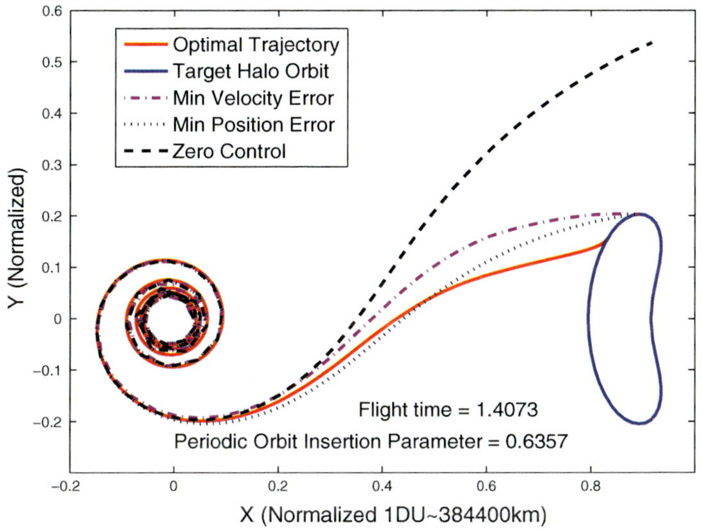


**Fig. 20** Optimal low-thrust trajectory to target orbit of amplitude $A_x = 3.0 \times 10^4$ km achieved using a succession of initial guesses

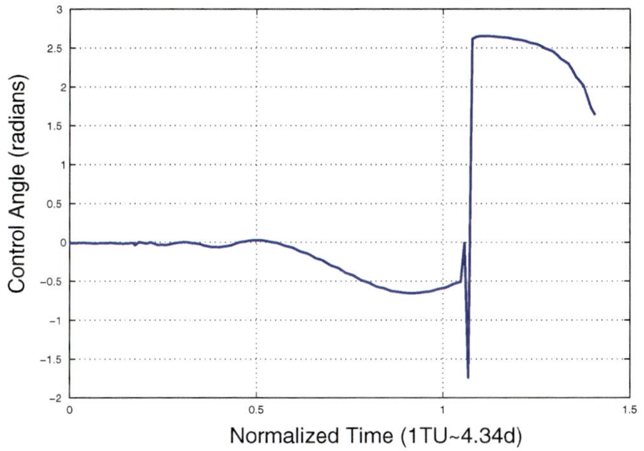


**Fig. 21** Thrust pointing angle time history for optimal low-thrust trajectory to a target orbit of amplitude $A_x = 3.0 \times 10^4$ km

To test whether the optimal trajectory found is the global minimum or merely a local minimum the optimization was repeated using a different initial guess for the insertion point. A position on the opposite side of the target periodic orbit was chosen, with $s = 0.25$. The initial guess for the optimal trajectory was then generated using the procedure outlined previously with the exception that the the optimal trajectory calculated previously (Fig. 20) was used as the initial guess for the position error optimization. The resulting time-minimizing trajectory is shown in Fig. 22.

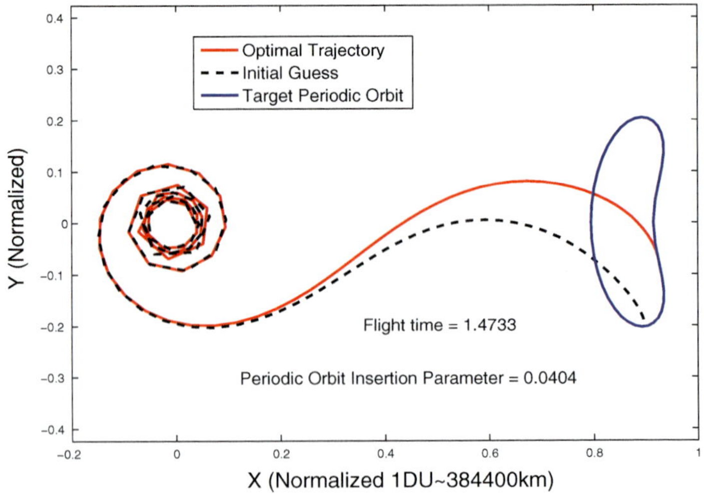


**Fig. 22** Locally optimal low-thrust trajectory to target periodic orbit with amplitude $A_x = 3.0 \times 10^4$ km from alternate initial guess

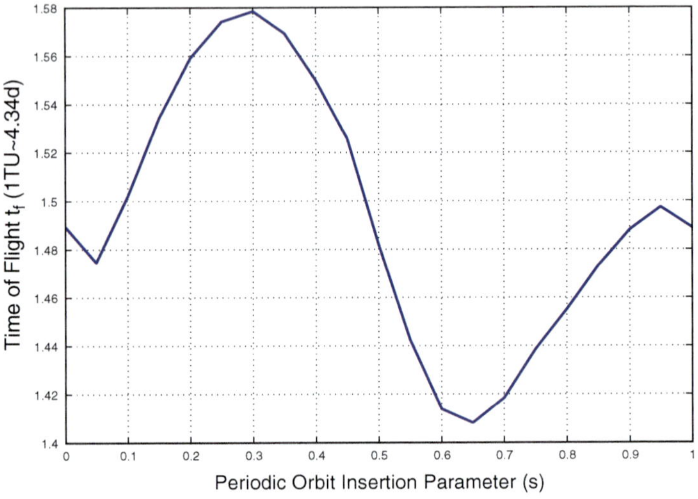


**Fig. 23** Time of flight for optimal trajectories to various points on a target orbit with amplitude $A_x = 3.0 \times 10^4$ km

The two trajectories found both represent local minima in the cost ($t_f$). To determine whether there are any more local minima it is useful to look at the solution space. Time minimizing trajectories were calculated for 21 evenly spaced insertion points around the periodic orbit $s = [0, 0.05, \ldots, 1]$. The results are shown in Fig. 23. There are only two local minima ($s = 0.0404, 0.6357$) therefore this procedure finding trajectories from two starting points on opposite sides of the orbit was able to locate all the minima. Note that there is a significant variation in time of flight as the insertion point on the periodic orbit is varied. The optimal point requires approximately 12% less flight time than the worst possible arrival point. Since the engine is operating continuously, entering the periodic orbit at the optimal point will save considerable propellant.

To test the procedure the process was repeated two different target orbits. One larger and one smaller, i.e. with amplitudes $A_x = 3.7 \times 10^4$ km and $A_x = 2.0 \times 10^4$ km respectively. Qualitatively similar results were obtained, two local minima were observed and the insertion point into the periodic orbit has a significant impact on the cost.

## 6 Conclusions

This work demonstrates that periodic orbits can be generated by posing the problem as an optimization problem and using direct transcription to find the solution. Periodic orbits with certain properties (amplitude, period, energy, etc...) can be generated by altering the constraints on the state variables. This method is able to produce periodic orbits quickly ($< 30$ s on a 2 GHz Intel Core 2 Duo), easily and reliably even from poor initial guesses.

Many current and proposed missions take advantage of Lagrange point orbits [13–15] and their manifolds so it is of great benefit to investigate how to get there cheaply. These trajectories are the "on-ramp" to the Interplantary Superhighway [2]. Optimal low-thrust trajectories were generated from Earth parking orbits to three example reference periodic orbits. The optimization method minimizes the time of flight whilst in the same computation also finding the optimal point to enter into the target orbit. This is necessary as the insertion point chosen and the path to get there both influence cost; i.e. the cheapest (closest) point to reach is not necessarily the cheapest point to insert into. Significant time (and fuel) savings can be realized by optimizing the insertion point in conjunction with the trajectory. Studies of time-minimizing trajectories to specified points around the orbit show that the optimization procedure successfully locates the optimal insertion points into the target periodic orbits.

The trajectories presented in this work are idealized. It is left to future work to generate a more realistic trajectory by, for example: extending the problem to three dimensions, including the change in mass of the spacecraft, using more realistic low-thrust engine performance characteristics, and using more sophisticated models of the solar system such as the JPL empheris. More efficient trajectories could be generated by allowing the optimizer to target the stable manifolds of the target periodic orbit in addition to the periodic orbits themselves, effectively adding a coast arc to the trajectory.

## References

1. Poincare, H., *New Methods of Celestial Mechanics. Volume II - Methods of Newcomb, Gylden, Lindstedt, and Bohlin.*, NTIS, United States, 1967, Translation.

2. Ross, S., *Cylindrical Manifolds and Tube Dynamics in the Restricted Three-Body Problem*, Ph.D. thesis, California Institute of Technology, 2004.
3. Tripathi, R. K., Wilson, J., Cucinotta, F., Anderson, B., and Simonsen, L., "Materials trade study for lunar/gateway missions," *Advances in Space Research*, Vol. 31, No. 11, 2003, pp. 2383 – 2388.
4. Richardson, D., "Analytic construction of periodic orbits about the collinear points," *Celestial Mechanics,* Vol. 22, No. 3, 1980, pp. 241 – 53.
5. Gomez, G. and Marcote, M., "High-order analytical solutions of Hill's equations," *Celestial Mechanics and Dynamical Astronomy*, Vol. 94, No. 2, 2006, pp. 197 – 211.
6. Guibout, V. and Scheeres, D., "Periodic orbits from generating functions," *Advances in the Astronautical Sciences*, Vol. 116, No. 2, 2004, pp. 1029–1048.
7. Gomez, G., Masdemont, J., and Simo, C., "Quasihalo orbits associated with libration points," *Journal of the Astronautical Sciences*, Vol. 46, No. 2, 1998, pp. 135 – 176.
8. Howell, K. and Pernicka, H., "Numerical determination of Lissajous trajectories in the restricted three-body problem," *Celestial Mechanics*, Vol. 41, No. 1–4, 1987–1988, pp. 107 – 24.
9. Rayman, M. D., Fraschetti, T. C., Raymond, C. A., and Russell, C. T., "Coupling of system resource margins through the use of electric propulsion: Implications in preparing for the Dawn mission to Ceres and Vesta," *Acta Astronautica*, Vol. 60, No. 10–11, 2007, pp. 930 – 938.
10. Russell, C., Barucci, M., Binzel, R., Capria, M., Christensen, U., Coradini, A., De Sanctis, M., Feldman, W., Jaumann, R., Keller, H., "Exploring the asteroid belt with ion propulsion: Dawn mission history, status and plans," *Advances in Space Research*, Vol. 40, No. 2, 2007, pp. 193 – 201.
11. Bryson, A. E and Ho Y. C, *Applied optimal control: Optimization, estimation and control*, Hemisphere Publishing Corporation, 1975.
12. Herman, A. L. and Conway, B. A., "Direct optimization using collocation based on high-order Gauss-Lobatto quadrature rules," *Journal of Guidance, Control, and Dynamics*, Vol. 19, No. 3, 1996, pp. 592 – 599.
13. Burnett, D., Barraclough, B., Bennett, R., Neugebauer, M., Oldham, L., Sasaki, C., Sevilla, D., Smith, N., Stansbery, E., Sweetnam, D., and Wiens, R., "The Genesis Discovery Mission: return of solar matter to Earth," *Space Science Reviews*, Vol. 105, No. 3–4, 2003, pp. 509 – 34.
14. Sabelhaus, P. A. and Decker, J., "James Webb space telescope: Project overview," *IEEE Aerospace and Electronic Systems Magazine*, Vol. 22, No. 7, 2007, pp. 3 – 13.
15. Schwarzschild, B., "WMAP spacecraft maps the entire cosmic microwave sky with unprecedented precision," *Physics Today*, Vol. 56, No. 4, 2003, pp. 21 – 24.
16. Tarragó, P., *Study and Assessment of Low-Energy Earth-Moon Transfer Trajectories*, Master's thesis, Université de Liége, 2007.

# On the Use of the Earth-Moon Lagrangian Point $L_1$ for Supporting the Manned Lunar Exploration

**Carlos Corral van Damme, Raúl Cadenas Gorgojo, Jesús Gil-Fernández, and Mariella Graziano**

**Abstract** Several space agencies are currently studying plans for returning humans to the Moon; ESA and the European industry, in particular, are conducting preliminary studies for the CSTS (Crew Space Transportation System) program. One of the scenarios under analysis involves a rendezvous between the crew transportation vehicle and the lunar lander (or a permanent space station) in the colinear Earth-Moon Lagrangian point $L_1$. Trajectories for transferring between the Earth and orbits around these points, as well as from them to the lunar surface and back, taking into account mission requirements such as lunar global accessibility, anytime return, and mission duration, are investigated in this paper. The merits of these trajectories are compared to a more conventional mission architecture where the staging node is placed on a low lunar orbit.

## 1 Introduction

Transportation systems able to support the future human exploration of the Moon are currently under investigation in Europe in the frame of the CSTS (Crew Space Transportation System) program. These systems involve typically at least two different vehicles: a crew transfer vehicle, in charge of bringing the crew from the Earth to the lunar vicinity and back, and a lunar lander and ascent vehicle, for transferring the astronauts to the lunar surface and from there back to the crew transfer vehicle. There exist basically two possibilities for the location of the staging node in the vicinity of the Moon where the crew transfer vehicle will remain and wait for the ascending lander from the lunar surface: it could be placed in low lunar orbit (LLO), as in the case of the Apollo missions, or it could be located in an orbit around one of the two colinear Lagrangian points in the Earth-Moon system $L_1$ or $L_2$ (EML1 or EML2). The latter option (in particular, a gateway station at $L_1$) has been often proposed [1, 2]. Compared to the Apollo solution, it would offer the following main advantages:

C.Corral van Damme (✉)
GMV, Tres Cantos, Madrid, 28760, Spain
e-mail: cfcorral@gmv.com

E. Perozzi, S. Ferraz-Mello (eds.), *Space Manifold Dynamics*,
DOI 10.1007/978-1-4419-0348-8_7, 

1) Continuous accessibility from the Earth: $L_1$ can be reached at any time, so there is no restriction on the launch window from the Earth or from a LEO.
2) Continuous accessibility to any location on the lunar surface: any landing site on the Moon can be reached at any time.
3) Continuous communications with the Earth and the near side of the Moon (the far side of the Moon could also be covered with relay satellites around $L_2$, as proposed in [3]).

A station in orbit around EML1 could also be used as hub for future missions to other bodies in the Solar System, such as Mars or the asteroids. In this paper, however, we only consider its potential use for manned missions to the Moon as an alternative to LLO. The main requirements for the manned exploration missions under consideration within the CSTS program are:

1) Global access to the lunar surface. This means that it shall be possible to land at any point on the lunar surface for any mission, independently of the launch date and the resultant geometry.
2) Any time return in less than 5 days. In case of contingencies, it shall be possible, at any time during the mission, to abort the mission and return the crew to the Earth in less than 5 days.

In this paper, we investigate the transfers between the Earth, the Moon and EML1, considering the requirements of manned missions and with the objective to compare the mission performance (total Delta-V, mission duration, etc.) with the LLO mission architecture.

## 2 Transfers Between the Earth and EML1

The transfers from the Earth (or from a LEO) to orbits around the co-linear Lagrangian point $L_1$ in the Earth-Moon system and back are necessarily different for manned and unmanned spacecraft. For a manned spacecraft, the transfers need to be *fast*. This does not mean inevitably as fast as possible, but there is clearly a limit for the maximum duration of these transfers. The transfers for the unmanned spacecrafts do not need to be fast; they may be designed to minimize the total required Delta-V, which normally implies long transfers times. We will concentrate in the following on computing fast transfers for manned missions.

### *2.1 From the Earth to EML1*

As commented before, for transfering a crew from the Earth to an orbit around EML1, the transfer duration should be limited to few days. For these short transfer durations, the trajectories to EML1 are very similar to the bi-impulsive,

Hohmann-like transfers in the two-body problem [4, 5]. The complex effects of the special dynamics around EML1 require lower relative velocities and longer time intervals to show up. For fast transfers, the spacecraft is injected into a high elliptical orbit, whose apogee altitude is close to the $L_1$ distance. Then, a manoeuvre is performed at apogee to modify the velocity of the spacecraft for the insertion in the corresponding orbit around the Lagrangian point.

The total Delta-V required for these fast transfers depends on several factors:

1) the position of the Moon along its orbit (that is, the launch date, with a period of some 28 days). The distance to EML1 varies because of the eccentricity of the lunar orbit ($e_{\text{Moon}} = 0.05$). The departure transfer Delta-V is highest when the trajectory arrives at the Lagrangian point when the Moon is at its apogee.
2) the altitude of the initial LEO. The higher the altitude of the initial LEO, the lower the required Delta-V.
3) the inclination of the transfer orbit with respect to the Moon orbital plane. In general, the transfer Delta-V will be higher for large inclinations with respect to the lunar orbital plane, although this will depend on the type of orbit around the Lagrangian points.
4) the type of orbit around the Lagrangian points. The transfer Delta-V will be different, for instance, for targetting a small amplitude Lissajous than for a halo orbit. The Delta-V could depend as well on the insertion point along that orbit (the phase).
5) the launch date may also affect the total Delta-V through the solar perturbation, whose effects depend on the relative geometry between the Sun and the transfer orbit. However, this effect is much smaller than the other factors.

In order to estimate the total Delta-V required for the fast transfers, the following parametric analysis has been run. From an initial 200 km altitude LEO, and an inclination of the transfer orbit of 28° with respect to the Moon orbital plane, two-impulse transfers to EML1 were computed as a function of the transfer time. The gravitational attractions from the Earth, the Moon, and the Sun (using numerical ephemeris from JPL ephemeris file DE 405) are considered. Note that, since the final orbit around EML1 is not known, a direct insertion into the EML1 has been assumed (that is, the second manoeuvre places the spacecraft directly into EML1, with zero velocity with respect to that point in the synodic frame). A direct EML1 insertion is representative also of the conditions for inserting into small-amplitude Lissajous orbits around EML1. These orbits are preferred for a gateway station over large amplitude Lissajous or halo orbits around EML1 because, in the latter case, the phasing in the orbit may compromise the continuous availability of launch windows from Earth and from the gateway station towards the Moon.

Figure 1 shows the results obtained, in terms of Delta-V as a function of the transfer time, for two positions of the Moon: at perigee (dashed lines), and at apogee (solid lines). The total Delta-V reaches a minimum for transfer times that are close to the times required for a Hohmann transfer (indicated with black lines). When the Moon is at perigee, the minimum Delta-V is 3.88 km/s, for a transfer time of 86 h

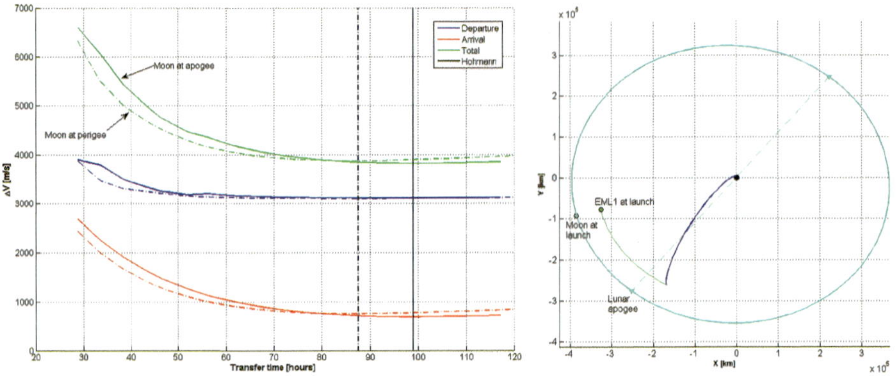


**Fig. 1** Fast transfers from a 200-km LEO to EML1. Delta-V requirements (*left*) and projection on the Earth equatorial plane (*right*)

(3.6 days), whereas when the Moon is at apogee, the minimum Delta-V is 3.83 km/s for a transfer time of 98 h. The typical aspect of these fast transfers is also illustrated in Fig. 1.

For unmanned spacecraft, the constraints on the maximum transfer duration do not necessarily apply. In some cases, it may be preferred to look for the minimum Delta-V transfers, even if they would imply very long transfers times. For slow transfers, the number of options increases significantly: it is possible to insert into the manifolds of the given orbits around EML1, to perform lunar swingbys to reduce the Delta-V, or to exploit the solar gravitational attraction for the same purpose, through excursions to the Sun-Earth SEL1 or SEL2 regions.

Contrary to the case of the Sun-Earth Lagrangian points, the unstable and stable invariant manifolds that depart from EML1 do not intersect the Earth. They generally pass at minimum distances of some 75,000 km or more with respect to the Earth. Hence, there is no free transport from an Earth parking orbit to the Lagrangian points in the Earth-Moon system. However, a transfer strategy may be considered where the spacecraft is first injected into an approximately Keplerian orbit that intersects the manifolds, and then, a second manoeuvre will be used to inject it into the manifold. These types of transfers have been investigated in [6, 7] (for halo orbits), and [8, 9] (including also Lissajous orbits), among others.

## 2.2 From EML1 to the Earth

Similarly to the outbound transfer from LEO to EML1, fast return transfers can be first approximated by the solution of the Lambert problem for the two-body motion, and then slightly modified for accounting for the corrections from the real dynamics (numerical ephemeris and Sun and Moon perturbations). Figure 2 illustrates the Delta-V requirements for these trajectories for transfer times up to 10 days. The

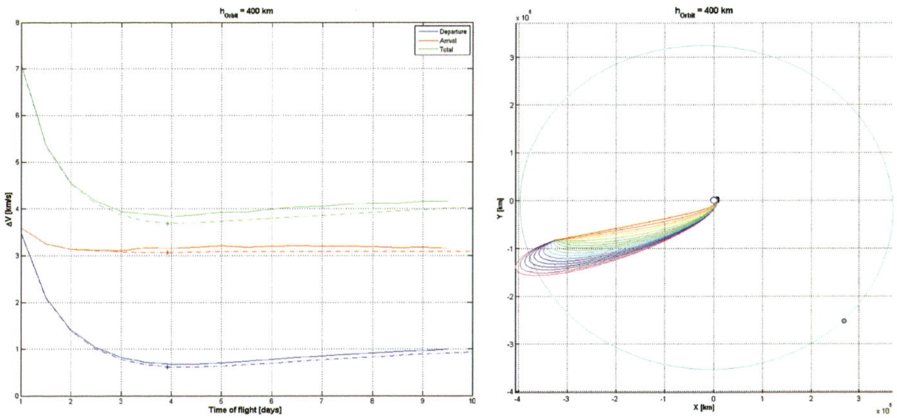


**Fig. 2** Fast transfers from EML1 to the Earth (LEO). Delta-V requirements (*left*) compared to the two-body dynamics assumption (*dashed*), and equatorial projection of trajectories for transfer times between 1 and 9 days (*right*)

Delta-V needed for the transfer using two-body dynamics is also shown. In general, the faster the trajectory, the better the approximation of the solution by the two-body Lambert problem (as for the outbound case). For longer trajectories, which have lower departure Delta-Vs from EML1, the dynamics are more complex and the two-body approximation is not that good. The appearance of these transfer trajectories is depicted also in Fig. 2.

When longer transfer times are allowed (at least 5 days), it may be advantageous to fly three-impulse trajectories for transferring between EML1 and a given LEO target orbit. The strategy is as follows: departing from EML1, a small $\Delta V$ is applied in the direction of the unstable manifold oriented to the Earth. Taking advantage of the non-linear dynamics of the EML1, this small $\Delta V$ places the spacecraft in an orbit with a perigee at an altitude of about 75,000–150,000 km, depending on the departure $\Delta V$ modulus, and on the angle of this $\Delta V$ w.r.t. the synodic plane. A second manoeuvre is applied at an intermediate point at the manifold to insert the spacecraft into an orbit with a perigee close to the target LEO orbit. This manoeuvre is required since, as already stressed, the manifolds from EML1 do not cross low altitude orbits. The time at which this manoeuvre is executed shall be optimized to get a total minimum $\Delta V$. The total $\Delta V$ will be the sum of the departure $\Delta V$, the intermediate $\Delta V$ to insert the spacecraft into a low perigee Earth orbit and the target orbit insertion $\Delta V$.

We have performed a parametric study on these types of trajectories by applying a $\Delta V$ at EML1 with two components: one in the direction of the unstable manifold ($\Delta V_{\text{unstable}}$), and one out-of-plane component ($\Delta V_{\text{per}}$). The resulting trajectory is propagated then until the first perigee. By perigee, it is understood the point where the velocity is perpendicular to the position vector. As it can be seen in Fig. 3, significant $\Delta V$ can be saved for small values of the initial $\Delta V$ and for long transfer

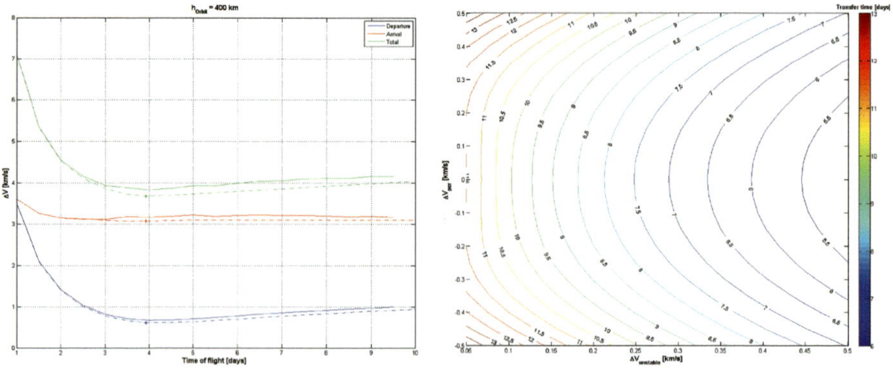


**Fig. 3** Total $\Delta$V (*left*) and transfer time (*right*) for the 3 impulse EML1-LEO transfers

times (11–13 days). As expected, the total DV and the total transfer time present a symmetry wrt the out-of-plane component of the velocity.

Two trajectories with different departure Delta-Vs at EML1 are plotted in Fig. 4 for illustration purposes. The trajectories are plotted until they reach the first point where the velocity is perpendicular to the position vector. At that point is where a minimum energy co-planar transfer is applied. The perigee radius is inversely proportional to the $\Delta$V applied at the unstable manifold.

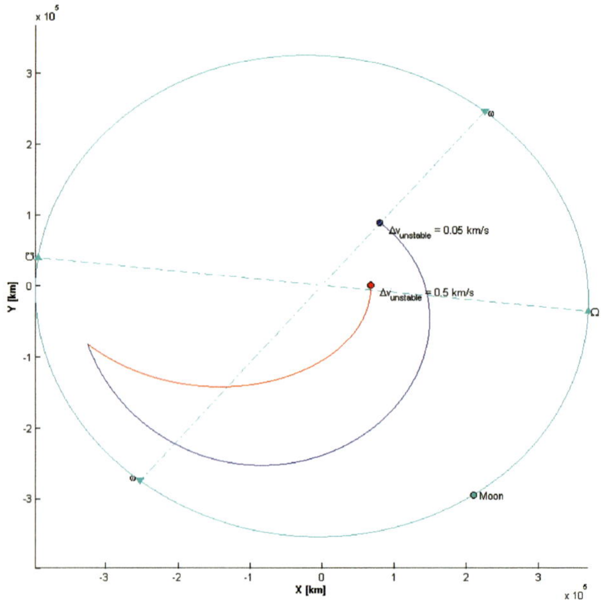


**Fig. 4** Trajectories for $\Delta V_{\text{unstable}} = 0.05$ (*blue*) and 0.5 (*red*) km/s

## 3 Analysis of Orbital Evolution of Orbits Around EML1

The two aspects of the dynamics of the orbits around EML1 with more relevance to the mission design are:

1) Orbit lifetime based on a free-drifting (uncontrolled) spacecraft.
2) Delta-V required to perform station keeping in a given Lissajous orbit or region around the libration points.

Concerning the first point, if the initial spacecraft state (position and velocity) is zero relative to the Earth-Moon $L_1$, the spacecraft would always drift from the $L_1$ position if no station keeping manoeuvres are performed. None of the free-drift orbits finalize in an Earth impact. They may finish on a lunar impact, or in a heliocentric or Earth high eccentric orbit (HEO) after one or several lunar swingbys. No deterministic pattern for those trajectories was found in the simulations performed for several starting epochs.

For the station-keeping allocation, several authors have given estimations on the cost of the control manoeuvres. These estimations range from around 100 m/s per year [10], to around 5 m/s in [11], depending on the assumed performances of the navigation system.

Simulations were performed to confirm these estimations with the complete dynamics and the following assumptions:

1) Only radial control (on position and velocity).
2) The navigation is simulated as the actual position plus the addition of a band limited white noise of known power.
3) The noise power for the range and range-rate measurements can be estimated as function of the range and range-rate variance respectively, the sample rate, the time interval for a single set of measurements and the time interval between sets of measurements. The values used were respectively:

    a. Time for a single set of measurements: 5 min,
    b. Time interval between sets of measurements: 1 day
    c. Sample rate: 1 min
    d. The control response is function of the error in radial position and velocity multiplied by a gain derived from the optimal solution of the linear problem [10]. In order to avoid controlling continuously, tolerances in the position and velocity errors of 1 m and 0.01 m/s, respectively, are allowed.

4) The acceleration control is then only applied when is larger than a certain minimum, which depends on the thruster. In the simulations a value of 0.01 m/s was used.

Figure 5 shows the results of the simulations for different assumptions on the accuracy of the navigation. These results are only preliminary. The problem of deter-

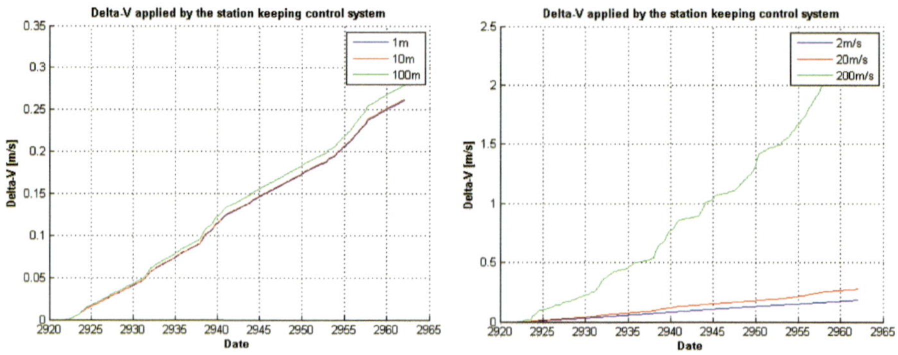


**Fig. 5** Station keeping Delta-V applied for a 40-day long propagation depending on the accuracy of the range system (*left*) and range-rate (*right*)

mining the optimal strategies for orbit maintenance at EML1 would require a deeper, dedicated analysis.

# 4 Transfers Between EML1 and the Moon

In this case, we are also interested in fast transfers between the Moon and EML1. In general, the method used to investigate the trajectories is based on solving the Lambert problem subject to the non-linearities of the full dynamics problem. However, as the transfer time increases, the trajectories deviate more from the two-body motion, and other methods are used, such as parametrically varying the initial conditions and just propagating.

For unmanned missions, as commented before, the transfer time is not a stringent constraint, and longer trajectories are usually allowed. In contrast to the case of the Earth, the unstable and stable manifolds that depart from both Lagrangian points do intersect the lunar surface.

## *4.1 From EML1 to the Lunar Surface*

The transfer trajectories from EML1 to the lunar surface have been studied through a parametric analysis. From EML1, an impulsive Delta-V is applied directly at EML1 and the trajectories are propagated considering the gravity of the Earth, the Moon and the Sun. The solar radiation perturbation has been neglected because of the expected high mass-to-area ratio of crew transfer vehicles. The impulsive Delta-V is varied in modulus (between 0.1 and 0.6 km/s), and direction (azimuth and elevation with respect to the synodic frame). Since the velocity in the synodic frame of a spacecraft in a small-amplitude Lissajous orbit is low, this analysis can be approximately extrapolated to any Lissajous orbit with low amplitudes. In the simulations performed, the transfer time is limited to a maximum of 3 days. With higher Delta-V it is also possible to obtain impact trajectories on the Moon surface, but the accessible region is not so wide. It was determined that, for this case, the elevation of the departure Delta-V on the synodic frame is limited to a range between –40 and 40° approximately. This would correspond to the region inside the unstable manifold that intersects the lunar surface.

Considering that the lunar surface is approximately an equipotential surface, the arrival velocity from EML1 is expected to be more or less constant at any point on the surface. The approximate value can be deduced from the conservation of the Jacobi constant of motion, from:

$$\frac{V^2}{2} = \frac{\mu_{\mathrm{moon}}}{R_{\mathrm{moon}}} - \frac{\mu_{\mathrm{moon}}}{r_{LL1}} \tag{1}$$

The above expression leads to an impact velocity on the Moon surface of $V = 2.37$ km/s. This value will vary slightly due to:

1) The eccentricity of the Moon, which causes the variation of the distance between EML1 and the Moon.
2) The surface of the Moon is not equipotential. There is a variation of this potential due to the own rotation of the Moon and its non-sphericity.
3) The value of the Jacobi constant varies with the value of the departure velocity wrt the synodic frame.
4) The perturbations from third bodies (Sun).

From all the trajectories computed with different departure Delta-V, only a subset impacts the Moon in the maximum allowed time (see Fig. 6). The maximum number of transfer trajectories that impact the Moon were found for a departure $\Delta$V between ~0.2 and 0.3 km/s, which is close to the value of the corresponding equivalent Hohmann transfer for an orbit whose apogee is located at EML1.

For lower departure $\Delta$Vs (0.1 km/s), no trajectories were found that impact on the Moon for the given transfer time of 3 days. There exist, however, trajectories for longer transfer times. The minimum transfer time required for this $\Delta$V is about 7 days. For $\Delta$Vs above 0.6 km/s, the number of impact trajectories is reduced. The higher the Delta-V, the more trajectories that reach hyperbolic conditions wrt the Moon, and do escape from its gravity field to follow geocentric or heliocentric trajectories.

Figure 7 represents the transfer times as a function of the Delta-V. There exist typically fast and slow transfers. As it can be seen, for $\Delta V = 0.3$ km/s the function is only continuous between 2 and 3-day transfer times for the departure epoch at the perigee of the lunar orbit, and the discontinuities increase as the epoch from perigee does. For higher $\Delta$V, the size of the discontinuity increases.

The results indicate that, indeed, any point on the Moon can be reached at any time if the departure $\Delta$V from EML1 is high enough for the maximum transfer time considered (see Figs. 8 and 9). The departure $\Delta$V determines the minimum transfer

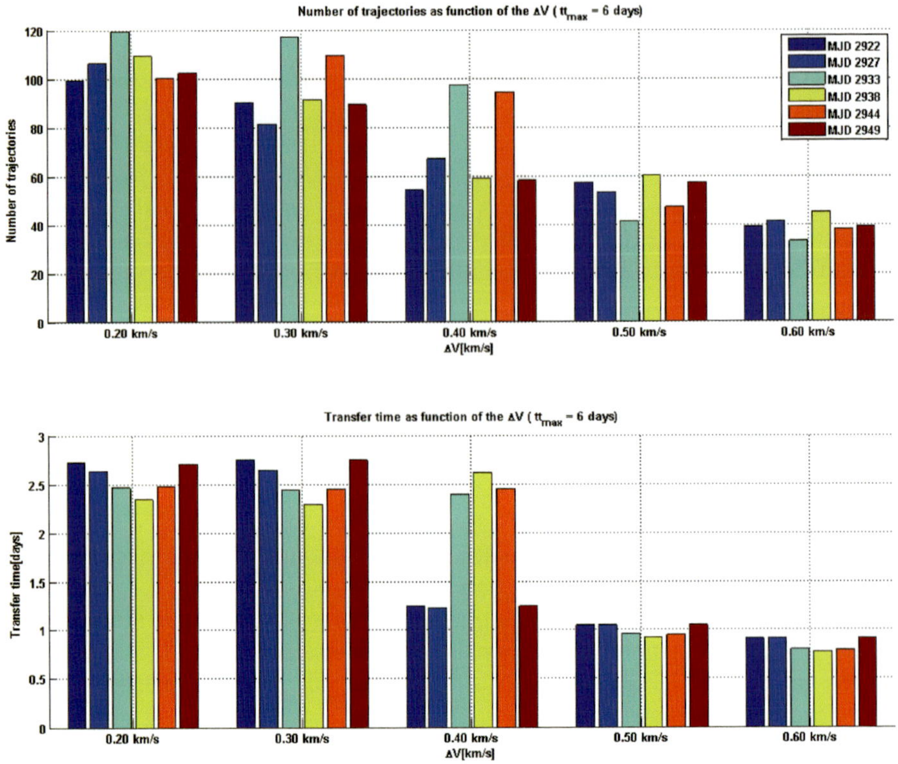


**Fig. 6** Number of trajectories and mean transfer time as function of the departure Delta-V

time to reach the lunar surface. Note that there is no additional cost in terms of $\Delta V$ for targetting any specific landing site. However, it must be noticed that the arrival velocity at the Moon for all these trajectories, for any point or transfer time, is high, $\sim$2.4 km/s.

Figure 10 illustrates the aspect of these trajectories for a departure $\Delta V$ of 300 m/s. The angle of the arrival velocity with respect to the local vertical is another important issue. This angle determines the direction in which the arrival $\Delta V$ must be applied. High angles with respect to the local vertical will mean trajectories very close to the Moon surface for the last part of the flight, with the associated risk. For most of the trajectories, it was found that this angle is close to 45° (as illustrated in Figs. 11 and 12). This angle could be also increased if an intermediate course manoeuvre is performed.

The operational constraints of all these trajectories are important. From the safety point of view, all the computed trajectories have the inconvenience of being direct impact trajectories (their pericentre is under the Moon surface). This implies that a failure in the execution of the final braking manoeuvre could lead to mission and crew loss. Clearly, for lunar surface missions, the final part of the trajectory must necessarily be an impact trajectory. Yet it is advisable to reduce as much as possible

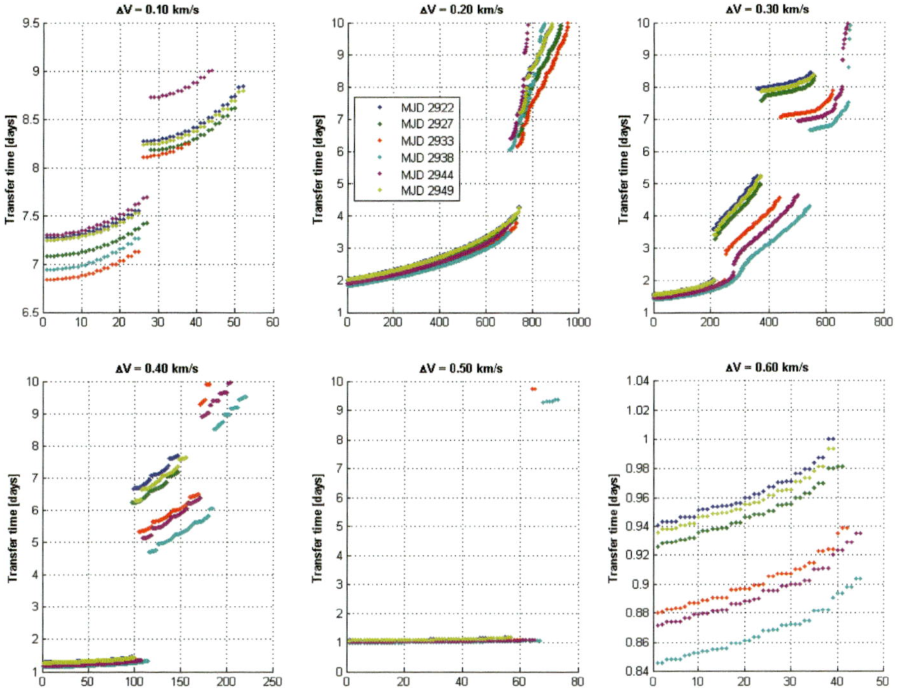


**Fig. 7** Number of transfer opportunities as function of the Delta-V

the duration of that part of the trajectory, with the objective to minimize the risks. It is indeed possible to target from EML1 higher pericentre altitudes (above the lunar surface), with the aim to insert into an intermediate low lunar orbit, and perform from there the descent to the surface. However, this would contraint the achievable range of landing sites, thus canceling one of the major benefits of the EML1 staging node (the capability to reach any location on the lunar surface).

## *4.2 From the Lunar Surface to EML1*

As in the previous section, a parametric analysis has been performed to estimate the $\Delta$V required for a two-impulse trajectory from any point on the surface of the Moon to the EML1 point. Numerical ephemeris and Sun perturbation are considered. The reference date for the departure point is taken with the Moon at its apogee. No gravity losses were considered.

The location of the departure point on the lunar surface has been varied through a grid of points in latitude and longitude. For convenience, we use a Moon-centered reference with the axes parallel to the synodic frame, so that the point with zero latitude and longitude corresponds to the surface point located just in the antipodes of EML1.

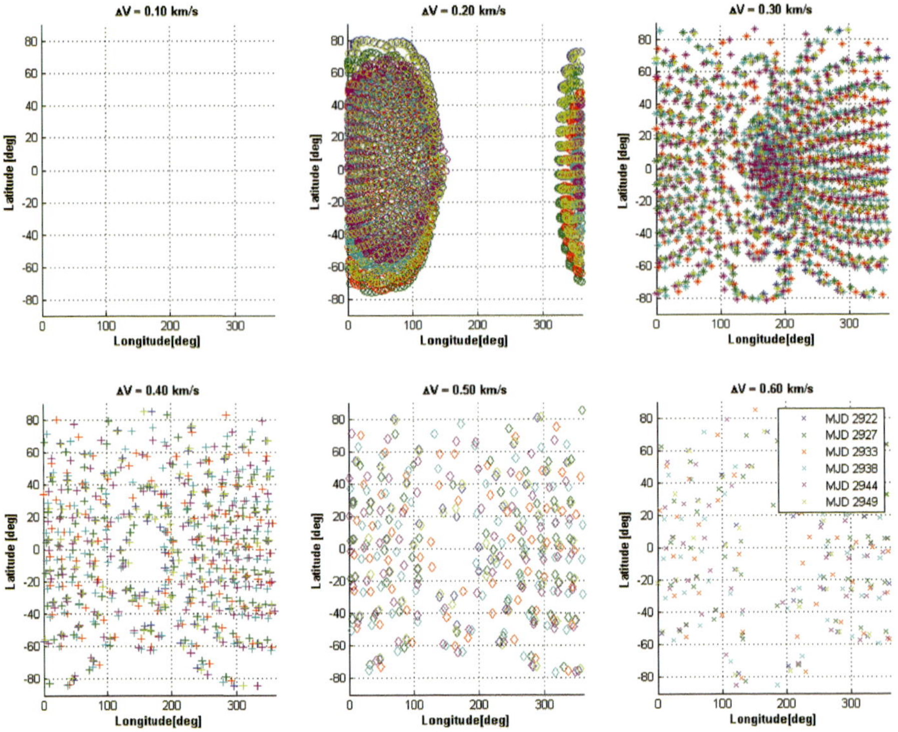


**Fig. 8** Transfer trajectories from EML1 to the Moon surface as function of the departure Delta-V for a maximum transfer time of 3 days

Figure 13 shows the results obtained for three different transfer times. The maximum ΔV at the departure and arrival corresponds to those points on the lunar surface located opposite to the direction of the EML1 point (in the far side of the Moon). The minimum ΔV at departure and arrival corresponds to the points located at the lunar equator and at some 75° of longitude. The ΔV contour lines converge to a region around 50° of longitude where the trajectories are splitted in direct and retrograde motion.

The differences in total ΔV for the different departure points are small (they do not exceed 20 m/s for transfer times lower than 3 days).

For mission safety considerations, we have investigated the evolution of the Moon-EML1 transfers for the contingency cases when the braking manoeuvre for inserting into an orbit around EML1 could not be performed. Figure 14 shows trajectories propagated 50 days after the arrival at the EML1 point without applying any manoeuvre at that point. For the shortest transfer times to EML1 (maximum arrival velocity), the subsequent trajectories are basically HEOs. The first perigee passage of those orbits is located at a distance of about half of the distance from

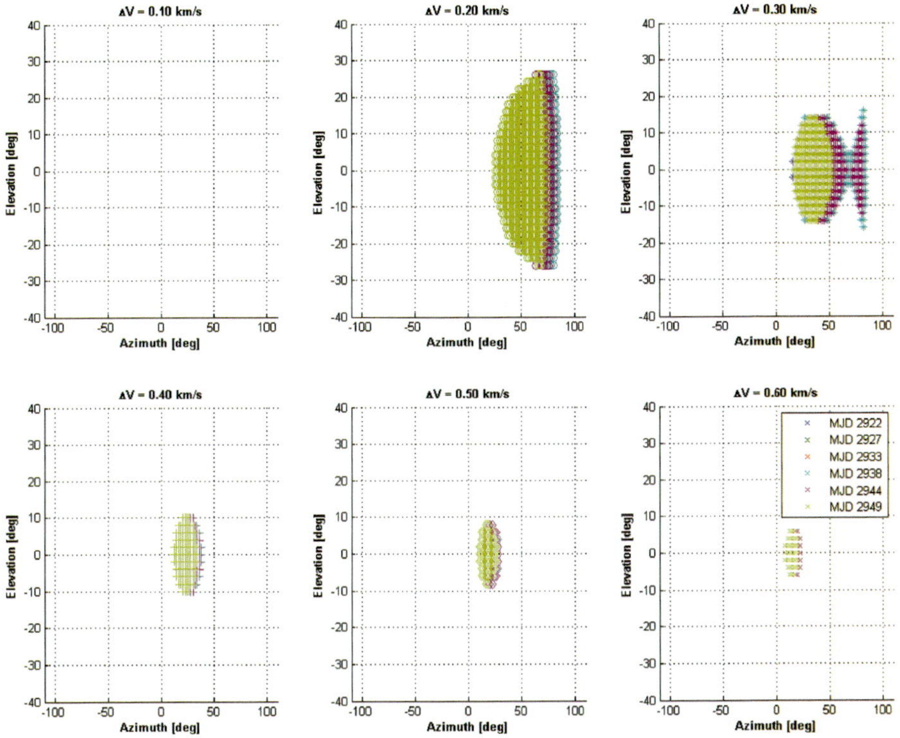


**Fig. 9** Direction of the impulsive departure Delta-V for a transfer from EML1 to the lunar surface for a maximum transfer time of 3 days

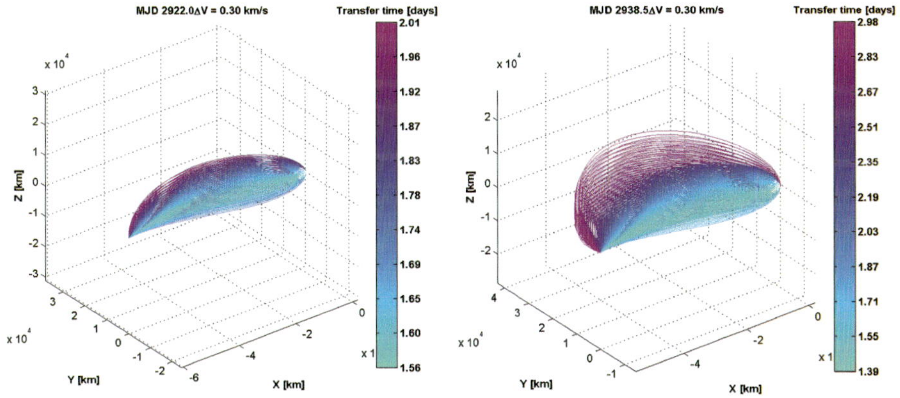


**Fig. 10** Transfer trajectories from EML1 to the Moon surface in the synodic frame as function of transfer time for Delta-V = 0.3 km/s (the Moon is at perigee in the *left* plot, at apogee in the *right*)

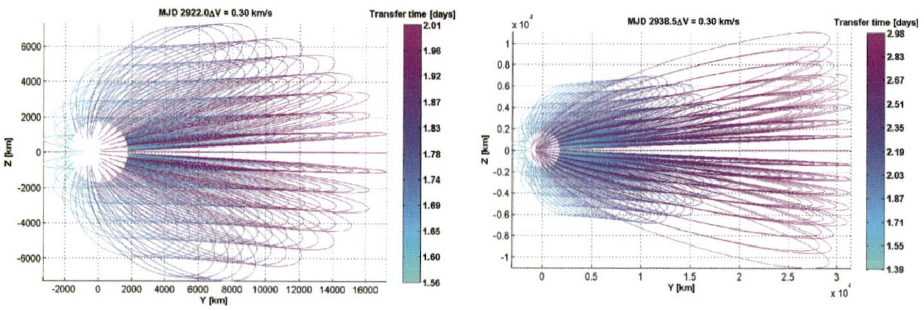


**Fig. 11** YZ-plane projections (synodic frame) of the trajectories from EML1 to the Moon

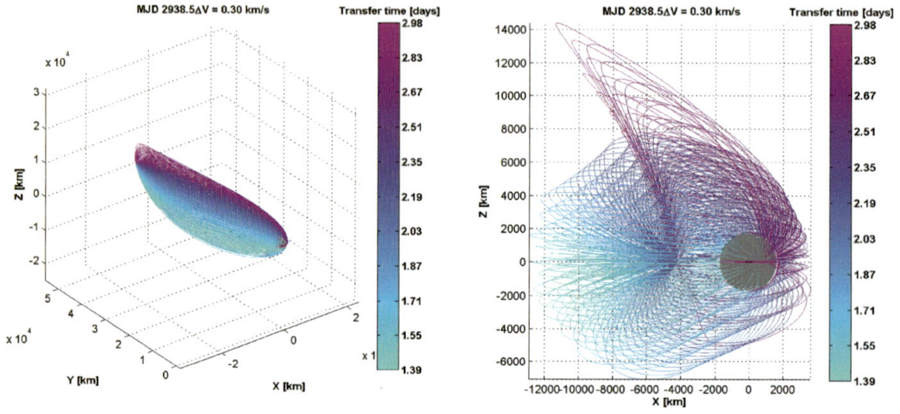


**Fig. 12** Transfer trajectories from the $L_1$ point to the Moon surface in the Moon rotating fixed reference in 3D (*left*) and XZ-projection (*right*)

the Earth to EML1. After the first perigee passage, the orbits either escape from the Earth following a heliocentric trajectory or describe a more eccentric HEO with perigee radius in the range of 10,000–70,000 km.

For the longest transfer times ($\sim$2.9 days), the evolution of the trajectories is more complex. Since the arrival velocity at EML1 is small, the Moon attracted again many of these trajectories (leading to lunar impact or to multiple swingbys and escape), while others were attracted by the Earth and kept in the cislunar region for longer times.

# 5 Summary

Table 1 summarizes the Delta-V requirements, the window opportunities and the mission duration for manned missions to the lunar surface with a staging node at either LLO or EML1. Although EML1 offers several advantages, in particular the continuous existence of transfer opportunities from EML1 to any point on the lunar

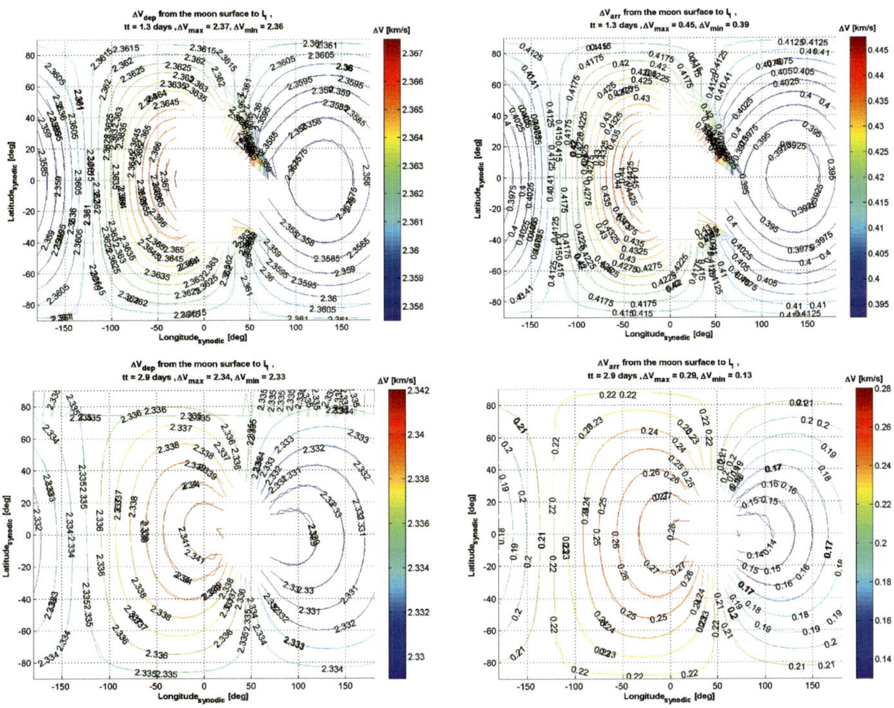


**Fig. 13** DV at the Moon surface (*left*) and EML1 point (*right*) for transfer times of 1.3 and 2.9 days as funcion of the location of the departure point at the lunar surface

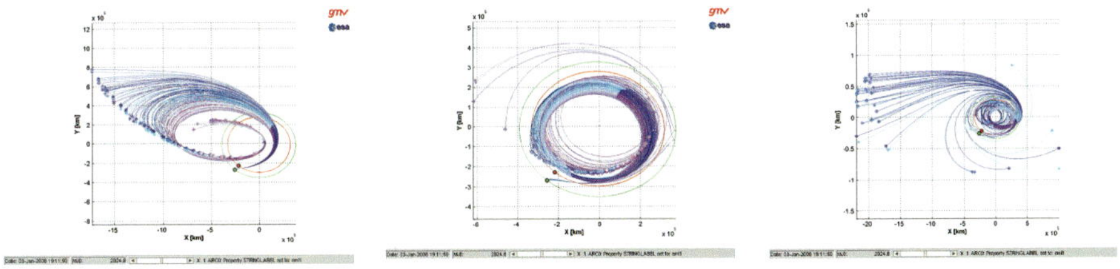

**Fig. 14** Trajectories from the Moon to EML1 propagated 50 days after the arrival at the EML1 point without applying any manoeuvre (the transfer times are 0.7, 1.7 and 2.9 days, respectively)

surface and to return to the Earth, the overall mission duration and associated Delta-V are higher than for LLO. The node at EML1 poses also some questions on the mission safety, since the trajectories to go from EML1 to the lunar surface are direct impact trajectories. Therefore, if the only objective is to locate a staging node in the vicinity of the Moon for supporting uniquely manned exploration missions to the lunar surface, the LLO alternative seems more advantageous.

For supporting longer Moon exploration architectures, including many cargo transfers and possibly also supporting missions to other destinations in the Solar

**Table 1** Summary of missions to the Moon through LLO and EML1

| Orbit / transfer | | LLO 100 km | EML1 |
|---|---|---|---|
| Transfer from LEO 200 km | Delta-V (km/s) | 4.0 (3.15 + 0.85) | [3.8–4.0] (3.1+[0.7–0.9]) |
| | Transfer duration (days) | ~ [4–5] | [3.6–4.1] |
| | Interval between transfer windows (days) | [8–10] (51° LEO); [11–14] (5° LEO) | [7–9] (51° LEO) [11–14] (5° LEO) |
| Transfer to/ from the lunar surface | Delta-V (km/s) | 1.71 | [2.54–2.64] ([0.2–0.3] + 2.34) |
| | Transfer duration | ~ 1 h | [1.4–3] days |
| | Interval between transfer windows | Depends on orbit incl. and lat. of landing site (from 1.5 h to 27 days) | Transfer possible at any time |
| Transfer back to Earth | Delta-V (km/s) | [0.8–0.9] (min. DV transfer window) Max 1.5 (anytime return) | ~ 0.7 |
| | Transfer duration | ~ [4–5] | [3.6–4.1] |
| | Interval between transfer windows | From 2 h to 14 days (min. DV transfer window) 2 h (anytime return) | Transfer possible at any time |

System, such as Mars or the asteroids, the trade-off would be different and the advantages of EML1 would probably be more decisive.

**Acknowledgments** The authors thank Dr. Ettore Perozzi for organizing the workshop on "Novel Spaceways for Scientific and Exploration Missions".

## References

1. Condon, J., Pearson, D., "The Role of Humans in Libration Point Missions with Specific Applications to an Earth-Moon Libration Point Gateway Station", Advances in the Astronautical Sciences: Astrodynamics 2001, Vol. 109, pp. 95–110.
2. Lo, M. W., Ross, S., "The Lunar L1 Gateway: Portal to the Stars and Beyond", AIAA Paper 2001–4768, August 2001.
3. Farquhar, R. W., "Lunar Communications with Libration-Point Satellites," *Journal of Spacecraft and Rockets*, Vol. 145, No. 24, 9 Dec. 1996, pp. 44–46.
4. Ranieri, C., et al. "Earth Moon Libration Point (L1) Gateway Station – Libration Point Transfer Vehicle Kickstage Disposal Options," *International Conference On Libration Point Orbits and Applications*, Parador d'Aiguablava, Girona, Spain, June 10–14, 2002
5. Prado, B. D. A., "Traveling Between the Lagrangian Points and Earth," *Acta Astronautica*, Vol. 39, No. 7, 1996, pp. 483–486
6. Parker, J. S., and Born, G. H., "Direct Lunar Halo Orbit Transfers", AAS 07-209, AAS/AIAA Spaceflight Mechanics Conference, Sedona, Arizona, January 2007.

7. Perozzi, E., Di Salvo, A.: Moon Harbor, the "Moon Base Gateway". In: Compagnone, F., Perozzi, E. (eds.), Proceedings of the Moon Base: a Challenge for Humanity. Donzelli Editore, Roma (2007)
8. Zazzera, F. B., et al., "Assessment of Mission Design Including Utilization of Libration Points and Weak Stability Boundaries.," Ariadna id:0./413, April 2004
9. Alessi, E. M., Gómez, G., and Masdemont, J. J., "LEO—Lissajous Transfers in the Earth-Moon System", IAC-08.C1.3.7, 59th International Astronautical Congress, Glasgow, Scotland, September 2008.
10. Farquhar, R. W., "The Control and Use of Libration-Point Satellites", Goddard Space Flight Center, Greenbelt, September 1970, NASA-TR=R-346.
11. Canalias, E., Gómez, G., Marconte, M., Masdemont, J. J., "Assessment of Mission Design Including Utilization of Libration Points and Weak Stability Boundaries", Department de Matematica Aplicada, UPC, Ariadna id: 03/4103.

# Manifolds and Radiation Protection

**Franco Rossitto, Vladislav M. Petrov, and Filippo Ongaro**

**Abstract** During the past 40 years humans have travelled beyond Earth's atmosphere, orbiting the planets for extended periods of time and landing on the Moon. Humans have survived this overwhelming challenge but to assure future exploration of space further expertise in the long term survival in space must be obtained. The International Space Station (ISS) provides this opportunity and allows space scientist to fine-tune their knowledge and prepare for even bolder human space missions. In this work we focus on the aspect of radiation, perhaps the most complex one from a physical and physiological perspective. Travel beyond the Earth's atmosphere and especially to Moon and Mars requires a precise consideration of the radiation environment as radiation exposure could be a show-stopper. At the moment scientists have not yet developed complete and reliable systems for radiation protection. Most likely an adequate level of protection will be reached through an integrated countermeasure system which could include: shields, monitoring of the environment, drugs to protect from damage, etc.

## 1 Introduction

During the past 40 years humans have travelled beyond Earth's atmosphere, orbiting the planets for extended periods of time and landing on the Moon. Humans have survived this overwhelming challenge but to assure future exploration of space further expertise in the long term survival in space must be obtained. The International Space Station (ISS) provides this opportunity and allows space scientist to fine-tune their knowledge and prepare for even bolder human space missions.

The environmental conditions found in space are not compatible with human life. The factors affecting human physiology are summarized below [1]:

F. Ongaro (✉)
ISMERIAN - Istituto di Medicina Rigenerativa e Anti-Aging
e-mail: filippo.ongaro@ismerian.com

E. Perozzi, S. Ferraz-Mello (eds.), *Space Manifold Dynamics*,
DOI 10.1007/978-1-4419-0348-8_8, 

1. Lack of atmosphere: Earth's atmosphere provides a combination of pressure, temperature and gas composition necessary for human survival.
2. Altered gravity environments: life is shaped by gravity and humans are exposed to the force of gravity throughout their entire existence. Travelling in space and especially from one planet to another will expose the body to different gravitational forces. During long term missions this affects especially the muscle-skeletal system inducing an important muscle and bone loss which might limit operational capabilities.
3. Radiation: Although the space environment is relatively devoid of matter it is very rich in energy. The atmosphere and the geomagnetic field protect the Earth from space radiation. But during a space mission crews are potentially exposed to electromagnetic waves, X-rays, protons, electrons, neutrons, alpha particles and heavy ions.
4. Thermal instability: the atmosphere stabilizes temperature on the Earth. In space radiation is the sole mode of energy transfer and thus heat depends directly on the distance from the Sun. Space crafts and crews are thus potentially exposed to extreme thermal stress. During EVA for example moving from sunlight to shade can provoke a temperature change of several hundreds degrees Celsius.
5. Space debris: micrometeorites and meteorites are a serious risk to space crews especially during an EVA. Moreover there is a large amount of orbital debris created by humans

In this work we focus on the aspect of radiation, perhaps the most complex one from a physical and physiological perspective. Travel beyond the Earth's atmosphere and especially to Moon and Mars require a precise consideration of the radiation environment as radiation exposure could be a show-stopper.

At the moment scientists have not yet developed complete and reliable systems for radiation protection. Most likely an adequate level of protection will be reached through an integrated countermeasure system which could include: shields, monitoring of the environment, drugs to protect from damage, etc.

## 2 Classification of Radiation

In the space environment there are 2 major types of radiation: ionizing and non-ionizing.

### *2.1 Ionizing:*

The have sufficient energy to knock material from atomic structures during a collision, which can release a cascade of electromagnetic or particle radiation. Astronauts report that with closed eyes it is frequent to see small flashes of light, evidence that the energy produced with nuclear collision is activating the visual receptors [2]. When ionizing radiation collides with cell nuclei a damage to the DNA can occur. This damage can either lead to cell death or to mutations able to initiate the development of a tumor. In this category we can recognize the following radiations:

- √ *Solar Particle Event:* natural in space or on planetary surface. This form of radiation is largely unpredictable and is produced by dynamic solar events with emission of electrons and protons.
- √ *Trapped Particles:* natural, localized outside the vehicle, constant with moderate intensity, within Earth's magnetic field (Van Allen's belts).
- √ *Galactic Cosmic Rays:* natural, ubiquitous and penetrating throughout universe, constant with low intensity.

### 2.2 *Non-Ionizing:*

This form of radiation does not have sufficient energy to directly damage genetic material but it is still very dangerous. Solar ultraviolet electromagnetic radiation, unattenuated by the atmosphere, can cause sever burns after only a few seconds. In this category we find the following radiations:

- √ *Technology Sources:* artificial, inside the vehicle, constant, low levels, derived from human made equipment or devices.
- √ *Solar UV***:** natural, outside vehicle and inside through windows, constant, high levels, produced by solar corona.

## 3 Radiation in Space Missions

Astronauts are professionally exposed to radiations and the ALARA principle (As Low As Reasonably Achievable) is applied to stay well below the dose limits. However there is a likelihood of high exposure especially in case of missions to the Moon or Mars. This exposure can increase the risk of cancer up to 3% and in acute conditions can lead to skin-reddening, nausea, vomiting and dehydration [3].

Damage can occur long after a mission is over in a manner proportional to the dose received [4]. On ISS career limits depend on gender and age at the beginning of exposure (Sv):

$$2.0 + 0.0075 \times (\text{age}-30) \text{ for males}$$
$$2.0 + 0.075 \times (\text{age}-38) \text{ for females}$$

The threat of radiation is different in different space environments inside or outside the Earth's magnetic field. A 2 and ½ years trip to Mars would expose an astronaut to nearly the maximum allowed life time limit of radiation. The Moon, with no atmosphere (and no magnetic field) is even more dangerous than the surface of Mars.

Mission planners at the Apollo time knew astronauts would be at great risk of a strong solar flare would occur during a mission. The short duration of the stay was paramount. The research on space radiation and its biological effects takes benefit from the ISS supporting activities in this field.

Particle radiation (GCR and SPE) goes through the human body and can tear apart stands of DNA. Damaged cells can lose their ability to perform normally and to repair themselves.

Electromagnetic radiation poses less problems and the effect of secondary neutrons is still unknown. For high energy particles emitted from the Sun during solar flares (SPE) the Earth's magnetic field is an effective shield. Radiation sickness from solar flare could kill an unprotected astronaut for example during an EVA.

Galactic Cosmic Rays (GCR) pose a greater long term risk of cancer, cataracts and other illnesses.

It is known that the Apollo astronauts absorbed highdoses (3 times the ISS) and 12 over 18 developed cataracts 4–10 years after their missions.

Exposure travelling through space is therefore twice compared to being on the lunar or Martian surface. An astronaut in a 6 month journey to Mars would be exposed to about 0.3 Sv or 0.6 Sv on a round trip. 18 months on the surface would add 0.4 Sv for a total of 1 Sv. Limits set by Space Agencies vary with age and gender but are in the range from 1 to 3 Sv.

## 4 Monitoring Systems on ISS

The are several systems on the ISS to measure and monitor radiation exposure. NASA's Phantom Torso is composed by 35 1 inch layers each carrying passive dosimeters. 2 dosimetry teloscopes are located in the US lab where streaming ions are measured. The DARA dosimetric mapping system in the US lab (ISS) measures the neutron dose by thermoluminescence devices.

NASDA's Bonner Ball Neutron Detector composed by 6 spheres filled by Helium 3 and covered by polyethylene. ALTEA, developed by ASI (Italian Space Agency), has been used to track HZE particles crossing the head of the astronaut and giving rise to retinal flashes. This device was previously on board MIR space station.

## 5 Protecting Space Crews

The future of space missions will depend to a high extent on the capability to protect crews from radiation exposure. Many strategies can be adopted, some concerning engineering and space craft design side and others medical applications.

For example, it is possible to use protective materials such as Polyethylene which absorbs 20% more cosmic rays than aluminium. Reinforced polyethylene developed at MSFC is 10 times stronger than aluminium and still lighter. Even if the full spaceship would not be built from plastic, this material could still be used to shield key areas like the crew quarters.

Fuel containing hydrogen is also a good shield. Wrapping the crew quarter with fuel tanks could be a valid alternative. Active shields can also be developed such as electrostatic (trapping) and/or magnetic (deflecting).

Other strategies would involve better medical monitoring (with direct measurement of DNA damage in astronauts for example) and intervention (drugs and nutrients specifically designed to help protect the genome from damaging radiations.

Specifically for the exploration of the Moon more data need to be collected. Apollo astronauts made some measurements of radiation on the Moon but the results are less complete than for Mars. It is certain though that the Moon would is worse than Mars and the ISS in terms of radiation exposure. The major concern is with Solar storms (SPE) for which the warning time can be as little as 18 h. Moreover the EVA suits have to be redesigned for long stay on the moon.

In conclusion: for the exploration missions and to ensure maximum protection. Go to moon first and practice safety.

## References


1. Waligora JM, editor. The physiological basis for spacecraft environmental limits. (NASA RP-1045). Washington DC: NASA Scientific and Technical Information Branch, 1979.
2. Lundquist CA, editor. Skylab's astronomy and space science. (NASA) SP-404). Washington DC: US Government Printing Office, 1979.
3. Boice JD jr, Blettner M, Auvinen A. Epidemiologic studies of pilots and aircrew. Health Phys 2000;79:576-584
4. Friedberg W, Copeland K, Duke FE et al. Radiation exposure during air travel: guidance provided by the Federal Aviation Administration for air carrier crews. Health Phys 2000;79:591–595

# Three-Body Invariant Manifold Transition with Electric Propulsion

**Pierpaolo Pergola, Koen Geurts, Cosmo Casaregola, and Mariano Andrenucci**

**Abstract** The advantageous combination of dynamical systems theory of three-body models with Electric Propulsion to design novel spacecraft mission in multi-body regimes has been investigated. Combining the advantages of Electric Propulsion with respect to propellant requirements and low-energy ballistic trajectories existing in the three-body model, multi-body planetary tours can be designed. The employment of power constrained Electric Propulsion at the solar distance of Uranus is enabled by the use of Radioisotope Thermoelectric Generators. This provides continuous availability of sufficient electrical power.

Not only a planetary tour of the Uranian system orbiting consecutively Oberon, Titania, Umbriel, Ariel and Miranda is designed, but also the required interplanetary trajectory transporting the spacecraft from the Earth to Uranus. Both the interplanetary trajectory and the planetary tour are computed in different three-body environments, where the start of the interplanetary phase is assisted by a high energy launch to limit the transfer time.

It is demonstrated that a feasible mission can be designed both in terms of transfer time and propellant mass requirement, with a scientifically interesting character. The spacecraft is unstable captured by the five moons for different periods of time, with a stable Uranian orbit as a final state.

## 1 Introduction

Previous studies have successfully employed coupled three-body environments to design multi-body tours using the low energy ballistic trajectories existing within this model. Significant savings in propellant mass have been obtained, together with a high scientific outcome, due to the multi-body character. Absent in these studies, however, is the interplanetary transfer preceding insertion into a planetary multi-body system. The current study not only presents a tour of the Uranian system, but also the required interplanetary trajectory.

---

P. Pergola (✉)
Alta S.p.A., Pisa, Italy
e-mail: p.pergola@alta-space.com

E. Perozzi, S. Ferraz-Mello (eds.), *Space Manifold Dynamics*,
DOI 10.1007/978-1-4419-0348-8_9, 

The tour within the Uranian system is designed in order to subsequently orbit Oberon, Titania, Umbriel, Ariel and Miranda. The spacecraft characteristics assumed are based upon previous studies [7] and correspond with technology currently available. Moreover, in the current study it has been assured that no impact trajectories exist by imposing a minimum orbital altitude above the surface of the moons.

The interplanetary transfer is assisted by a high energy launch, providing the spacecraft with a high departure velocity to ensure acceptable transfer times. Following launch the EP is used to decelerate the spacecraft in addition to trajectory maneuvering [4, 3, 5, 6].

In previous studies technical aspects such as the availability of sufficient electrical power at the solar distance of Uranus and the design of a small spacecraft maintaining the stringent total mass constraints have been investigated and discussed. These studies demonstrated the feasibility of designing relatively small spacecraft incorporating an RTG power source, Electric Propulsion and all required subsystem [3]. Based on these findings the technical aspect of the mission discussed in this study is omitted, where the main focus will be on the trajectory design.

## 2 Mathematical Fundamentals

The mathematical principles of the three-body model, the transformations and the optimization schemes are discussed, where these are applied to obtain the solutions discussed in the next section.

### 2.1 Circular Restricted Three-Body Model

The dynamical environment is governed by the circular restricted three-body model equations as given by Szebehely[10]:

$$\begin{aligned} \ddot{x} - 2\dot{y} &= \Omega_x \\ \ddot{y} + 2\dot{x} &= \Omega_y \\ \ddot{z} &= \Omega_z \end{aligned} \tag{1}$$

where the subscripts denote the partial derivatives and $\Omega$ is the function:

$$\Omega(x,y,z) = \frac{1}{2}(x^2 + y^2) + \frac{1-\mu}{r_1} + \frac{\mu}{r_2} + \frac{1}{2}\mu(1-\mu) \tag{2}$$

The $x$ and $y$ coordinates and their derivatives are computed with respect to a non dimensional, barycentric, synodic reference frame formed by the two main attractors and rotates with their relative angular velocity. The *mass parameter* ($\mu$) of the system is the only parameter necessary for the characterization of the specific three-body system. In this study this is used to differ between both the interplanetary phase as well as the different planet - moon systems. The distances between the two primaries, $m_1$ and $m_2$, and the spacecraft are given by $r_1$ and $r_2$, respectively:

$$\begin{aligned} r_1^2 &= (x+\mu)^2 + y^2 + z^2 \\ r_2^2 &= (x-1+\mu)^2 + y^2 + z^2 \end{aligned} \tag{3}$$

The system possesses one first integral of motion, the *Jacobi integral*, given as:

$$C = 2\Omega(x,y,z) - (\dot{x}^2 + \dot{y}^2 + \dot{z}^2) \tag{4}$$

where the Jacobi integral is related to the spacecraft energy by:

$$E = -C/2 \tag{5}$$

The system furthermore contains five equilibrium points that can be found by equating the partial derivatives of (1) to zero, which has three unstable collinear solutions ($L_{1,2,3}$) and two stable solutions ($L_{4,5}$) which form two equilateral triangles with the two primaries. The equilibrium points are critical points of the effective potential, $\Omega$, and the level surface of the Jacobi constant represents the *energy surface*, which is, in a planar environment, an invariant 3-dimensional manifold in a 4-dimensional phase space:

$$M(\mu,C) = \{(x,y,\dot{x},\dot{y}) \mid C(x,y,\dot{x},\dot{y}) = constant\} \tag{6}$$

where its projection onto position space gives the *Hill region* which confines the spacecraft's motion for the possessed energy:

$$M(\mu,C) = \{(x,y) \mid \Omega(x,y) \geq C/2\} \tag{7}$$

The spacecraft is confined to Hill's region as the kinetic energy would otherwise assume a negative value. The boundaries of this surface are the zero velocity curves, which cannot be traversed without a modification of the spacecraft's energy. For low energy values the spacecraft is confined in circular realms around either the primary or secondary body. With increasing energy the circular regions open up enabling transition of the spacecraft between the two primaries. A minimum energy value associated with $L_1$ permits transition between the two primaries, whereas a further increase in energy opens up the realm containing the secondary body where an energy of $E > E(L_2)$ permits the presence of the spacecraft in the outer region.

This principle is exploited to design the transition between two different Uranus - moon systems. Two different three-body systems are coupled, where the energy of the spacecraft with respect to the inner system must be high enough to open the neck region around $L_2$. This permits transition between the two systems provided that the dynamics are correctly matched.

In order to maintain the lowest energy levels possible, only the 1-dimensional manifolds associated with the libration points are evaluated and not the manifolds associated with periodic orbits around the libration points. These 1-dimensional manifolds are given by the stable and unstable eigenvalues of the coefficient matrix of the linearized dynamics as in (8). The manifold associated with the stable eigenvalues computes a ballistic trajectory towards the libration point, whereas the unstable eigenvalues compute ballistic trajectories away from the libration points.

The linearized equations of motions with the Jacobian matrix $A$ can be written as:

$$\dot{\mathbf{x}} = A \cdot \mathbf{x} \tag{8}$$

If $\lambda_s$ and $\lambda_u$ represent the stable and the unstable eigenvalues, respectively, ($\lambda_s < 0$ and $\lambda_u = -\lambda_s$) and $v_s$ and $v_u$ are the associated eigenvectors, the computation of the manifolds associated with the point in which the linearization has been done, only requires the propagation of a small perturbation in the direction of $v_s$ and $v_u$, as discussed by Zazzera [11]. If $x_0$ represents the state of the libration point of interest and $d$ the small perturbation, the manifolds associated with $L_1$ and $L_2$ can be obtained by propagating:

$$\begin{aligned} x_0^s &= x_0 \pm d\, v_s \\ x_0^u &= x_0 \pm d\, v_u \end{aligned} \tag{9}$$

The first of (9) must be propagated backward and the second forward, where there are two legs for each manifold.

## 2.2 Reference System Transformation

Consideration of both the interplanetary and planetary phase of the mission requires an appropriate reference system transformation. The planetary tour is computed subsequently considering planar models formed by Uranus and one of its moons, where the initial state must be computed with respect to the outermost moon Oberon.

The Uranian tour initiates on the exterior invariant manifold leading toward the second libration point in the Uranus - Oberon system. This initial state in the Uranus - Oberon system is transformed to the Sun - Uranus system and forms the boundary condition for the interplanetary trajectory optimization. The transformation provides for a decoupling of the two systems with the advantage that both mission phases can be computed and optimized independently.

The high inclination of Uranus' spin axis and its moons having approximately equatorial orbits, results in an offset of the orbital plane of the moons with respect to the fundamental Sun - Uranus plane [9]. This is schematically shown in Fig. 1. A generalization is applied where Uranus' axis of rotation is assumed to coincide with the orbital plane of Uranus. Moreover, the moons are all considered to have circular, equatorial orbits, thus their orbital planes being exactly perpendicular to the Sun - Uranus plane. The system of reference transformation takes into account both this rotation of the principal axes, in addition to the velocity conversion. Due

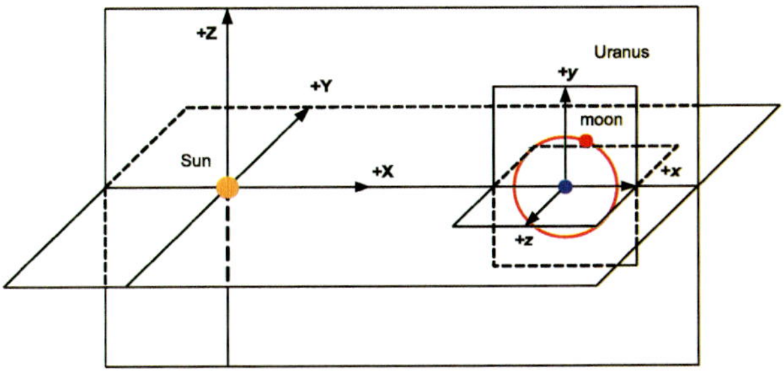


**Fig. 1** Schematic overview of the two principal reference systems

to this rotation the initial conditions of the planetary tour have components of the position and velocity only in the Sun Uranus $x$-$z$ plane.

## *2.3 Thrust Acceleration*

In order to compute the thrust acceleration the components of the thrust vector along each fundamental axis have been included in the equations of motion. The thrust is approximated by the available power of 1 kW and specific impulse of 3200 s, using the relationship:

$$T = \frac{2\eta_T P}{g I_{sp}} \tag{10}$$

where a constant thrust efficiency $\eta_T = 0.5$ is assumed. The thrust thus has a constant modulus where the individual components orientating the thrust vector are computed by the optimization processes. Moreover, an equation for the propellant mass consumption is required:

$$\begin{aligned} \ddot{x} - 2\dot{y} &= \Omega_x + a_x \\ \ddot{y} + 2\dot{x} &= \Omega_y + a_y \\ \ddot{z} &= \Omega_z + a_z \\ \dot{m} &= -|\mathbf{T}|/(g I_{sp}) \end{aligned} \tag{11}$$

where $\mathbf{a} = \mathbf{T}/m$.

# 3 Interplanetary Phase

The initial conditions for the interplanetary phase are derived from the position and velocity that correspond with a 1 AU circular orbit around the Sun. The position of departure on the circular orbit, together with the excess energy, has a strong effect on the shape of the trajectory. Erroneous selection of initial position or excess energy might render the transfer impossible or excessively long in duration. A previous study [4] identified the angular region leading to minimized transfer times, moreover, it showed a low sensitivity to excess energy magnitude for the range investigated. This angle has been adopted in the current study and equals 230° degrees from the $x$-axis in anti-clockwise direction.

Following launch the EP must modify the velocity in all three dimensions in order to adhere the imposed final conditions: the initial state is in the $x$-$y$ plane, whereas the final state which corresponds with the start of the Uranian tour is in the $x$-$z$ plane. The optimization algorithm therefore computed both in- and out of plane thrust components, while minimizing the required time to arrive at the final state.

## *3.1 Trajectory Optimization*

The Earth to Uranus transfer has been computed using two different, sequential, optimization techniques. Initially a gradient method was applied to generate an accurate initial guess for a subsequently used forward shooting method, both based on the calculus of variations. The latter method requires a precise initial guess of the Lagrange multipliers at $t = t_0$, which is obtained by the former, more robust, optimization scheme as discussed by Bryson[1].

Both techniques are based on the calculus of variations where a control vector $\mathbf{u}(t)$ is obtained that minimizes the functional $J$. To maintain feasible mission durations the transfer time is subjected to minimization, where a limit on the final mass after the interplanetary transfer is imposed indirectly by ensuring departure with an initial mass that results in $m_f \geq 500\,\text{kg}$, this to ensure sufficient spacecraft and propellant mass for the planetary tour. This mass constraint is respected by applying a numerical scheme that ensured an initial mass sufficient to arrive at the final state with the required mass.

## *3.2 Trajectory Conjunction*

The interplanetary trajectory is connected with the Uranian tour by the previously discussed transformation of reference system, however, in addition a conjunction phase is required to dissipate the spacecraft's energy in order to satisfy the tours initial conditions. The conjunction phase resembles a classical low-thrust planar orbit transfer, where a spiraling motion is performed gradually closing in on Uranus. The duration of the conjunction phase was arbitrarily chosen as the time required to reach a distance from Uranus equal to the sphere of influence (SOI).

The thrust vector orientation during this phase is opposing the velocity vector, thus decreasing the velocity magnitude and consequently approaching Uranus. The conjunction phase is computed by a backward integration in the Sun - Uranus system starting from the tour's initial conditions until the SOI radius. This state is taken as the boundary condition for the interplanetary optimization code.

# 4 Uranian Tour

The second part of the study investigated the possibilities to continue the interplanetary transfer with a tour of the Uranian system, advantageously exploiting the three-body low energy ballistic trajectories. This results in a tour with a scientifically interesting character as many different environments are experienced and technically interesting as the propellant requirements are limited.

## *4.1 Approach*

The state obtained by the spacecraft after the execution of the interplanetary transfer and the subsequent conjunction phase is the intersection between the exterior, stable manifold of Oberon and the first Poincare section. This stable manifold will transport the spacecraft ballistically to the second libration point, $L_2$, in the Uranus - Oberon system. Moreover, more generally, the manifolds associated with the libration points of each moon are computed in the relative synodic frame and subsequently translated and scaled to the Uranus - Oberon system, which is chosen as the main system of reference for the tour construction as it is the outermost moon considered.

The transformation between the two systems takes into account the initial phase of the moons and the associated non-autonomous phase difference during the entire transfer. A manifold of a generic moon, in this system, appears as a trajectory that flows from a radius greater than the radius of the circular orbit of the moon, wraps around the moon's orbit and finally arrives inside the moon's circular orbit. In the main reference frame the manifolds of Oberon are time independent, whereas the manifolds associated with the other moons are time dependent (periodic).

It is worth noting that the manifold used for the construction of the capture arc of each moon is the stable manifold associated with $L_2$, this as it is the ballistic trajectory that leads the spacecraft towards the moon from the outer realm. Its computation requires a propagation of the initial conditions (9) for a time span that must begin at the same final time as the powered phase of the previous step. The propagation is performed backward for a time span that identifies the time duration for which the spacecraft lies on the stable manifold. The duration of this time span ($t_{man}$) and the initial position of the relative moon ($\theta$) are terms of the control vector. Furthermore, the exit time from the previously considered unstable manifold of $L_1$, ($t_0$), is also considered a term of the control vector.

As opposed to the interplanetary trajectory, optimization with respect to the required transfer time is not applied anymore. Time optimization results in a continuous, maximized thrust modulus, whereas this is not necessarily the case when minimizing the required propellant mass as applied for the planetary tour optimization scheme. An appropriate thrust law, based on ($\alpha$,$\tau$) which are the thrust angle and modulus, respectively, is required in order to establish the connection between the final conditions of the propulsion phase and the insertion conditions on the manifold

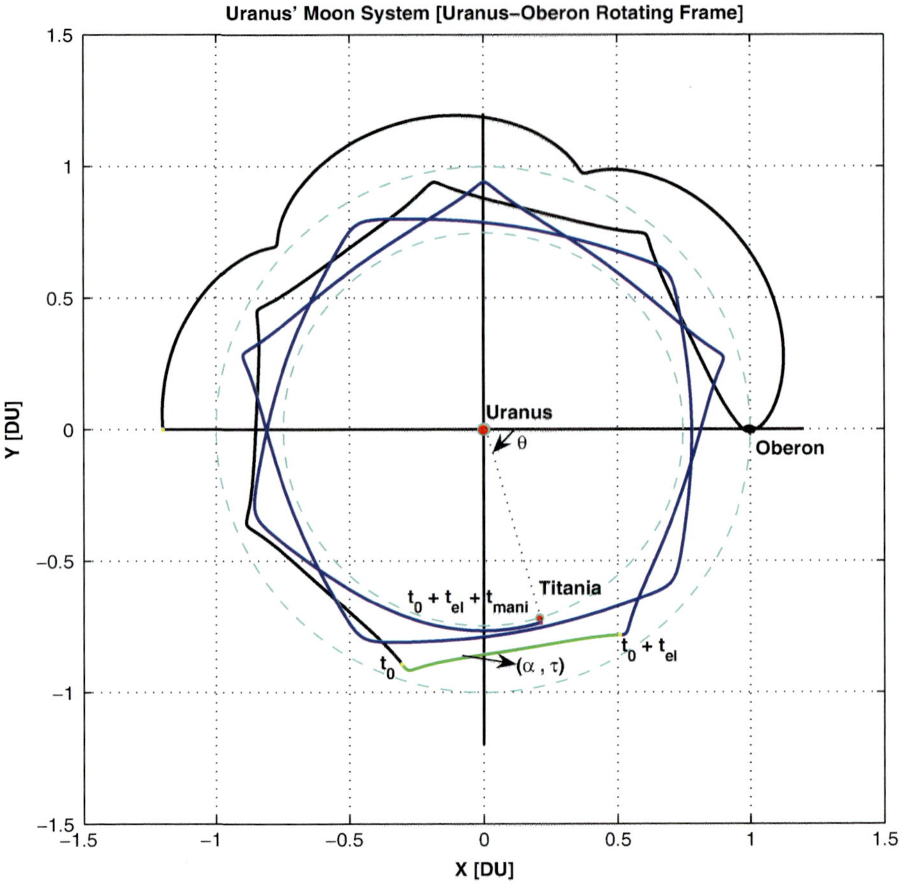


**Fig. 2** Optimization parameters

of the target moon. The thrust must be considered for a time span to be determined, being ($t_{EP}$), where these parameters are determined by the optimization scheme. The definition of the control vector elements for the first passage are shown in Fig. 2.

The complete control vector (**u**) used for each passage of the tour is:

$$\mathbf{u} = \{t_0, t_{EP}, t_{man}, \alpha, \tau, \theta\} \tag{12}$$

The control vector elements are determined by an optimization process that computes the passage with minimum propellant mass, subjected to the constraint that the final state of the propulsion phase must match with the initial state of the $L_2$ stable manifold of the target moon. It must be noted that the stable manifold associated with the second libration point of the specific moon, when propagated for a ballistic time greater than $t_{man}$, performs various closed orbits around that moon after which it passes onto the unstable manifold of $L_1$ of the same moon. This transition is called a *heteroclinic connection* [8] and is used to obtain the starting conditions for the next passage. In fact, $t_0$ is the exit time from the unstable manifold associated with the first libration point of the previous moon considered.

## 4.2 Tour Optimization

The problem is stated as a constrained minimization problem with equality constraints on the final state function of the control vector and with inequality constrains on the elements of the control vector, which has an upper and a lower bound ($u_{ub}$, $u_{lb}$), these identify the feasibility envelope ($U$) for $\mathbf{u}$.

$$\begin{aligned} &\min_{U} f(\mathbf{u}) \textit{ subjected to:} \\ &c_{eq}(u) = 0 \\ &u_{lb} \leq u \leq u_{ub} \end{aligned} \tag{13}$$

This is a nonlinear programming problem with only active constraints [2]. The functional to be minimized, $f(\mathbf{u})$, is the required propellant mass during the propulsion phase, which is a nonlinear objective function with multiple nonlinear constraints.

A sequential quadratic programming technique has been implemented to find the optimal solution. This technique converts the objective function in a quadratic form and linearizes the constraints. Moreover, at each iteration an approximation of the Hessian of the Lagrangian is made using a Quasi-Newton updating method. This type of optimization process is strongly dependent on the quality of the initial guess and possibly results in a high computational load due to a poor initial guess or when approaching the feasibility region boundary.

Using this method the thrust law $(\alpha, \tau)$ is included in the control vector by a time discretization of the propulsion phase. It has been divided into $N$-mesh points and at each point the thrust modulus and angle have been considered as elements of the control vector. So the total dimension of $\mathbf{u}$ equals: $2 \times N + 4$. The thrust law between two consecutive mesh points has been linearly interpolated.

Due to the extreme sensibility of the three-body system to the initial conditions this kind of approach is not sufficient to assure the passage between two manifolds. In fact, the chaotic dynamics of the model lead to completely different solutions even for very similar initial conditions.

In order to improve the precision of the conjunction points a further optimization process has been implemented starting from the solution of the nonlinear programming problem. In this second step the function to be minimized is only the distance in the phase space between the end point of the propelled phase and the initial condition required for the insertion onto the stable $L_2$ manifold. A simplex algorithm has been used taking the output of the previous step as initial guess. This approach assures a local solution that requires approximately the same propellant mass as given by the minimization process.

The value of the small perturbation introduced in (9) has been subjected to a numeric iteration in order to obtain a value that corresponded with a minimum altitude not less than 50 km above the surface.

## 5 Results

An overview of the input characteristics of the spacecraft and trajectory are given in Table 1:

**Table 1** Transfer inputs

| Input | Unit | Value |
|---|---|---|
| Power | W | 1000 |
| Specific impulse | s | 3200 |
| Thrust efficiency | – | 0.5 |
| C3 energy | $km^2/s^2$ | 130 |

The excess velocity corresponding with the excess energy has been summed to the Earth's velocity on a 1 AU circular orbit in the Sun - Uranus system. The location of departure on the circular Sun orbit was selected as 230° degrees, as detailed in the previous section.

### *5.1 Interplanetary Trajectory*

Figure 3 presents the interplanetary transfer shown in the Sun - Uranus synodic system. The visualization in the *x*-*z* plane shows a large excursion below the *x*-*y* plane, however, it must be noted that the scaling along the *z*-axis is two orders of magnitude smaller than along the *x*-axis.

Figure 4 shows the instantaneous values for the radius as measured from the Sun, the velocity with respect to the synodic reference system and the mass decrease during the transfer. Results are shown both for the solution obtained by the gradient method as well as the forward shooting method, represented by the blue and red lines, respectively. The continuation represented by the black line is the conjunction phase, which is shown only for the more precise forward shooting solution. The spiraling motion can be observed from the subfigure showing the spacecraft velocity, where the black line oscillates. This oscillation is also present in the Sun radius, however, due to the scale this is less pronounced. The propellant mass consumption equates to a linear decrease in spacecraft mass as no coasting phases are considered by the optimization scheme.

The conjunction phase is shown in more detail in Fig. 5 where the final state of the forward shooting optimization is continued by the conjunction phase shown in black. During the conjunction phase the spacecraft energy gradually decreases leading to a closure of the Hill region. In addition, the spacecraft is approaching Uranus

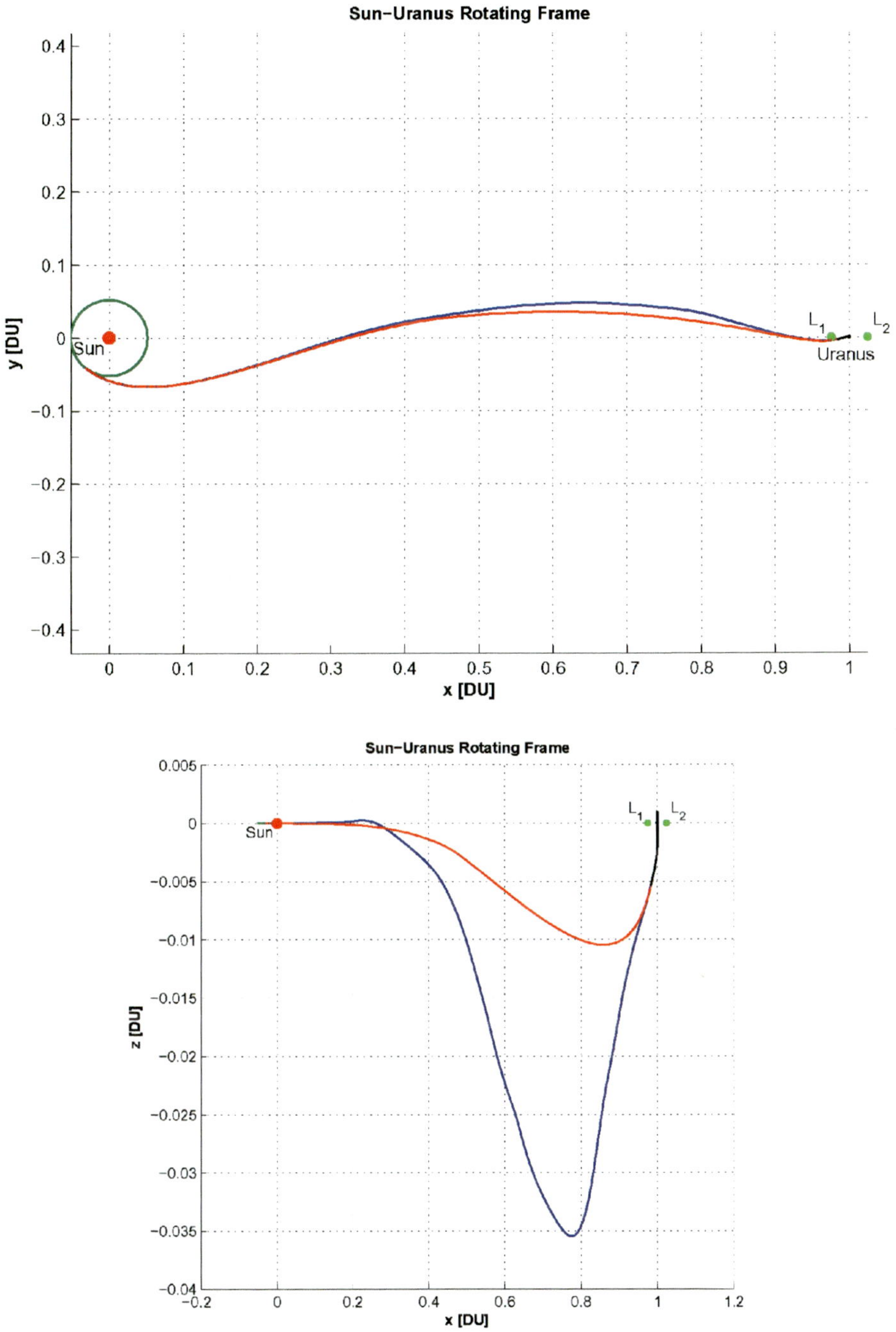


**Fig. 3** Overview of the interplanetary trajectory. **a** xy-plane. **b** xz-plane

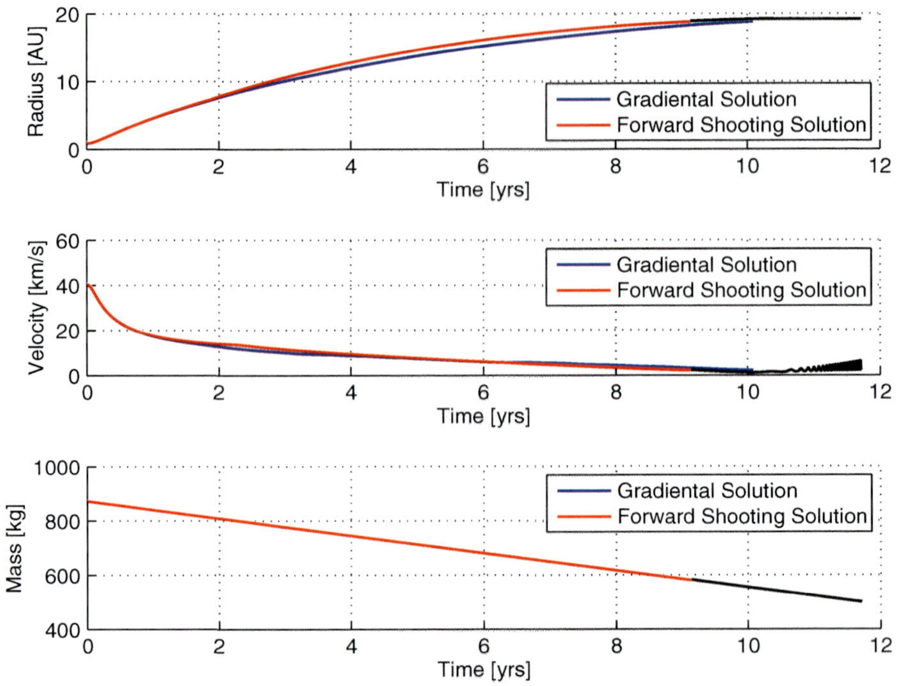


**Fig. 4** Instantaneous radius, velocity and mass

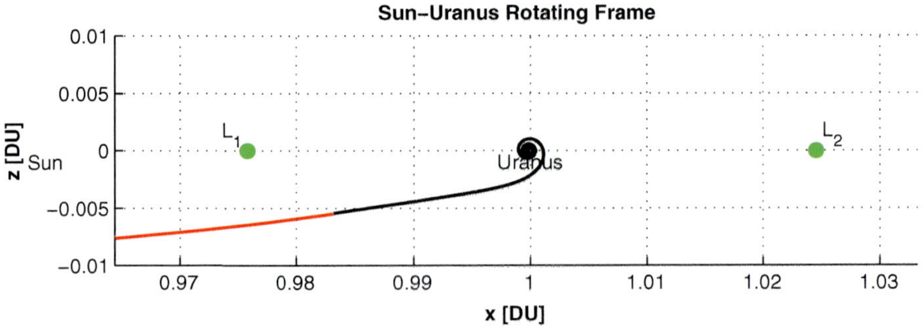


**Fig. 5** Detail of the conjunction phase

aligning the position and velocity to the requirements imposed by the planetary tour, as discussed in the next section. The duration equals approximately 2.56 years with a mass consumption of 83 kg. This in addition to the 9.14 years required to reach the SOI of Uranus with a mass consumption of 292.7 kg.

Figure 6 shows the in- and out-of plane thrust angles computed by the forward shooting optimization scheme. It is seen that after approximately 2.35 years the thrust angles demonstrate a rapid variation of the thrust direction. The $\alpha$-angle represents the in-plane thrust angle measured positive in anti-clockwise direction from

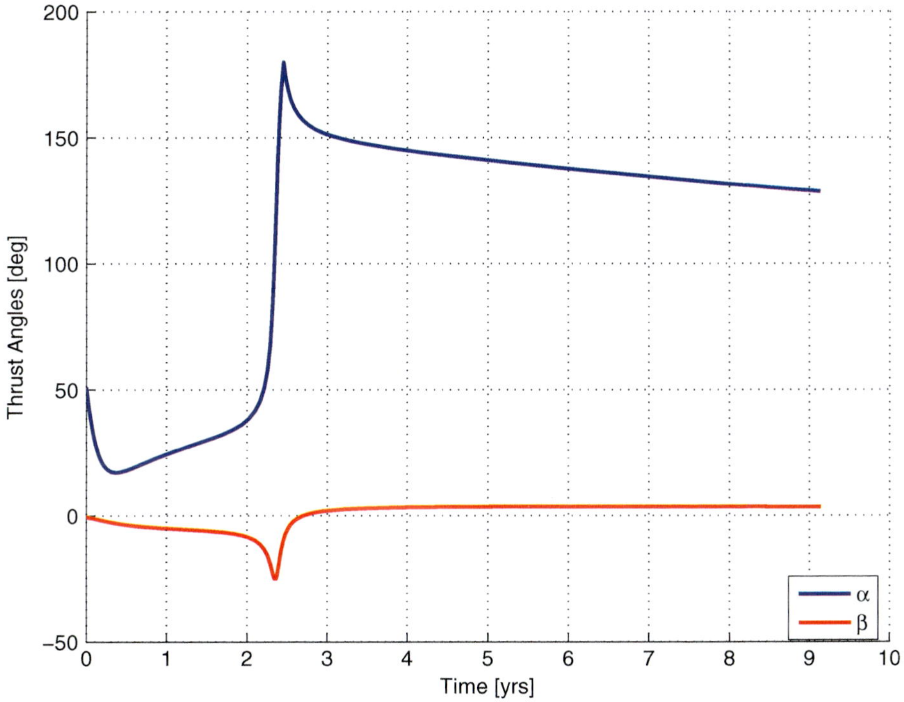


**Fig. 6** Time variation of the thrust angles by the forward shooting method

the $x$-axis, whereas the $\beta$-angle represents the out-of-plane thrust angle measured positive along the positive $z$-axis. Before the variation in direction the $\alpha$-angle comprises an accelerative component in the positive $x$-direction, where this changes into a decelerative behavior afterward. The $\beta$-angle demonstrates only a slight out-of-plane excursion, which is explained by the fact that the velocity component along $z$ at the final state is relatively small and the total distance of 19.2 AU is available to achieve this.

The general output of the interplanetary phase are summerized in Table 2:

**Table 2** Transfer outputs

| Output | Unit | Value |
|---|---|---|
| Interplanetary transfer time | yrs | 11.7 |
| Initial mass | kg | 875 |
| Propellant mass | kg | 375 |
| Fuel mass fraction | – | 0.417 |
| Transfer $\Delta$ V | $km^2/s^2$ | 17.57 |

**Table 3** Tour outputs

| Output | Unit | Value |
|---|---|---|
| Tour transfer time | days | 957.6 |
| Tour mass fraction | – | 0.070 |
| Propellant mass | kg | 35 |
| Tour $\Delta$ V | $km^2/s^2$ | 2.26 |

After some numerical iterations the initial spacecraft mass at launch has been determined at 875 kg, this, coupled with the assigned excess energy of $C_3 = 130\,km^2/s^2$, corresponds with the Atlas launcher performance when equipped with the Star 48 V upper stage.

## 5.2 Uranian Tour

The computation of the different legs of the planetary tour commenced with the determination of the optimized thrusting law as described in the previous section. For all cases, that is, all manifold transitions, the initial guess constituted an angle of approximately 180° with respect to the velocity vector. This corresponds with an anti-tangential thrust dissipating the spacecraft's energy, justified by the fact that all transitions performed corresponded with a decrease in orbital altitude with respect to Uranus. The starting point of the tour, the position on the exterior $L_2$ Uranus - Oberon manifold, has been arbitrarily fixed on the intersection with the $x$-axis, where it is considered that $t_0 = 0$. A fixed number of mesh points, $N = 10$, has been arbitrarily chosen for normal transfers where this is doubled for long propulsion phases to limit the computational time.

A sequential quadratic programming scheme has been applied to compute the solution, corresponding with convergence of the relative error of the equality constraints. In order to constrain the computational time, a tolerance of 5% has been imposed on the phase-distance of the conjunction states.

The main system of reference to which all other, coupled systems are scaled is the Uranus - Oberon system. Figure 7 shows the entire tour with respect to this reference frame, together with a transformation to the inertial reference frame.

In the figure the propulsion phases are represented by green lines and the ballistic arcs by blue lines. Maintaining this convention the position, velocity and mass are shown in Fig. 8. It is noted that the velocities are with respect to the principal reference frame where the velocity of Oberon equals zero. The passage of a moon is indicated by the rapid oscillations in the velocity and position plots.

The transition from the Oberon to the Titania system requires only a relatively short propulsion phase of approximately 56 days due to the close proximity of the two moons and their similar physical conditions. This with respect to the transition to Umbriel and Miranda, requiring approximately 150 and 128 days respectively. This is explained by the large physical and radial difference with respect to the preceding system.

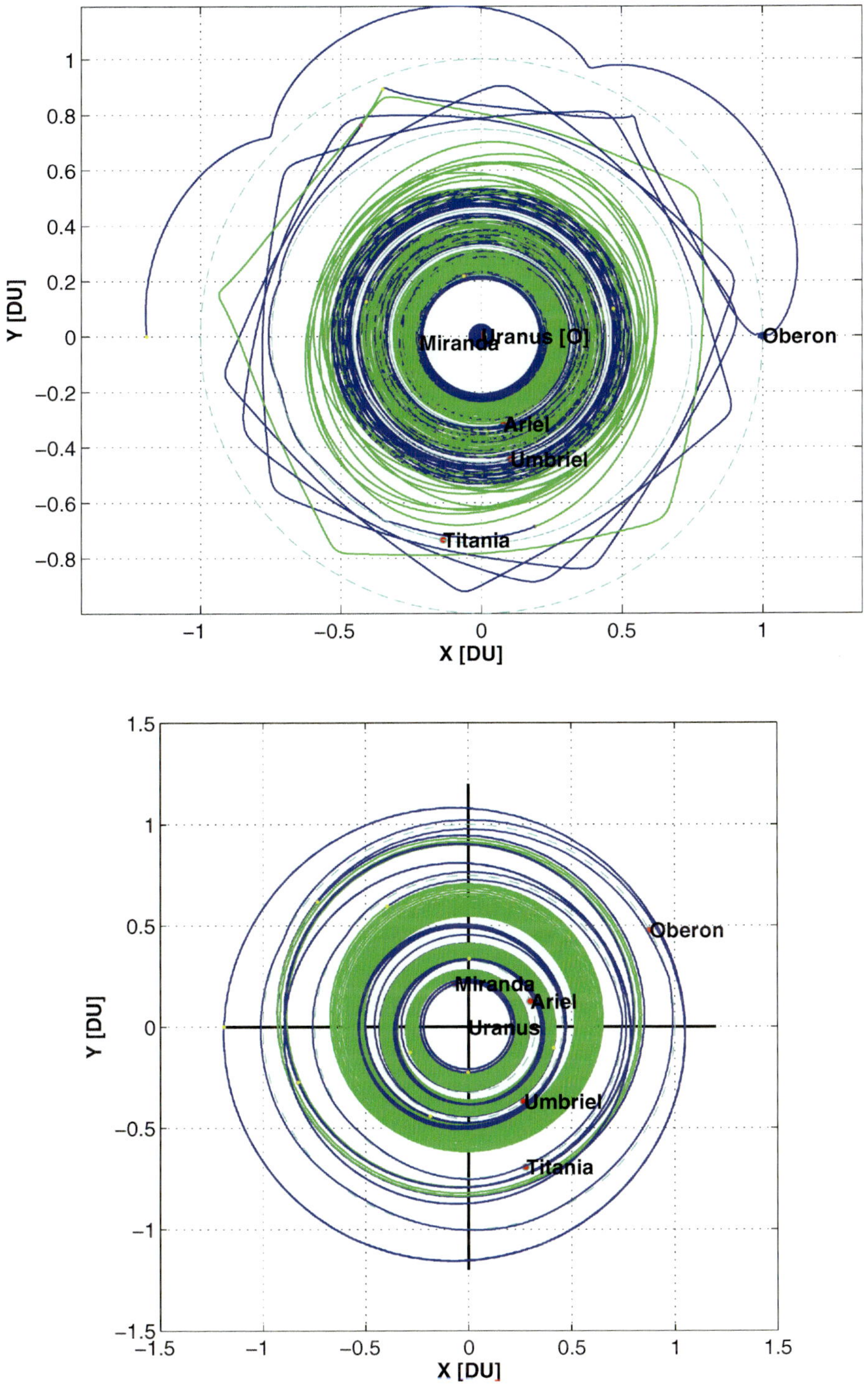


**Fig. 7** The planetary tour. **a** Uranus - Oberon synodic system. **b** Uranus inertial system

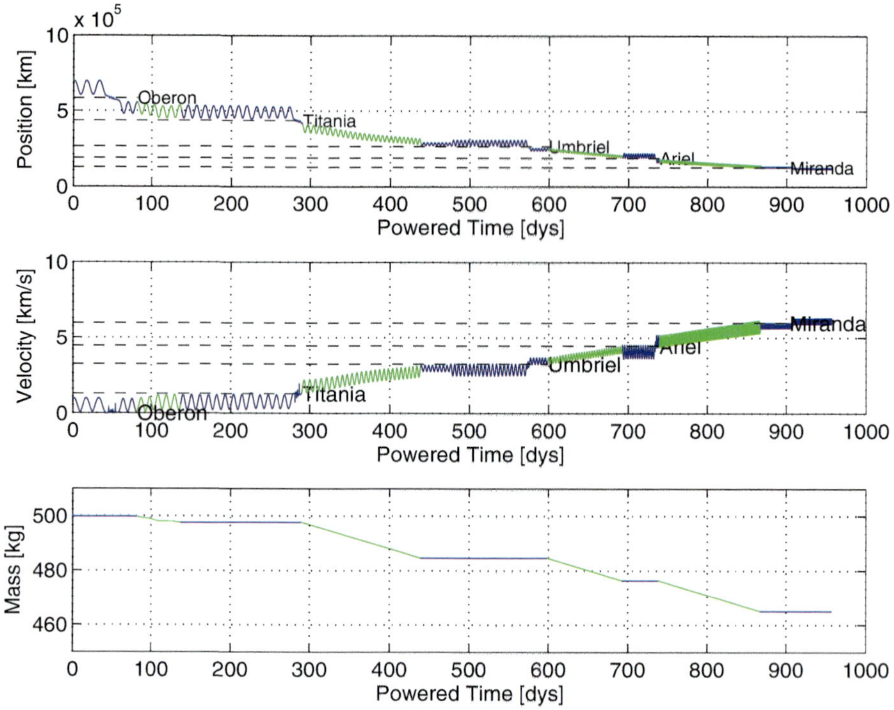


**Fig. 8** Instantaneous position, velocity and mass during the tour

The transitions from the outer to the inner Lagrange point, $L_2$ to $L_1$ respectively, for a certain system, correspond to the so called heteroclinic connections between the two manifolds associated with the Lagrange points. These connections are shown in Fig. 9. The figures display the ballistic trajectories within the appropriate synodic system and the Hill region associated with the $L_2$ energy, where they are propagated for the effective time that the spacecraft follows the heteroclinic connection. The minimal orbit altitude for each moon is set to be equal or greater than 50 km. This is achieved by numerical iteration of the small perturbation factor, $d$, discussed in the previous section.

Moreover, the number of closed orbits around each moon and the associated duration of the unstable ballistic capture are also strongly dependent on the value of the perturbation used to compute the manifold. Neither the number of closed orbits nor the capture duration have been parameters of optimization in this study, merely the constraint of minimum altitude has been imposed and obeyed. This resulted in a capture duration ranging from several days to almost a month.

Figure 9 shows a zoom of the closed orbits performed around the different moons, together with the incoming and outgoing ballistic arcs. The trajectories go from right to left where in the neck regions the spacecraft closely passes the libration points.

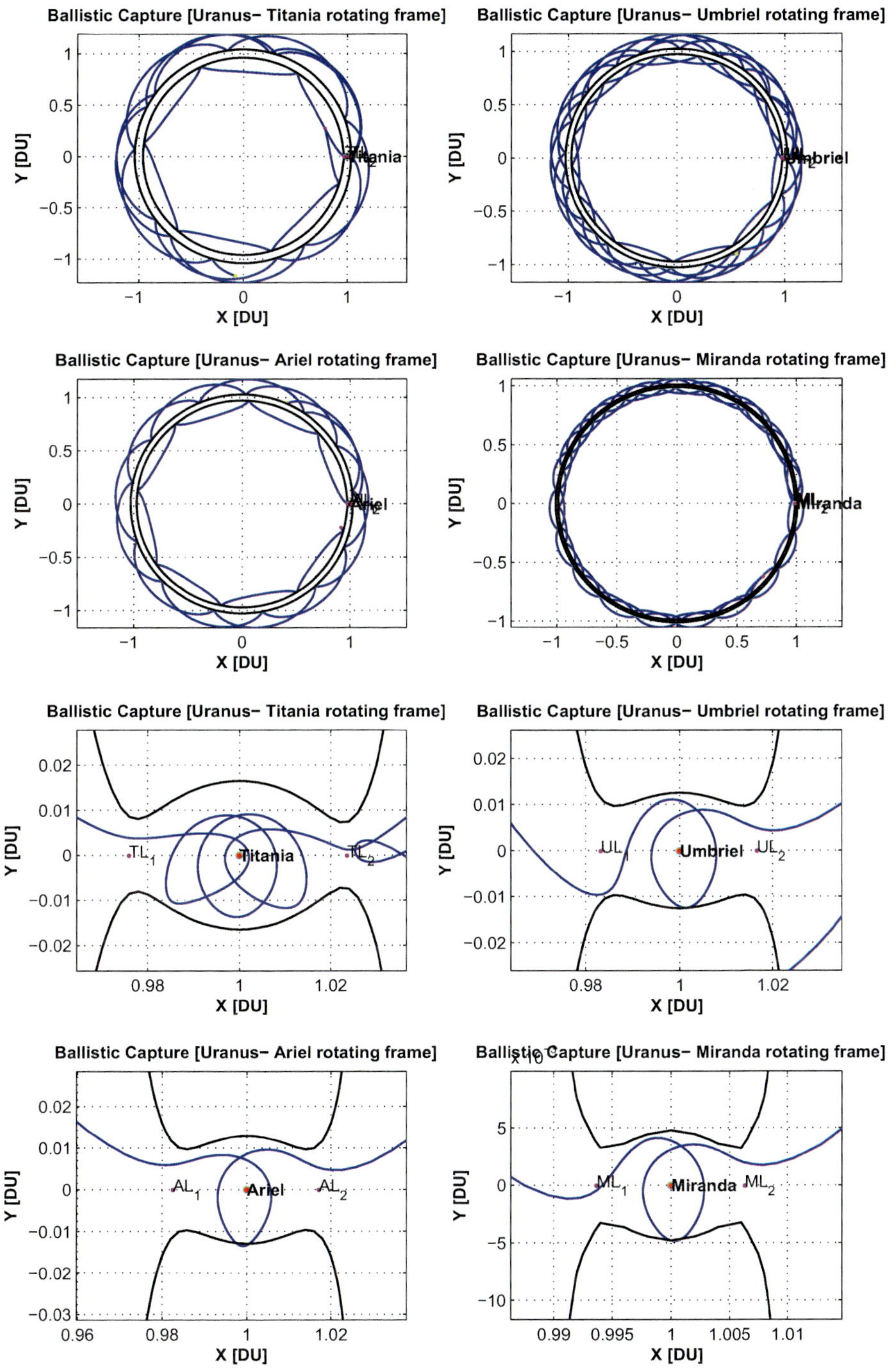


**Fig. 9** Detail of the heteroclinic connections. **a** Manifold transitions. **b** Uranus inertial system

In Fig. 10 the variation of the energy and the Jacobi constant are shown for the entire tour. The energy is shown with respect to the Uranus - Oberon system, where the energy on both the stable and unstable manifolds associated with Oberon remains constant. This behavior is also maintained during the passage through the moon realms, whereas for the other systems these passages result in a modification of the energy. Furthermore, the energy on the manifolds associated with the other systems also remains approximately constant.

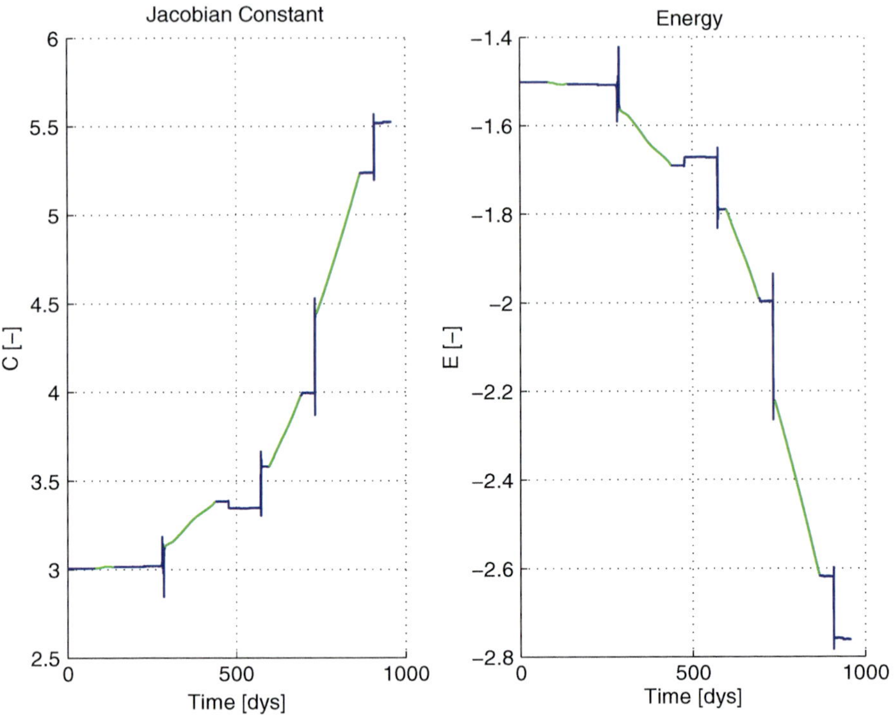


**Fig. 10** Value of the energy and Jacobi constant

## 6 Conclusions

The work presented in this study demonstrates that by a combination of coupled three-body models and Electric Propulsion very interesting scientific missions to Uranus can be designed. Moreover, it is shown that inclusion of the interplanetary trajectory does not render the mission infeasible neither in terms of mission duration nor with respect to the mass budgets. Many aspects do impose very high technological demands such as the EP propulsion operational time, however, for the current study this is omitted for simplicity reasons.

The interplanetary trajectory presents a time minimized solution adhering the appropriate conditions for the planetary tour to start. The optimization scheme computed a solution modifying the spacecraft's state in all 6 dimensions, with in addition a conjunction phase that dissipates the excess energy in order to start the planetary tour.

The planetary tour combines the advantages of dynamical systems theory within the three-body model and the use of EP, which opens a wide range of possible mission scenarios. The tour performs transitions between five different planetary three-body systems, establishing unstable captured orbits at each moon where the spacecraft is guided into a stable, circular orbit around Uranus after departure from the last moon considered.

Future developments could invoke the evaluation of manifolds associated with periodic orbits instead of the libration point, giving more freedom in the manifold intersections and therefore the number and duration of closed orbits around the moons. Moreover, increasing the envelope of available ballistic trajectories could decrease the propellant mass requirements even further. In addition, considering a three-dimensional environment for the Uranian tour would present an even more realistic analysis.

## References


1. Arthur E Bryson, J.: Dynamic Optimization, 1st edn. Addison Wesley Longman Inc., California (1999)
2. Betts, J.: Survey of numerical methods for trajectory optimization. Journal of Guidance, Control and Dynamics **21**(2), 193–207 (1998)
3. Casaregola, C., Geurts, K., Pergola, P., Andrenucci, M.: Radioisotope low-power electric propulsion missions to the outer planets. In: AIAA-2007-5234. 43rd Joint Propulsion Conference (2007)
4. Geurts, K., Casaregola, C., Pergola, P., Andrenucci, M.: Exploitation of three-body dynamics by electric propulsion for outer planetary missions. In: AIAA-2007-5228. 43rd Joint Propulsion Conference (2007)
5. Noble, R.J.: Radioisotope electric propulsion of science-craft to the outer solar system and near-interstellar space. In: FERMILAB-Conf-98/231. Proceedings 2nd IAA Symposium on Realistic Near-Term Advanced Scientific Space Missions (1998)
6. Oleson, S.R., Benson, S., Gefert, L., Patterson, M., Schreiber, J.: Radioisotope electric propulsion for fast outer planetary orbiters. In: AIAA-2002-3967. 38th Joint Propulsion Conference (2002)
7. Pergola, P., Casaregola, C., Geurts, K., Andrenucci, M.: Three body invariant manifold transition with electric propulsion. In: IEPC-2007-305. 30th International Electric Propulsion Conference (2007)
8. Ross, S.D., Koon, W.S., Lo, M.W., Marsden, J.E.: Heteroclinic connections between periodic orbits and resonance transitions in celestial mechanics. Journal of Chaos **10**(2), 427–469 (2000)
9. Seidelmann, P.K.: Explanatory Supplement to the Astronomic Almanac, 1st edn. University Science Books, California (2006)
10. Szebehely, V.: Theory of Orbits: The Restricted Problem of Three Bodies, 1st edn. Academic Press Inc., New York (1967)
11. Zazzera, F.B., Topputo, F., Massari, M.: Assessment of mission design including utilization of libration points and weak stability boundaries. Ariadna Study, ESA (2003)

# From Sputnik to the Moon: Astrophysics and Cosmology from Space

**Carlo Burigana, Nazzareno Mandolesi, and Luca Valenziano**

**Abstract** The launch of the Sputnik in October 1957 signed the beginning of space era. Just after few years Bruno Rossi opened the era of astrophysics observations. The Italian community has played, since then, an important role in the space community. After the success of many satellite missions, such as BeppoSAX, XMM-Newton, INTEGRAL, now the next frontier could be pioneering the scientific activities on the Moon. The absence of atmospheric emission and telecommunication interferences joined to the possibility of locating scientific instrumentation of relevant size and adaptive in time makes the Moon an ideal astronomical site for many branches of the modern astrophysics and cosmology and for dedicated fundamental physics experiments. Accurate measurements of the cosmic microwave background (CMB) radiation and of the radio sky at extremely long wavelengths could take a great advantage from the opportunity of observations from the Moon. In this context, we discuss here some aspects of particular interest: the CMB anisotropy in polarization and total intensity (at very small scales), the CMB spectrum. Some guidelines for future experiments from the Moon are presented.

## 1 The Birth of Space Astrophysics

Space astrophysics starts from the challenge in sending a probe outside the atmosphere. Unexpectedly the Soviet Union sent on October, 4th 1957 the first spacecraft in orbit around our planet. Presented as the USSR contribution to the International Geophysical Year, it was a sphere of 0.5 m, weighing about 80 kg, with an orbit of 96.3 min, while it was sending the famous "bip-bip" signal. Radio-amateurs all around the world caught the signals from the Sputnik, spreading the information of the first man-made object in space across the continents.

After the first exciting days, a paranoid race started in the USA, trying to overcome the Soviet result. "For the USA establishment, the launch of Sputnik was quite

C. Burigana (✉)
INAF-IASF Bologna, I-40129 Bologna, Italy
e-mail: burigana@iasfbo.inaf.it

E. Perozzi, S. Ferraz-Mello (eds.), *Space Manifold Dynamics*,
DOI 10.1007/978-1-4419-0348-8_10, 

a shock, producing a quasi hysterical reaction in universities and research institutions, beside naturally the impact on the defence and industrial outfits", Livio Scarsi, recalls in his memories. "First in space. First in everything. Second in space, second in everything. At MIT the activities stirred up by the event were particularly noticeable: The President of MIT was nominated special adviser to the White House for space policy".

The USA decided that in a very short time they should have sent their satellite in space. Unfortunately, at the first attempt, largely announced by media, the rocket was not able leave the launch pad and felt down after few meters.

Then USA involved the German team from Peenemunde, who has developed the V2 rockets during the second world war. Head of these "tiger team" was the famous engineer Wernher von Braun. In about one year, they succeeded in realizing a rocket and sending the first satellite, Explorer 1, into space.

During the same period, a young Italian scientist, Bruno Rossi, who had escaped for racial reasons from Italy, was settled at the MIT. He has been working with E. Fermi at the Manhattan Project, as responsible of detector group. He soon became a member of NASA.

Rossi had hint that, outside the Earth atmosphere, some high energy radiation of cosmic origin should be present. Livio Scarsi, who at that time had joined the Bruno Rossi team at MIT, recalls:

> I remember to have been enrolled in a "crash course" of technical Russian language, mandatory for all MIT staff members and research fellows. Practically all the Rossi Research Group (4th and 5th floor of building 26 in the Campus of Cambridge) was re-directed to a new space oriented programme.

The budget of NASA raised from 400 M USD in 1959 to 4000 M USD in 1963 and its personnel from 8000 to 30,000 people.

B. Rossi (MIT) in 1961 thought about the possibility of exploring the sky in X rays: he proposed his idea to AS&E (American Science and Engineering Inc.). He convinced a young Italian scientist, Riccardo Giacconi, to work on a rocket experiment for the detection of X rays from the moon. The project was accepted and flown in 1962. After two failures, at the third attempt the unexpected Sco X-1 source was discovered

The fathers of space astrophysics in Europe are Edoardo Amaldi and Pierre Auger. These two eminent scientists met in April 1959 in Paris, Jardin du Luxemburg, to discuss a scientific space policy in Europe and start up an organization on the model of CERN. In the same year, Amaldi writes "Space Research in Europe". The following year, two organizations were finally agreed: ESRO (European Space Research Organization) and ELDO (European Launcher Development Organization). Eventually, in 1971, ESA (European Space Agency) was born from the fusion of ESRO & ELDO.

During the 1960's, Bruno Rossi, stimulated some Italian colleagues, among who Domenico Brini at the Università di Bologna, to develop small balloon experiments to detect X-rays from cosmic origin from the high atmosphere.

While the Giacconi's experiment signed the start of the space astrophysics, the activity in Bologna represented the beginning of the X-ray astronomy and of Space research in Italy.

From the first balloon launches from the Physics Department in Bologna, the tradition after Domenico Brini lead to the launch of the BeppoSAX satellite, named after Giuseppe (Beppo) Occhialini, in 1996. This mission had an enormous success during the six year of operations, being appointed of the Bruno Rossi Prize, considered the Nobel prize for astrophysics.

The Italian scientist have been participating, often with leading roles, up to the present to the most important scientific space missions in astrophysics, such as XMM-Newton, INTEGRAL, Planck, AGILE, ISO and many others.

## 2 From Planck to the Moon

In 1992, an Italian team proposed a mission in response of an ESA call for mission ideas. It was named COBRAS (COsmic Background RAdiation Satellite). The scientific goal was an accurate map of the cosmic background radiation anisotropies at radio frequencies. For the first time, the needs for an extremely accurate control of systematic effects led to an orbit far from the Earth. "Preferred orbit: libration point of the Sun-Earth system $L_2$, in order to minimize earth radiation and optimize the thermal performance of the payload", quoting from the original proposal.

This payload, later merged with SAMBA (Satellite to Measure the BAckground Anisotropies), is now ready to fly as Planck Low Frequency Instrument. Planck will image the sky in 9 different bands at millimetre and sub-millimetre wavelengths with an accuracy limited only by astrophysical foregrounds.

The new frontier for space astrophysics could take advantage of the human settlement on the Moon. While satellite are a cost-effective way to explore the sky, large and heavy instruments cannot be flown on spacecrafts. On the contrary, the Moon offers an opportunity for such projects.

## 3 Why from the Moon

The absence of atmospheric emission and telecommunication interferences joined to the possibility of locating scientific instrumentation of relevant size and adaptive in time makes the Moon an ideal astronomical site for many branches of the modern astrophysics and cosmology and for dedicated fundamental physics experiments. For example, regarding ultraviolet (UV), optical and infrared (IR) astronomy, 1–2 m, 4–8 m, or, finally, 50–100 m (OverWhelming Large telescope) class telescopes could be dedicated respectively to solar observations aimed to solar magnetism and spectropolarimetry, to the observation of the IR radiation from primeval stars and galaxies, and to wide field imaging of large scale structures with unprecedented resolution, i.e. up to the diffraction limit of ~ 0.0005 arcs.

The possibility of greatly improving the gravity theories, and in particular of testing the Einstein's general relativity principles, through a new generation of Lunar Laser Ranging (LLR) array aimed to overcome the problem of geometric librations [11] will certainly represent a major contribution also for general relativity astrophysics and cosmology other than a remarkable example of fundamental physics experiments from the Moon.

Among the various astrophysical branches, accurate measurements of the cosmic microwave background (CMB) radiation, a field in which the interesting signal is so small to call for a great suppression of any (even very small) potential systematic effect, and of the radio sky at very long wavelengths, where the atmosphere prevents the sky observation, could take a great advantage from the opportunity of observations from the Moon. We discuss here four aspects of particular interest in this context: CMB polarization anisotropies, CMB total intensity anisotropies on very small scales, CMB absolute temperature measurements (i.e. CMB spectrum).

## 4 CMB Anisotropies

The state of the art of CMB anisotropy observations at angular scales larger than ~15 arcmin is represented by the second release of the measures of the NASA WMAP[1] satellite [27, 26, 40, 54] that greatly improved the fundamental results of COBE/DMR [53] and balloon-borne and ground-based experiments (e.g. [14, 25, 24, 33, 37, 22, 1]).

In general, the exploitation of the cosmological information encoded in the microwave anisotropy maps requires to separate the astrophysical and cosmological signal from the various classes of spurious systematic effects and then to separate the CMB cosmological signal from the astrophysical contributions, of local nature (i.e. produced in the Solar System, in particular the diffuse contribution represented by the Zodiacal Light Emission) and of Galactic and extragalactic origin. The astrophysical and cosmological signals can be distinguished on the basis of their different spatial correlations and frequency dependences. The latter property calls for multi-frequency observations of the microwave sky possibly complemented by radio and far-IR observations.

The CMB anisotropy is analyzed in terms of the two-point angular correlation or in terms of its transform, the *angular power spectrum* (APS), $C_l$, as function of the *multipole*, $\ell$ [inversely proportional to the angular scale, $\ell \sim 180/\theta$ (deg)], widely used in the comparison with theoretical predictions. WMAP obtained a very accurate measure of the CMB APS in a wide range of multipoles ($\ell$ between 2 and ~ 800). It allowed to greatly improve the accuracy of our knowledge about a wide set of cosmological parameters, essentially confirming the so-called cosmological concordance model. WMAP obtained a high-precision measure not only of the total intensity (i.e. temperature, TT mode) APS but also an accurate measure of the APS of the cross-correlation between the total intensity and the polarized intensity

[1] http://lambda.gsfc.nasa.gov/product/map/current/

(TE mode) and a first measure of the low $\ell$ tail of the polarization APS (namely, the EE mode) both for the CMB component and, in particular, for the polarized foregrounds. The forthcoming ESA Planck[2] satellite (e.g. [58, 41]) will allow to greatly improve in the next years the WMAP measure of the TT, TE, and EE APS thanks to the high sensitivity of its two instruments, the *Low Frequency Instrument* ([36]) and the *High Frequency Instrument* [44] that respectively represent the state of the art of radiometer and bolometer technology and cover a wide frequency range, from 30 to 857 GHz, of extreme relevance for an accurate separation of the CMB signal from the astrophysical one.

## 5 CMB Polarization Anisotropy Beyond Planck

In spite of the great cosmological opportunities opened by Planck, the full exploitation of the scientific information encoded in the CMB polarization anisotropies [28, 29, 51] requires a new generation of experiments and a numerous set of projects and proposals are currently under realization or study (see e.g. the B-Pol[3] project as an example of an ESA free-flight mission; [15]). In particular, the observation of the weak signal of the so-called BB mode of the CMB polarization anisotropy is of extreme interest for cosmology and fundamental physics: it likely represents a unique way to identify and study the primeval gravitational wave background (e.g. [47, 23, 55, 46, 17]) associated to the tensor perturbations. The amplitude ratio (T/S) between tensor and scalar perturbations proves the inflation energy scale. On the other hand, the study of this weak signature requires an overall sensitivity significantly better than that achievable by Planck (that could allow only a mere detection, under the hypothesis of extremely high values of T/S close to the limit of compatibility with current data) and, in particular, a much better comprehension of the polarized foreground contamination. Figure 1 illustrates this concept for two different T/S ratios considering the same sky areas adopted by the WMAP team in the polarization data analysis. As evident, the multipole region about $1 \sim 100$ (i.e. at angular scales $\theta \sim 2°$) exhibits a broad peak of the primeval BB mode that makes these scales particularly favourable for the study of the primeval gravitational wave background. Also, another relevant broad peak, produced during the cosmological reionization epoch, appears at $\ell \leq 10$–$20$. Since its amplitude depends on the T/S ratio, its observation provides a relevant consistency test of the T/S ratio derived from the primeval BB mode amplitude bump at $\ell \sim 100$. These arguments prove the relevance of observations of large sky areas (up to the all-sky) with an angular resolution (FWHM) of about 0.5° and with a frequency coverage suitable to separate the CMB from the foreground.

Another fundamental consideration derives from the presence of the BB mode induced by gravitational lensing [30], an integrated effect associated to the deflection of the CMB photons from the cosmic structures at relatively late epochs, that

---

[2]http://www.rssd.esa.int/planck

[3]http://www.b-pol.org/index.php

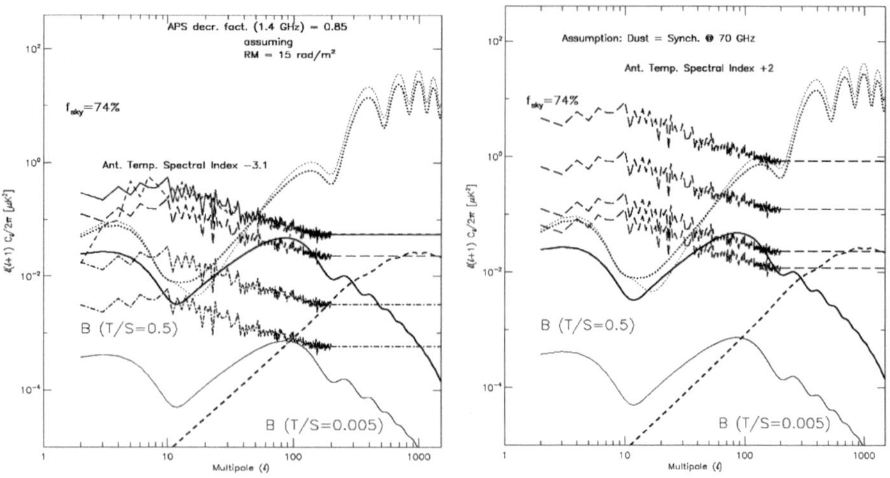


**Fig. 1** Comparison between the CMB anisotropy polarization APS and the Galactic foreground APS (synchrotron emission, left panel, dust emission, right panel). The polarized synchrotron model reported here (at 60.3, 70, 100, and 143 GHz from the *top* to the *bottom*) is based on a joint analysis of radio surveys and WMAP data (see [5] for further details); at $l > 200$ a flattening of the Galactic polarized APS is assumed ($C_l \propto l^{-2}$). The polarized dust model reported here (at the same above frequencies but from the *bottom* to the *top*) is based on the assumption to have similar power contributions from synchrotron and dust at ~70 GHz, i.e. close to the frequency of minimum polarized foreground. On dedicated (cleaner) sky areas the foreground contamination can be significantly smaller (by about one or two orders of magnitude in terms of $C_l$ [34, 10]). Note the oscillating nature of the CMB EE mode APS (*dots*) and the two broad bumps of the CMB BB primordial mode here reported for cosmological parameters consistent with WMAP and two values of T/S (0.5, thick lines, 0.005, thin lines). Note also the lensing contribution (dashes) peaking at $\ell \sim 1000$. See also the text

peaks at $\ell \sim 1000$. The amplitude of this term is relatively well defined, differently from the primeval BB mode largely dependent on T/S. As evident in Fig. 1, in spite of its dominance at multipoles much larger than those of the bump of the primeval BB mode, the lensing relative contribution to the global BB mode is not negligible also at smaller multipoles, depending on the T/S ratio. Although a certain separation of lensing and primeval BB modes is feasible in principle only with measures at $\ell$ less than about 200, it is clear that an accurate measures of the BB mode up to $\ell \sim$ 1000–2000, an intrinsically interesting scientific theme carrying information on the late stages of the Universe evolution during its cosmological constant or dark energy dominated dynamical evolution, greatly helps also the separation of the lensing contribution to the global signal at lower $\ell$'s and then the accurate measure of the primordial BB mode. This calls for accurate polarization measures with an angular resolution of about 10 arcmin, i.e. for relatively large size projects.

Given the wide range of values theoretically permitted for the T/S ratio, related also to the variety of inflation models, and then the possibility to miss the observation of the primeval BB mode (because of realistic experiment sensitivity limitation and foreground and lensing "contamination"), it is very interesting to note that a very accurate measures of the CMB polarization is in any case of high scientific interest (not only for the upper limit analyses on the primordial BB mode). A remarkable example is represented by the study of the formation and evolution of the cosmic structures and of the early stages of star formation in connection with the physical process of cosmological reionization, related also to the nature of the dark matter (see e.g. [39, 43, 6, 9] and references therein). The details of this process leave small features in the EE polarization mode (see Fig. 2), observable with high accuracy polarization experiments with an excellent rejection of the various classes of systematic effects and an extreme control of the foreground contamination.

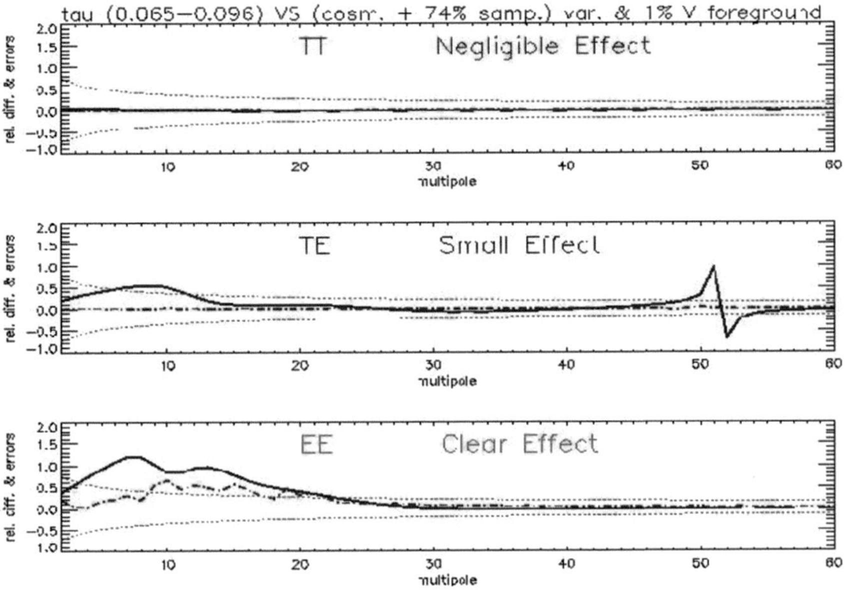


**Fig. 2** Relative difference (*solid lines*) between the TT, TE, EE CMB APS for cosmological parameters and two different values (0.065 and 0.096) of the Thomson optical depth, τ, corresponding to different cosmological scenarios, compatible with WMAP compared with the cosmic and sampling variance (*dots*) and with a residual (1%) foreground contamination (*dot-dashes*) in the WMAP V band with the same polarization sky mask adopted by the WMAP team. Note the remarkable dependence of the EE mode on τ

# 6 Small Scale Anisotropy Generated by Sunyaev-Zeldovich Effects

A further interesting field of the CMB cosmology is the study of the anisotropies at very small scales. The inverse Compton scattering of CMB photons on warm or hot electrons can generate a global distortion of the CMB spectrum (i.e. a Comptonization spectrum, see next section) as well as a local distortion towards a cluster of galaxies because of the presence of hot intergalactic gas. The latter case is the well known thermal Sunyaev-Zeldovich (SZ) effect ([57]; see also [45]), largely proved through radio, microwave, and X-ray observations.[4] Other than a local signature, this effect is also a source of small scale (i.e. large multipole) anisotropy (e.g. [35]) generated by the ensemble of SZ effects towards populations of galaxy clusters. There is also another kind of SZ effect, of Doppler nature, produced by the overall motion of a given galaxy cluster, the so-called kinetic SZ effect. Analogously to the case of the thermal effect, it also produces a small scale anisotropy signal because of the combined effects from many clusters. The two kinds of SZ effect show a different frequency behaviour and a different power at the various multipoles, the thermal one peaking at $l \sim$ few $\times$ $10^3$ the kinetic one peaking at $l \sim 10^5$, i.e. at multipoles where the primordial CMB anisotropy APS power is largely smoothed out by the diffusion damping [52] effect. Clearly, their study requires to achieve resolution and sensitivity much better than those suitable for studying the CMB primary temperature anisotropy. In particular, it calls for a resolution of ~0.1–1 arcmin at various frequencies about 217 GHz, where the thermal effect drops out, to be able to recognize the frequency signatures of such effects. The accurate study of the SZ anisotropy and the dedicated mapping of galaxy clusters provide crucial information on the evolution of cosmic structures, the dark energy and the physical processes in galaxy clusters.

## 7 A Concept Proposal for a Moon-Based Project Dedicated to CMB Polarization Anisotropy

While a free-flight space mission is suitable for a very clean imaging of CMB polarization anisotropies at angular scales larger than ~0.5°, a Moon-based project represents a very promising scientific opportunity thanks to the possibility to jointly achieve a wide frequency coverage and the $\approx$ arcmin angular resolution.

The main scientific objectives of a such project can be summarized as it follows: accurate measure of the EE mode essentially free from systematic effects; measure of the BB mode at the best achievable instrumental level (note that the theoretical BB mode level is not determined by theory); resolution suitable to reach multipoles $\ell \sim 1500$ in order to clearly distinguish the primordial BB mode from the lensing BB mode; precise mapping of polarized foregrounds ($\sim 0.1\ \mu$K) in order to control the systematic effects related to the microwave sky and to perform accurate astrophysical studies.

The 1st Level Requirements of such a project include: experimental site in a crater to prevent Sun illumination and located on the far side of the Moon, possibly close to the Moon equator to allow all-sky mapping; frequency coverage from ~10 to ~500 GHz for a very precise mapping of the various classes of astrophysical polarized foregrounds suitable for their study and separation from the

[4] It could be also present at smaller scales towards active and primeval galaxies (e.g. [42, 16]).

CMB component; nearly full sky coverage; the use of two receiver (solid lines); polarimetric technologies with a common frequency band to check systematic effects.

The 2nd Level Requirements include the following guidelines. Sensitivity better than 0.1 mK $s^{-1/2}$ to reconstruct the BB mode at least up to $\ell$ ~ 500 for $(\ell(\ell+1)C^{BB}{}_{\ell})^{1/2} > 0.1$ μK or at least up to $\ell$ ~ 200–300 for $(\ell(\ell+1)C^{BB}{}_{\ell})^{1/2} > 0.01$ μK at least in the cosmological frequency channels (~50–150 GHz). FWHM resolution of $\approx$1 arcmin at ~100 GHz; lifetime appropriate to allow at least four full sky surveys. Number of receivers > $N\times1000$, where $N$ is the number of frequency bands. HEMT technology at $\nu \leq$ 70–100 GHz and bolometer technology at $\nu \geq$ 70–100 GHz. Cryogenics technology dependent on the adopted detector technology (type, number, dissipation) and project (lifetime, experiment, etc.), with a possible cryo chain combination of the following systems: from 300 to ~50 K: ~3 passive radiators, or their combination with active coolers, mechanical (low vibration) or sorption, to reduce number/dimensions of passive stages; from ~50 to ~20–30 K (operating T for HEMT and pre-cooling stage for lower T refrigerators): sorption/JT cooler with $H_2$ or Ne, or mechanical cooler (pulse tube, low vibration level), or solid/liquid $H_2$ or Ne storage; from ~20–30 to ~5 K (pre-cooling for 0.1 K cooler): sorption/JT $H_2$cooler, or LHe storage, or mechanical cooler (pulse tube, vibration level); from ~5 to 0.1 K: dilution cooler. OMT based architecture for an accurate separation of the polarized signals.

In order to achieve the resolution discussed above, the typical telescope size could be different at the various frequencies. For example, assuming an edge taper of −30 dB (similar to that adopted for WMAP and Planck) a telescope size of ~3.5, 6.5, 10 m is respectively necessary to reach at 20 GHz a multipole l ~500, 1000, 1500. Going to 100 GHz (500 GHz) these sizes become respectively ~0.7, 1.3, 2 m (~0.1, 0.3, 0.4 m). Therefore, a possible alternative to the huge complexity of a single very large telescope with a very complex focal unit common to all detectors can be represented by a set of simpler/smaller optical systems designed to optimize the project synergy. For example, one (or two, given the large size of feeds at low frequencies) HEMT-based telescope(s) at $\nu \leq$ 70–100 GHz, one bolometer-based telescope at cosmological frequencies (from ~70–100 to ~150 GHz), one bolometer-based telescope at $\nu \geq$ 200 GHz.

## 8 CMB Spectrum

The accurate measure of the absolute temperature of the CMB on a wide frequency range means to observationally study the possible deviations of the CMB spectrum from a blackbody spectrum.

The main problem in this kind of measurements does not particularly relies on statistical sensitivity but is represented by the necessity of an extremely accurate control of systematic effects, and in particular of the *absolute joint calibration* of the data at the various frequencies, and of an *absolute* separation of the CMB signal from the atmospheric and astrophysical foreground (see [48] for a recent discussion and references therein). For these reasons the COBE/FIRAS [18, 38] space measures of the CMB spectrum, carried out at $\lambda \leq 1$ cm, still represent the state of the art in this field while measurements at $\lambda \geq 1$ cm currently exhibit remarkable uncertainties (see Fig. 3, bottom panel).[5]

Various classes of physical, astrophysical and cosmological processes could have imprinted signatures in the CMB spectrum at different cosmic epochs (see e.g. [12] and references therein). The details of these features depend on global cosmological parameters and, in particular, on the (possible wide) set of parameters that characterizes the specific considered process. However, two process parameters mainly determine the spectral shape of such spectral distortions: the epoch (or redshift, $z_h$) at which the considered process occurred and the fractional (i.e. with respect to the radiation energy density in the absence of this process) amount of energy exchanged in the primeval plasma, $\Delta\varepsilon/\varepsilon_i$, ([2] and references therein). In Fig. 3 (bottom panel) we show some kinds of spectral distortions, widely considered in the literature, assuming energy exchanges compatible with the limits set by current measures (mainly by COBE/FIRAS, see [49]). Note that late Comptonization like [59, 60] and free-free distortions as those shown in Fig. 3 are predicted on the basis of the astrophysical processes during the early stages of cosmic structure formation even in relatively standard scenarios while early Bose-Einstein like [56] distortions could have been produced in some fundamental physics models, for example those involving some particle or scalar field decay with radiative channels or energy exchanges coupled to the primeval plasma. In this regard, the development of very accurate CMB spectrum measurements represents a crucial probe for current cosmology, providing a relevant information complementary to that contained in the CMB anisotropy pattern. While the good agreement of the CMB spectrum with a blackbody spectrum has represented a fundamental verification of the standard cosmological model, the possible lack of detection of small spectral distortions such those observable with experiments 10–100 times more sensitive than COBE/FIRAS would represent a problem for our understanding of the evolution of the universe as it would have been if the anisotropies were missed by COBE/DMR.

## 9 CMB Spectrum Beyond COBE/FIRAS

Obviously, there is room for measurement improvements at both $\lambda > 1$ cm and $\lambda < 1$ cm (see e.g. [3, 4, 8] and references therein). While measurements at $\lambda < 1$ cm with an extremely high accuracy and sensitivity (much better

[5] Improvements in absolute spectrum measurements have been recently achieved with the NASA ARCADE balloon-borne experiment [19, 31] (see http://arcade.gsfc.nasa.gov/) and announced during the conclusion of the writing of this contribution; they suggest a signal excess at 3.3 GHz [20, 50, 32].

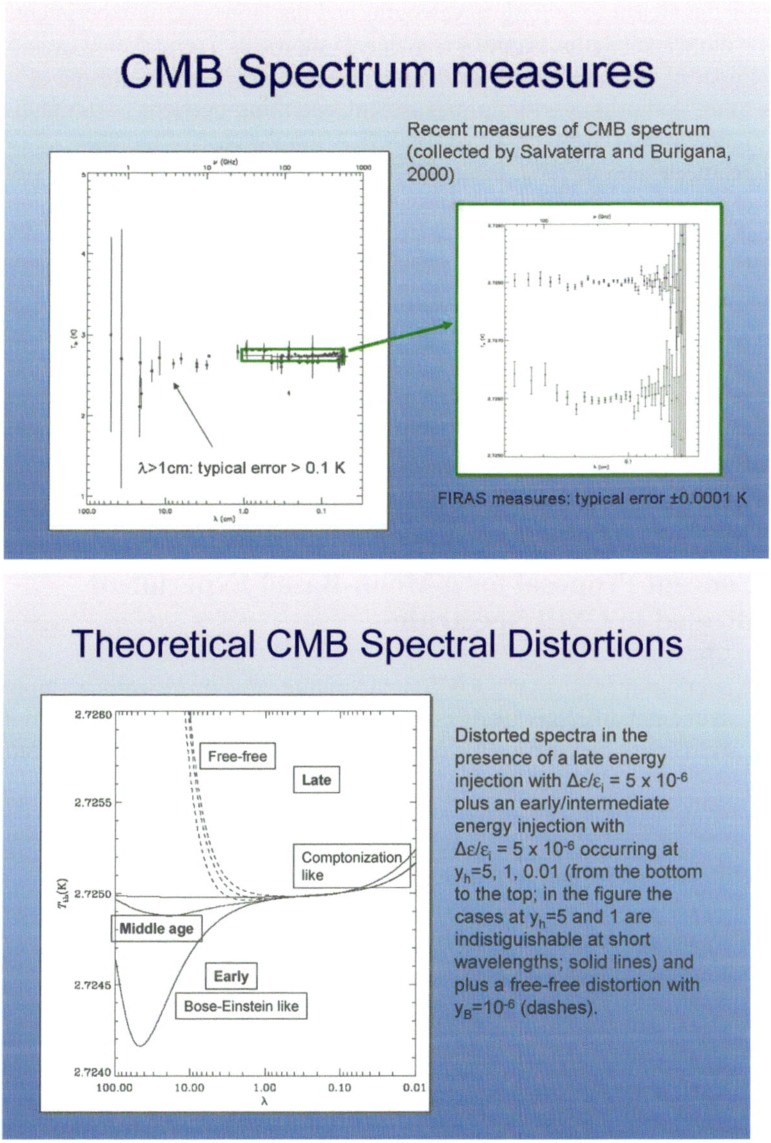


**Fig. 3** *Top panel:* state of the art of the measures of the CMB absolute temperature (but see also footnote 5). Note the huge difference between the precision of COBE/FIRAS (at $\lambda \leq 1$ cm) and of the measurements at $\lambda \geq 1$ cm. *Bottom panel:* typical distorted spectra predicted in the case of energy dissipation processes at different cosmic times for distortion parameters compatible with current observational limits

than FIRAS) could likely take advantage from a free-flight (possibly in $L_2$) satellite dedicated project [21] thanks to the smaller size necessary for instruments operating at shorter wavelengths, measurements at $\lambda > 1$ cm are of particular interest for a Moon-based project because of the evident necessity of improving the measures at long wavelengths and the large size of experiments operating at ~ dm wavelengths. Figure 4 (top panel) shows, as a remarkable example, the improvement of the constraints on the energy exchanges in the plasma at various cosmic times achievable with an experiment operating between ~ 0.5 and 20 cm with a precision similar to that of COBE/FIRAS in the case of the absence of spectral distortion detection: note, in particular, the improvement of a factor ~ 50 with respect to the current limits for processes possibly occurred at early or intermediate epochs. Figure 4 (bottom panel) also shows the reconstruction of the energy exchanges in the plasma at the various cosmic epochs in the case of the observation of deviations from a Planckian spectrum with an instrument with the above properties. Finally, we note that extending the observational range to $\lambda \sim 1$ m, as possible with relatively large size equipments more suitable for a Moon-based project than for a free-flyer, would allow to identify also the long wavelength raising (see Fig. 3, top panel) of the absolute CMB temperature in the case of a Bose-Einstein like early (or intermediate) spectral distortion.[6]

## 10 A Concept Proposal for a Moon-Based Experiment Dedicated to CMB Spectrum

Absolute measurements of the CMB temperature are intrinsically complex, the major experimental difficulty likely coming not from the sensitivity but from the calibration. We discuss here below the main specifications of an experiment dedicated to complement the COBE/FIRAS information on CMB spectral distortions.

The operation frequency range should be between ~ 0.4 and ~ 50 GHz (i.e. between ~ 75 and ~ 0.6 cm), in order to reach the low frequency raising of a Bose-Einstein like distortion and to match the lowest COBE/FIRAS frequencies, with a frequency spectral resolution not so high (about 10%) but suitable to appropriately cover the frequency range in order to distinguish the CMB (spectral distortion) signal from the foreground. The angular resolution requirement is not critical in this context (7°–8°) although a significantly better sky sampling is likely very useful for several data reduction aspects. A wide sky coverage (larger than $10^4$ degree$^2$ towards relatively foreground clean regions) is necessary although an all-sky coverage is certainly the best solution. The final sensitivity (E.O.L.) should be better 0.1 mK per each resolution element (both on each frequency map pixel and for each spectral bin).

Concerning the optics, the following requirements should be satisfied: low secondary lobes, ground shielding to reject ground signal and assure thermal stability, pointing accuracy better than ~1 arcmin. In order to maintain an appropriate thermal

[6]In particular, this would permit the determination of the frequency location of the minimum of the distortion [13], so independently proving the product of the baryon density with the square of the Hubble constant.

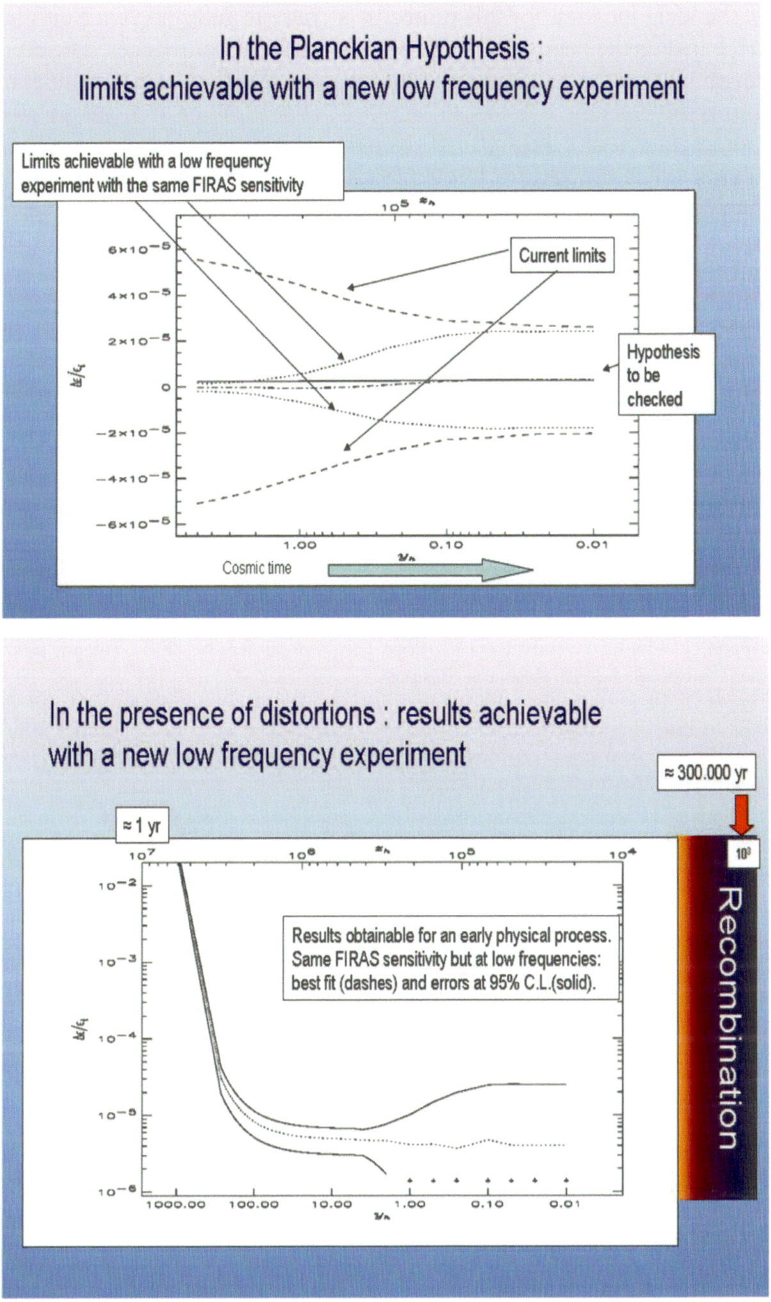


**Fig. 4** Improvement of the knowledge on the energy exchanges in the primeval plasma at various cosmic epochs achievable with an experiment with a sensitivity similar to that of COBE/FIRAS but at $\lambda > 1$ cm. *Top panel:* limits achieved in the case of the observation of a Planckian spectrum. *Bottom panel*: constraints (upper/lower limits – *solid lines* – and *best fit* – *dots*) obtained in the case of the observation of a distorted spectrum. Limits at 95% CL. See [7]

stability, the ideal location for this project is a crater to fully prevent Sun illumination and located on the far side of the Moon to avoid RF disturbances. The cryogenic calibrator should be cooled down to a temperature close to the CMB temperature of ~ 2.7 K plus the Galactic one ($\rightarrow$ to $\approx$ 3–5 K, depending on frequency). It should have a return loss less than $-60$ dB over the whole frequency range. The inter-calibration between the various frequency channels should be at an accuracy level better than 30 $\mu$K with a thermal stability better than 1 mK and a well sampled temperature monitoring (and a temperature determination accuracy better than 10 $\mu$K). Note that this experiment is based on radio receiver technology, with a stability defined by the above characteristics, and a cryogenic calibrator as key element.

**Acknowledgment** It is a pleasure to thank S. Dell'Agnello, F. Finelli, L. La Porta, G. Morgante, L.A. Popa, P. Procopio, R. Ragazzoni, R. Salvaterra, and F. Villa for collaborations and constructive discussions.

## References

1. Benoit A., et al., 2003, A&A, 399, L19
2. Burigana C., De Zotti G., Danese L., 1995, A&A, 303, 323
3. Burigana C., De Zotti G., Feretti L., 2004, New Astron. Rev., 48, 1107
4. Burigana C., Finelli, F., Salvaterra R., Popa L.A., Mandolesi N., 2004, Recent Res. Devel. Astron. Astrophys., 2, 59, Research Signpost
5. Burigana C., La Porta L., Reich W., Reich P., Gonzalez-Nuevo J., Massardi M., De Zotti G., 2006a, in Proc. of *"CMB and Physics of the Early Universe"*, 20–22 April, 2006, Ischia, Italy, Editorial board: G. De Zotti et al., PoS(CMB2006)016 (http://pos.sissa.it//archive/conferences/027/016/CMB2006_016.pdf), astro-ph/0607469
6. Burigana C., Popa L.A., Finelli F., Salvaterra R., De Zotti G., Mandolesi N., 2006b, in Proc. of JENAM 2004 meeting *"The many scales in the Universe"*, 13–17 September, 2004, Granada, Spain, Eds. J.C. del Toro et al., Springer, Dordrecht, The Netherlands, astro-ph/0411415
7. Burigana C., Salvaterra R., 2003, MNRAS, 342, 543
8. Burigana C., Salvaterra R., Zizzo A., 2004, in Proc. of *"Plasma in the Laboratory and in the Universe: New Insights and New Challenges"*, 16–19 September 2003, Como, Italy, AIP Conference Proceedings, Vol. 703, p. 397, Eds. G. Bertin et al., Melville, NY
9. Burigana C., Popa L.A., Salvaterra R., Schneider R., Choudhury T.R., Ferrara A., 2008, MNRAS, 385, 404
10. Carretti E., Bernardi G., Cortiglioni S., 2006, MNRAS, 373, L93
11. Currie D.G., et al., 2006, *A Lunar Laser Ranging Array for the 21ST Century*, Submitted Proposal, see Cantone C., et al., 2006, LNF-06/28 (IR), November 1, 2006
12. Danese L., Burigana C., 1993, in Proceedings of *"Present and Future of the Cosmic Microwave Background"*, 28 June–1 July, 1993, Santander, Spain, Lecture in Physics, Vol. 429, p. 28, Eds. J.L. Sanz et al., Springer Verlag, Heidelberg, FRG
13. Danese L., De Zotti G., 1980, A&A, 84, 364
14. de Bernardis P., et al., 2000, Nature, 404, 955
15. de Bernardis P., Bucher M., Burigana C., Piccirillo L., for the B-Pol Collaboratio., 2003, Experimental Astronomy, 23, 5
16. De Zotti G., Burigana C., Cavaliere A., Danese L., Granato G.L., Lapi A., Platania P., Silva L., 2004, in Proceedings. of *"Plasma in the Laboratory and in the Universe: New Insights and New Challenges"*, 16–19 September 2003, Como, Italy, AIP Conference Proceedings, Vol. 703, p. 375, Eds. G. Bertin et al., Melville, NY
17. Fabbri R., Pollock M.D., 1983, Phys. Lett. B, 125, 445

18. Fixsen D.J., et al., 1996, ApJ, 473, 576
19. Fixsen D.J., et al., 2004, ApJ, 612, 86
20. Fixsen D.J., et al., 2009, ApJ, submitted, astro-ph/0901.0555v1
21. Fixsen D.J., Mather J.C., 2002, ApJ, 581, 817
22. Grainge K., et al., 2003, MNRAS, 341, L23
23. Grishchuk L.P., 1975, Sov. Phys., JEPT, 40, 409
24. Halverson N.W., et al., 2002, ApJ, 568, 38
25. Hanany S., et al., 2000, ApJ, 545, L5
26. Hinshaw G., et al., 2007, ApJS, 170, 288
27. Jarosik N., et al., 2007, ApJS, 170, 263
28. Kaiser N., 1983, MNRAS, 202, 1169
29. Kamionkowski M., Kosowski A., Stebbins A., 1997, Phys. Rev. D, 55, 7368
30. Knox L., Song Y.-S., 2002, PRL, 89, 011303
31. Kogut A., et al., 2004, ApJS, 154, 493
32. Kogut A., et al., 2009, ApJ, submitted, astro-ph/0901.0562v1
33. Kovac J., et al. 2002, Nature, 420, 772
34. La Porta L., Burigana C., Reich W., Reich P., 2006, A&A, 455, L9
35. Ma C.-P., Fry J.N., 2002, PRL, 88, 211301
36. Mandolesi N., et al, 1998, *Planck* LFI, A Proposal Submitted to the ESA
37. Mason B.S., et al., 2003, ApJ, 591, 540
38. Mather J.C., et al., 1999, ApJ, 512, 511
39. Naselsky P., Chiang L.-Y., 2004, MNRAS, 347, 795
40. Page L., et al., 2007, ApJS, 170, 335
41. Planck Collaboration, 2006, *The Scientific Programme of Planck*, astro-ph/0604069
42. Platania P., Burigana C., De Zotti G., Lazzaro E., Bersanelli M., 2002, MNRAS, 337, 242
43. Popa L.A., Burigana C., Mandolesi N., 2005, New Astron., 11, 173
44. Puget J.L., et al, 1998, HFI *for the Planck Mission,* A Proposal Submitted to the ESA
45. Rephaeli Y., 1995, ARA&A, 33, 541
46. Rubakov V.A., Sazhin M.V., Veryaskin A.V., 1982, Phys. Lett. B, 115, 198
47. Sachs R.K., Wolfe A.M., 1967, ApJ, 147, 73
48. Salvaterra R., Burigana C., 2000, Int. Rep. ITeSRE/CNR 270/2000, astro-ph/0206350
49. Salvaterra R., Burigana C., 2002, MNRAS, 336, 592
50. Seiffert M., et al., 2009, astro-ph/0901.0559v1
51. Seljak U., Zaldarriaga, M., 1997, PRL, 78, 2054
52. Silk J., 1968, ApJ, 151, 459
53. Smoot G.F., et al., 1992, ApJ, 361, L1
54. Spergel D.N., et al., 2007, ApJS, 170, 377
55. Starobinksy A.A., 1979, JEPT Lett., 30, 682
56. Sunyaev R.A., Zeldovich Ya.B., 1970, Ap&SS, 7, 20
57. Sunyaev R.A., Zeldovich Ya.B., 1972, Comm. Astrophys. Space Phys., 4, 173
58. Tauber J.A., 2003, Adv. Space Res., 34, 491
59. Zeldovich Ya.B., Sunyaev R.A., 1969, Ap&SS, 4, 301
60. Zeldovich Ya.B., Illarionov A.F., Sunyaev R.A., 1972, Zh. Eksp. Teor. Fiz., 62. 1216 (Sov. Phys. JEPT, 35, 643)

# Space Exploration: How Science and Economy may Work Together

Silvano Casini

**Abstract** Space Exploration will last for a very long period of time: every single parameter impacting on that venture will dramatically evolve for generations. What follows is meant to be both an example of what could be done, and a possible way to start the process. It is also hoped to trigger detailed and constructive reflections in the vast community of Space Exploration potential stake-holders.

## 1 Why Space Exploration?

To explore, discover and then settle in new territories is typical of the mankind. It is what has been going on for thousands of years on Earth, resulting in human settlements everywhere on the globe, even in regions characterized by harsh environmental conditions.

The U.S. Presidential Directive of January 4th, 2004 has added a new dimension to exploration. No more confined to terrestrial continents, exploration is now targeting bigger and further continents, namely the celestial bodies of the solar system ... and beyond.

This is not just a cultural revolution. The Presidential Directive in fact states: "The fundamental goal of this vision is to advance U.S. scientific, security and economic interests through a robust space exploration program." [1] The points is that this statement about U.S. interests applies also to any other advanced country, and more so when one pays attention to two major and innovative concepts well rooted in the Presidential Directive: Space Exploration has to be the tool to give birth to a new economy and, as such, has to count on the participation of private investors.

Having in mind the scientific, security and economic interests, which are or should be the celestial bodies to be explored? As Space Explorations is about "the solar system, and beyond", the exploration of the solar system means, for the next few generations, robotic missions to Near Earth Objects, Jupiter's Icy Moons, Saturn's Satellites, as well as manned missions to the Moon, and later to Mars.

S. Casini (✉)
DdeB – Domaine de Beauregard Sarl, Houston, TX, USA
e-mail: silvano.casini@gmail.com

E. Perozzi, S. Ferraz-Mello (eds.), *Space Manifold Dynamics*,
DOI 10.1007/978-1-4419-0348-8_11, 

"Beyond" means that robotic and manned expansion into the solar systems will provide capabilities and infrastructures enhancing the study of the mysteries of the deepest universe, up to imaging exoplanets . . . and beyond!

## 2 Space Exploration as a Challenge and an Opportunity for Europe

Europe has not the U.S. experience and capabilities to start by itself the exploration of the solar system. Therefore, to protect its long term interest Europe has to define a strategy complementary to the U.S. one, and apt to federate other space-faring countries: the major ones (Russia, Japan, India, and China) and the developing ones (of whom Brazil is a good example).

Moreover, Europe has to add to its limited institutional space resources the contributions coming from non-space stakeholders (such as institutions dedicated to medicine, pharmacy, biology, physics, chemistry, material technology, nanotechnology, life cycle support, waste management, civil engineering, . . .), and from a strong participation of private investors.

By respecting these assumptions, Europe could assume the role of "co-leader" in the Space Exploration endeavour. [7]

## 3 The Importance of Public Awareness

Space Exploration is a venture which will go on for generations, and for the first 20–25 years will mostly be financed by public institutions (Space Agencies et similia). Provided that the institutional Space Exploration programmes will develop the necessary technologies (e.g. for economic and safe space transportation and life support systems) that the private companies will then adapt to their needs, private funding will accelerate to the point of rapidly taking over public funding in the subsequent years (this has been already done in several instances: the most appropriate one to be here mentioned is the U.S. government support to the very young aircraft industry by paying for airmail service for a decade or two, to permit the aircraft technology to develop to the point where it was relatively safe to fly). This means that in order to be successful, Space Explorations has to have for the next 25 years a continuous vivid impact on public opinion so to assure the required regular flow of public and private funds. In other words, a positive public awareness is a must and as such has to be pursued. This is a difficult task but not an impossible one, because Space Exploration has intrinsic elements striking the public imagination.

One set of such elements is related to the "adventure" aspects. The manned exploration of difficult areas has been, since Jason's expedition, always fascinating the mankind. The heroes that conquered the North and South poles of Earth in the first part of the XX century, and often lost their life in that attempt, went there for scientific, political, or strategic reasons, but certainly to also face and overcome extreme and even unknown difficulties.

Their fate stimulated the vivid interest of the media and by consequence of the worldwide public opinion. This resulted in a long and expensive series of expeditions dedicated to such conquests.

It is reasonable to assume that the exploration of the solar system will even be a better subject to forcefully promote similar emotions.

A spectacular Earthrise image sent by the HDTV payload on board the Japanese Kaguya lunar mission (Courtesy JAXA)

In addition to the exploration events, other associated activities may stimulate the Media, and therefore the public opinion. Tourism, sport and contests will be the natural follow-up of any advancement in the exploration of the solar systems. We have already enough elements (e.g. the demand for manned sub-orbital flights as well as manned flights to the International Space Station) comforting those who think that space tourism will quickly develop. The fact that private investors are now developing in-space infrastructures to receive tourists (the investment by Mr. Bigelow in the development and deployment of an inflatable Space Station in LEO for private astronauts is a good example for that) is even a reason to think that the space tourism age has already started.

Sport activities and contests will be a natural follow-on of space tourism, taking advantage from the proliferation of in-space and in-situ (e.g. on the surface of the Moon) Space Exploration infrastructures. The availability of propellant depots in orbit will induce the organization of in-space speed racing as well as endurance competitions, just as we know them on Earth. The building of bases on the Moon should facilitate the set up and the perpetuation of something like a lunar Paris-Dakar challenge. Similarly, the same kind of people who devote themselves to climb the Everest mountain will be excited by the idea of climbing lunar craters, as most probably will be their great grandsons at the idea of climbing the highest mountain on Mars: Olympus Mons volcano, 21200 m high.

Last but not least, even the building of large innovative and challenging infrastructures in specific in-space points (e.g. Lagrangian points) or on other celestial bodies (like on the Moon) will draw the interest and the participation of the public.

All mentioned achievements will be not only important because of their impact on public opinion, and by consequence in assuring a continuous and planned availability of institutional funds. They, in fact, will generate by themselves an important flow of private funds, which ultimately will cover most of the running costs of Space Exploration infrastructures.

So said, they will not however be sufficient to assure public awareness and support to Space Exploration in the initial phase (20–25 years), and without that support Space Exploration would abort.

## 4 The Impact of Science

Science will be the real triggering factor. Almost all disciplines will take advantage from Space Exploration, and probably new ones will be generated by it.

The improved knowledge of the solar system and its celestial bodies and that of the deep universe will certainly enjoy a quantum leap. Discoveries about the origin of the solar system and about the origin of life, the detailed mapping of the solar system celestial bodies, the discovery and the analysis of their marvellous natural phenomena, will cause emotion in the public opinion. The same will come out from the upgrading of our understanding of the nature of the Universe and from the improved capability in studying and eventually imaging exoplanets.

But not only basic Science will contribute to this popular movement. Applied Science too will play an important role. To implement the required infrastructures, to perform exploratory missions, and to settle temporarily or permanently on other celestial bodies or in-space large platforms (also known as safe havens, or space harbours) will demand new technologies and techniques, such as reliable and environmentally friendly sources of energy, efficient methods of waste recycling, economical systems for life cycle support, . . . .

Nearly every one of these technologies and techniques will have a practical application on Earth. Just as an example, let us consider the impact of new sure, reliable and clean sources of energy. They could radically solve environmental and geopolitical problems that we are nowadays experiencing on Earth by using fossil fuels.

Medicine too will have to be more and more preventive, to protect astronauts from serious disease such as blindness, cancer, osteoporosis, psychical and neuronal problems, so paving the way for more efficient prevention and treatment of diseases on Earth.

It clearly appears from the mentioned examples that science applied to Space Exploration will also be instrumental in solving such problems as increasing energy demand, better environmental protection, and preventive medical care for a growing and aging population.

It is therefore evident that Science (fundamental Science, because of its intrinsic importance and natural glamour on public imagination, and applied Science, because of the interest in its new discoveries and inventions as tools to solve major societal problems of the Globe) will be, as mentioned, the triggering factor to implement and develop a strong public movement in favour of Space Exploration in all space-faring countries.

## 5 Manned Space Exploration Constraints

While robotic exploration of the solar system can be a "go as you pay" programme, manned exploration will require large front investments in space transportation systems, in-space platforms and, generally speaking, enabling technologies and techniques.

Such an investment will be so huge that even the USA will have problems in implementing it beyond filling the gap, after the Shuttle retirement, about an autonomous capability to send American astronauts in Low-altitude Earth orbits (LEO). [8]

To make it affordable it is mandatory to get Space Exploration organized as a global endeavour or, in alternative, as the venture of no more than two complementary groups of space-faring countries. In this latter case there will be a few leading or co-leading countries, and not just a single country going to monopolize it.

Once affordability of Space Exploration is assured, a much bigger problem has to be resolved: how to assure its sustainability. In fact, as the Apollo programme has demonstrated, a big programme may become affordable on the basis of strategic interests, but not sustainable on the long run (unless one implements a kind of a "war economy" – but also in this case the situation remains not sustainable at longer term, as demonstrated by the collapse of Soviet Union).

The only way to make Space Exploration sustainable, and assure the deployment of human beings throughout the solar system, is to consider Space Exploration as the past explorations and discoveries were considered: a way to create a new and bigger wealth, namely additional and completely new segments of the economy.

This can only be achieved if the Space Exploration programme and initial infrastructures will be conceived in such a way to be friendly to private investors from the very beginning.

## 6 Economy and Science

The new economy originated by Space Exploration will have a twofold aspect. As mentioned in previous paragraphs, nearly all new scientific and technological achievements will have a direct impact on Earth economy by their capability to solve major worldwide problems.

The new ways for producing and transferring electrical energy, as imposed by Space Exploration environment and requirements, could in fact revolutionize the way of doing it also on Earth, so (i) allowing elimination of pollution, elimination of dependency from countries rich on fossil fuels politically unstable, elimination of most of high voltage power lines, and (ii) lowering the overall energy bill. Just these outcomes would free an enormous amount of financial resources. Soon however, Space Exploration will generate, as for past explorations, new markets and a new economy by itself.

Just to remain focused on the energy problem, let us mention that the possibility of solar thermal or photovoltaic giant power plants in geostationary orbit (GEO), with energy transmitted by micro-wave or laser systems either to Earth or other space-based plants, is seriously considered, even by NSSO (U.S. National Security Space Office).

It is therefore an option worth to be examined very closely and considered very carefully, and assessed in all its details. In any case, to have such big plants built in GEO with components and systems produced on Earth would be foolish, due to the enormous costs of getting them out from the Earth gravity well. Fortunately, more than 95% of the total mass of those plants could be produced on the Moon, at an overwhelming cost advantage versus the same items lifted from the Earth's surface. This approach requires the deployment of robotic/automatic mining and production plants on the Moon, and the presence of a large space station in the Earth-Moon Lagrange point $L_1$ (where to receive materials and components from the Moon, to be assembled into major systems to be robotically shipped to GEO orbit). In turn, the operability of these assets imposes the availability of routine space transportation systems for two-ways flights from Earth to LEO, from LEO to $L_1$, from $L_1$ to the Moon. This means the deployment of in-space fuel depots, advanced telecommunications, search and rescue capability, water depots, general logistic support, . . . . . ... In other words, an entirely new sector of the economy.

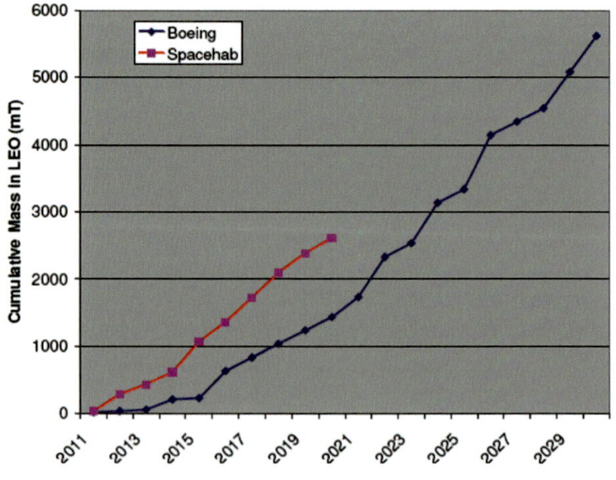


Cumulative mass in LEO, in the time frame 2011–2030, as derived from the studies made by Boeing [2] and Spacehab [3]. Space Exploration will induce an increased traffic requirement from Earth to LEO and this will ultimately favour the development of reusable systems

Moreover, the infrastructures so deployed will also be utilized to assure a quick start of Space Tourism and Ventures, as well as Media activities. In turn, the operators of these activities will soon invest (and, as mentioned, they are already doing that now) in dedicated facilities and specific logistic support, so adding momentum to the development of the in-space economy.

More could be said about Space Exploration as a tool to renovate our global economy. But it is time to come back to the relationship among Science, Space Exploration, and Economy.

First of all, growth of the population in orbit, in space-to-space travelling, and in-situ on the Moon will require a perfect understanding of the solar system weather and its sudden perturbations. Moreover, efficient and active protection systems against solar radiations will be necessary. Last but not least, new celestial mechanics principles, such as those pertaining to space dynamic manifolds, have to be implemented to make affordable and sustainable deep space transportation.

These are just three examples about the advancement of a mix of fundamental and applied Science, as requested by Space Exploration. But even more important is the point about the availability of large structures in space, and production capabilities on the Moon. All those assets, while deployed for economical or strategic reasons, will be available to Science too. As already mentioned this could eliminate some of the limitations that astronomy is facing today. By exploiting these assets, it would be possible to conceive very large telescopes to be assembled in Earth-Moon $L_1$ and robotically shipped to the Sun-Earth $L_2$ (about 14 days away). Those telescopes could be larger than any telescope possibly produced and launched from Earth. Moreover they could be called back for repairing, overhauling, refuelling and upgrading, to Earth-Moon $L_1$ where the existing space harbour i.e. a large operational platform, will assure all the necessary jigs, tools, facilities, and a reasonable comfort to human operators, as well as the availability of a park of free-flying cooperating robots.

In this way, the large investments required to produce and deploy those giant telescopes would be better justified.

In a similar way, in Earth-Moon $L_1$ (which is a convenient gateway to interplanetary space [4]) one could assemble heavy satellites to be shipped towards other celestial bodies (Mars, Icy Moons of Jupiter, Saturn's Moons) or even to GEO, to accomplish valuable robotic missions. This approach would in fact allow more sizeable scientific results and a better return on investment than those which could be allowed by robotic missions directly launched from Earth (a typical case is the hinted Mars Sample Return mission which at an outrageous cost would bring back a ridiculous quantity of Mars soil specimen).

A specific area of interlaced interests between Science and Economy is the robotic exploration of Near Earth Objects (NEO) population. The scientific importance of NEOs is well known [6]. Less known is the fact that some NEOs are rich in rare materials which are more and more scarce on Earth, or are in regions with very unstable political situations. For these reasons, the interest on NEOs by major worldwide mining companies is now taking shape. Clearly robotic mining missions could

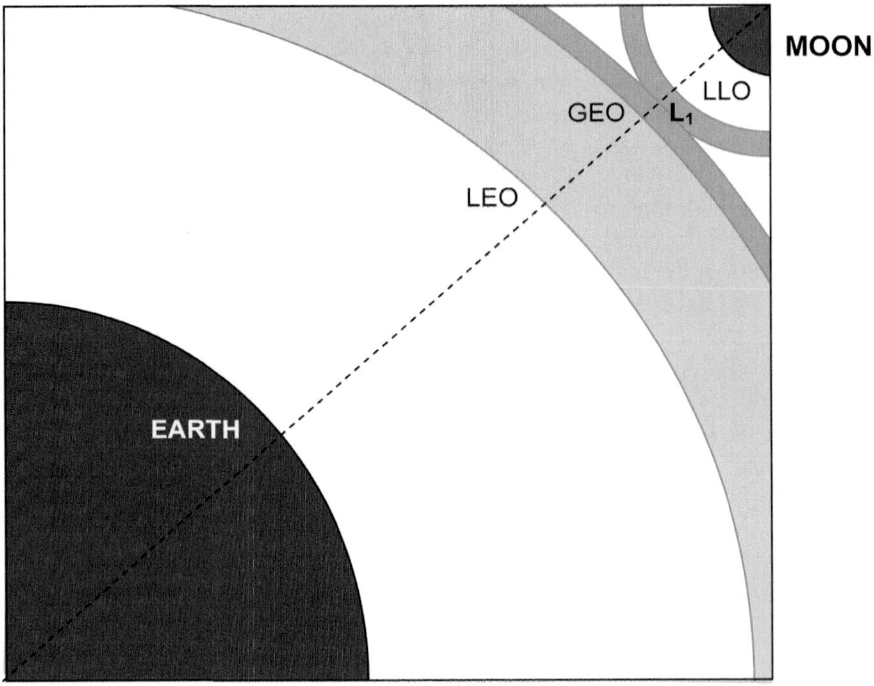


Earth-Moon Transportation Energy drawn after reference [5]. The diagram represents, in a linear scale, the change in velocity required to reach circular orbits of increasing distance from the Earth and from the Moon (LLO – typical low altitude lunar orbits) thus clearly showing the favourable location of the Earth-Moon lagrangian point $L_1$

also provide elements of interest to the scientific community, and a NEO scientific mission could similarly provide information valuable to better define and organize mining missions. Moreover, robotic exploration and mining of NEOs might also be a strong motivation for development of space cooperating robots and preparation for cooperating robots to travel throughout the Solar System.

## 7 Conclusions

Space Exploration is not just another bigger and more difficult space programme. Strictly speaking it is not even a space programme. It is the most audacious exploration endeavour even conceived, and it will stay well beyond the 21st century.

Achieving human presence in space and on other celestial bodies of the solar system is a step that has the potential of creating sustained prosperity and hope for every human being. It will also stimulate great works and efforts and will expand the borders of science and technology. Space Exploration will also be a cultural revolution as soon as the mankind will realize that it is no more confined to Earth. It will require, to be sustainable, the exploitation of Moon natural resources and huge in-space infrastructures (in LEO and in Earth-Moon $L_1$). More than that, Space Exploration can be deployed only if routine Earth-to-LEO transportation systems and new (very efficient) space-to-space ships will be developed.

Astronomy, as well as all other involved disciplines, could take advantage of the overall Space Exploration logistic capability. In turn this would provide better and quicker return from the overall investment. In a similar way, the dynamic mapping of Space manifolds and a complete understanding of their mechanics and environment could be the key to implement an affordable space-to-space transportation system.

Space Exploration is therefore a unique opportunity to combine science and economy, as it is an overall unique opportunity for mankind. Therefore the scientific international community and, in particular, representatives of the astronomical branch should propose advanced programmes based on the principle of taking advantage of Space Exploration infrastructures (free flying cooperating robots included) and logistic support which will be available at the time of the injection into orbit of these new astronomical missions. Specific programmes to quickly improve the understanding of space dynamic manifolds should also be rapidly submitted: there is a desperate need for solutions of the space-to-space transportation cost conundrum.

The scientific international community has now a unique opportunity to get extra funds. Astronomy could lead the way: by defining advanced projects interlaced with overall Space Exploration investments, and by being assertive in programming them, many of present dreams could become a reality in the next twenty-twenty five years.

Money is not always the major problem. The most horrible problem for the decision making people is the lack of vision and coordination in what is proposed to them for selection and priority assignment.

## References

1. *The Vision for Space Exploration.* NASA, February 2004.
2. *Boeing SoS Concept Design*, CE&R CA-1 Final Review, March 2005
3. *Spacehab Concept Exploration and Refinement of the Lunar Exploration System of Systems*, Final Report, March 2005
4. *The Lunar L1 Gateway : Portal to the Stars and Beyond* - Martin W. Lo, Shane D. Ross – AIAA Space 2001 Conference
5. *Space Resource Economic Analysis Toolkit: The Case For Commercial Lunar Ice Mining*, Brad R. Blair, Javier Diaz, Michael B. Duke, Elisabeth Lamassoure, Robert Easter, Mark Oderman and Marc Vaucher, Final Report to NASA Exploration Team, December 20, 2002
6. *Cosmic Vision. Space Science for Europe 2015–2025.* ESA BR-247, October 2005.
7. *Space Exploration: a Challenge and an Opportunity for Europe.* ESTEC Contract Final Report, June 2006.
8. *Does The Nasa Constellation Architecture Offer Opportunities To Achieve Space Science Goals In Space?* Harley A. Thronson, Daniel F. Lester, Adam F. Dissel, David C. Folta, John Stevens, Jason Budinoff. IAC-08-A5.3.6, 2008

# Index



E. Perozzi, S. Ferraz-Mello (eds.), *Space Manifold Dynamics*,
DOI 10.1007/978-1-4419-0348-8, 

GPSR Compliance

*The European Union's (EU) General Product Safety Regulation (GPSR) is a set of rules that requires consumer products to be safe and our obligations to ensure this.*

*If you have any concerns about our products, you can contact us on ProductSafety@springernature.com*

In case Publisher is established outside the EU, the EU authorized representative is:

Springer Nature Customer Service Center GmbH
Europaplatz 3
69115 Heidelberg, Germany

**Batch number: 10756376**

Printed by Printforce, the Netherlands